Green Reaction Media in Organic Synthesis

Edited by

KOICHI MIKAMI
Professor of Synthetic Organic Chemistry
Department of Applied Chemistry
Tokyo Institute of Technology
Japan

Blackwell
Publishing

Editorial Offices:
Blackwell Publishing Ltd, 9600 Garsington Road, Oxford OX4 2DQ, UK
Tel: +44 (0)1865 776868
Blackwell Publishing Professional, 2121 State Avenue, Ames, Iowa 50014-8300, USA
Tel: +1 515 292 0140
Blackwell Publishing Asia, 550 Swanston Street, Carlton, Victoria 3053, Australia
Tel: +61 (0)3 8359 1011

First published 2005 by Blackwell Publishing Ltd

Library of Congress Cataloging-in-Publication Data

Green reaction media in organic synthesis/edited by Koichi Mikami.
p. cm.
Includes bibliographical references and index.
ISBN-13: 978-1-4051-3402-6 (acid-free paper)
ISBN-10: 1-4051-3402-X (acid-free paper)

1. Solvents – Environmental aspects. 2. Organic compounds – Synthesis – Environmental aspects. 3. Green products. I. Mikami, Koichi.

TP247.5.G67 2005
660′.2844 – dc22

2005003242

ISBN-13: 978-1-4051-3402-6
ISBN-10: 1-4051-3402-X

A catalogue record for this title is available from the British Library

Set in 10/12 pt Times
by Newgen Imaging Systems (P) Ltd, Chennai, India
Printed and bound in India
by Replika Press Pvt, Ltd, Kundli

The publisher's policy is to use permanent paper from mills that operate a sustainable forestry policy, and which has been manufactured from pulp processed using acid-free and elementary chlorine-free practices. Furthermore, the publisher ensures that the text paper and cover board used have met acceptable environmental accreditation standards.

For further information on Blackwell Publishing, visit our website:
www.blackwellpublishing.com

Contents

Contributors

Dennis P. Curran	Department of Chemistry, Chevron Science Center, 219 Parkman Avenue, Pittsburgh, PA 15260, USA
Charlotte Emnet	Institut für Organische Chemie, Friedrich-Alexander Universität Erlangen-Nürnberg, Henkestrasse 42, 91054 Erlangen, Germany
John A. Gladysz	Institut für Organische Chemie, Friedrich-Alexander Universität Erlangen-Nürnberg, Henkestrasse 42, 91054 Erlangen, Germany
Jonathan J. Jodry	Chemical Research Center, Central Glass Co., Ltd., 2805 Imafuku-Nakadai, Kawagoe-shi, Saitama-ken 350-1151, Japan
Hiroshi Matsubara	Department of Chemistry, Osaka Prefecture University, Gakuen-cho, Sakai-shi, Osaka 599-8531, Japan
Koichi Mikami	Department of Applied Chemistry, Tokyo Institute of Technology, O-okayama 2-12-1-S1-29, Meguro-ku, Tokyo 152-8552, Japan
Yutaka Nakamura	Niigata University of Pharmacy & Applied Life Sciences, 265-1 Higashijima, Niitsu 956-8603, Japan
R. Scott Oakes	Department of Chemistry, University of Leeds, Leeds, LS2 9JT, UK
Christopher M. Rayner	Department of Chemistry, University of Leeds, Leeds, LS2 9JT, UK

Ilhyong Ryu Department of Chemistry, Osaka Prefecture University, Gakuen-cho, Sakai-shi, Osaka 599-8531, Japan

Toshiyasu Sakakura National Insititute of Advanced Industrial Science and Technology (AIST), Tsukuba 305-8565, Japan

Seiji Takeuchi Niigata University of Pharmacy & Applied Life Sciences, 265-1 Higashijima, Niitsu 956-8603, Japan

Hiroyuki Yasuda National Insititute of Advanced Industrial Science and Technology (AIST), Tsukuba 305-8565, Japan

Preface

Green, sustainable chemistry involves the designing of chemical processes with a view to reducing or even eliminating the use and production of hazardous materials. Recent endeavours have focused on limiting the use of organic solvents and replacing them with new, environmentally benign media. The chemical industry is interested in these cost-effective alternative solvents and processes.

This book provides a broad overview of the three most commonly used green reaction media. Directed at researchers and professionals in academic and industrial laboratories, it will also serve as a textbook for graduate courses on green chemistry. Successful green reactions are considered, and experimental sections at the ends of the chapters provide important practical details, with illustrations of potential applications. Sufficient information is included to allow selection of the most appropriate medium. Extensively referenced, the volume offers a point of entry into the detailed literature.

I wish to thank Professor Jim Coxon of the University of Canterbury, New Zealand, for his continual encouragement and editorial support, and for his original suggestion that we construct a book on this topic with a limit of 200 pages. I am grateful to Professors Ryu and Sakakura for their helpful discussions, and to all the authors for bringing this project to life. Finally, I wish to thank Dr Jonathan Jodry, previously a post-doctoral fellow in my group and now a research chemist at Central Glass Company, not only for his support as a co-author but also for his editorial and graphical assistance.

Koichi Mikami

1 Introduction

Koichi Mikami

1.1 Green reaction media

Recent years have witnessed a major drive to increase the efficiency of organic transformations while lowering the amount of waste materials. Catalysis plays a central role in these efforts: many industrial catalytic reactions have been developed to work with as low as 0.1% of catalyst loading, with extremely large substrate-to-catalyst ratios in some cases. However, even with such developments in catalysis, the weight ratio of waste material to desired products in chemical and pharmaceutical processes is still between 5 and 100, largely as a result of the use of organic solvents. The US environmental protection agency (EPA) reports that around 100 billion tons of industrial wastes are produced each year.[1] The annual use of organic solvents – worth more than US$5 billion – has a significant environmental impact.[2] Many organic solvents are hazardous and can be deleterious to human health. They are volatile and cause an environmental threat by polluting the atmosphere. A typical example of the problem is the hole in the ozone layer, which has been attributed to the release of CFCs (chlorofluorocarbons) in the atmosphere. Until recently, few efforts had been made to limit the use of organic solvents, to replace them with new green reaction media or to eliminate them completely (solvent-free conditions).[3] New international treaties such as the Montreal Protocol are beginning to increase pressure to minimize the use of solvents. Industries themselves are interested in cost-effective alternatives. It is expected that, in the coming decades, this will lead to a complete remodelling or modification of the infrastructure of many industrial processes currently in use, since new green reaction media might at least partially replace traditional organic solvents.[4] The term *green chemistry* has been proposed for new chemical processes that reduce or eliminate the use and production of hazardous materials.[5,6]

This book reviews reactions in which ionic liquids, fluorous media and supercritical CO_2 are used, as these solvents are the most promising new types of green reaction media.[7] Sufficient details are provided to allow researchers to explore the use of these solvents in specific reactions. Typical examples of reaction conditions and workup procedures are included at the end of each chapter to allow chemists to utilize these new technologies with confidence, and extensive references to the literature are listed. Other standard green reaction media such as water,[8–10] ethanol, aqueous surfactant micelles and polymers,[11] as well as solvent-free conditions,[3] are outside the scope of this book.

1.2 Ionic liquids

Fluorous media and supercritical CO_2 are, like water and all organic solvents, molecular reaction media. In contrast, ionic liquids are salts formed by mismatched associations of cations and anions. It is usually assumed that salts are solid under normal conditions and that they melt only at very high temperatures; for example, NaCl has to be heated to 803°C before it becomes liquid. However, it has been discovered that special pairings of cations and anions can lead to salts that are liquid around room temperature. These salts, called ionic liquids, constitute very promising media as they display many interesting properties.[12,13]

- *They have no effective vapor pressure.* Since ionic liquids are salts, they cannot evaporate. For this reason, they cannot contaminate the atmosphere and are a medium of choice for green chemistry. Reaction products can be separated by simple distillation.
- *They are outstandingly good solvents.* Ionic liquids can dissolve a wide variety of organic, organometallic, inorganic and polymeric compounds. The high solubility of reagents and catalysts in ionic liquids allows reactions to be performed in concentrated solutions, and hence small reactors. In addition, gases are usually more soluble in ionic liquids than in classical organic solvents.
- *They are stable at a wide range of temperatures.* Being liquid over a range of 300°C is usual for ionic liquids, and melting points can be as low as −96°C. Compared with traditional organic solvents and water, much higher kinetic control can be attained in such media.
- *They are immiscible with many organic solvents.* This immiscibility extends their use to biphasic systems. Furthermore, this property is extremely valuable for catalytic reactions because the product can be extracted from an ionic liquid using organic solvents, whereas the catalyst remains in the ionic liquid and can be directly recycled and reused.
- *Their synthesis can be carried out easily and reasonably inexpensively.* Ionic liquids remain expensive, even though they are now widely available commercially. Furthermore, a huge number of derivatives can be synthesized, making ionic liquids the most tunable type of solvent.

For all these reasons, ionic liquids are attracting a growing interest in both academic and industrial research. They are obtained by informed combination of a cation and an anion with mismatched shapes. Some of the most common are represented in Figure 1.1.

The advantages and disadvantages of these solvents are shown in Table 1.1. A number of interesting and useful properties of ionic liquids can be exploited, and a summary of the most important physical and chemical properties of ionic liquids, along with their applications in synthetic chemistry, is shown in Table 1.2. Ionic liquids are sometimes described in the literature using other terms, including

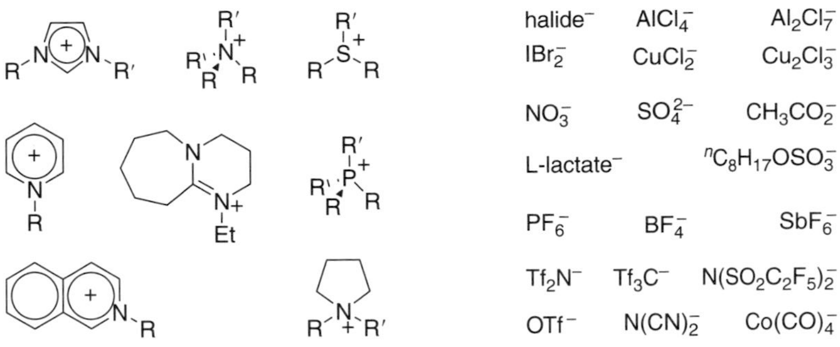

halide$^-$ $AlCl_4^-$ $Al_2Cl_7^-$
IBr_2^- $CuCl_2^-$ $Cu_2Cl_3^-$
NO_3^- SO_4^{2-} $CH_3CO_2^-$
L-lactate$^-$ $^nC_8H_{17}OSO_3^-$
PF_6^- BF_4^- SbF_6^-
Tf_2N^- Tf_3C^- $N(SO_2C_2F_5)_2^-$
OTf^- $N(CN)_2^-$ $Co(CO)_4^-$

Figure 1.1

Table 1.1

Advantages	Disadvantages
No vapor pressure	Purification difficult, quality variable
Non-flammable	High viscosity
Highly tunable	Commercially expensive
Easy separation of products and recycling of catalysts	Toxicity not fully understood
Potential for high solubility of gases	Reacts with strong nucleophiles

Table 1.2

Property	Applications
Saline structure	New reactions and selectivities possible
High solubility of organic reagents and catalysts	Reactions performed in concentrated conditions
High solubility of gases	Reactions involving gases can be accelerated
Immiscibility with several organic solvents	Catalysts can be recycled and reused; efficient biphasic reactions
High tunability	Solvents can be designed for specific applications

room-temperature molten salts, *ambient-temperature molten salts*, *ionic fluids*, *liquid organic salts* and *organic ionic liquids (OILs)*.

1.3 Fluorous media

Perfluorocarbons (PFCs) came to public notice in the 1960s with the publication of an account of a mouse breathing in this medium. As well as the fact that they dissolve gases (such as oxygen), PFCs have the important property of being easily separable from their hydrocarbon analogues. Thus, PFCs are considered the solvents of choice for gas/liquid reactions, liquid/liquid biphasic separations and the purification of products and catalysts.[14] The first use of PFCs as reaction media was reported by Zhu in 1993.[15] One year later, Horvárth and Rábai coined the term "fluorous" by analogy with "queous phases".[16,17] They also used a fluorous biphasic system (FBS) to carry out easy phase separation of an organic product and recycling of a fluorinated catalyst.

The use of fluorous solvents has proliferated as a result of their attractive features.

- *Gases have good solubilities in fluorous solvents.* In contrast, many organic materials have much lower solubility in fluorous media than in standard organic solvents. The result is that organic materials are easily separable from fluorous solvents.
- *PFCs are immiscible with many organic solvents.* This extends their use to biphasic systems. Furthermore, this property is extremely valuable for catalytic reactions, as the product can be extracted from the fluorous media using organic solvents, whereas the fluorous catalyst remains in the fluorous media and can be directly recycled.
- *They are chemically stable.* Compared with traditional organic solvents, fluorous media are inert in common reaction conditions.
- *They have low viscosity.* For this reason, they are media of choice for smooth passive transport of reagents.
- *They are non-toxic, have zero ozone depletion potential and have significantly low green house potential.* For these reasons, they are friendly to life and the environment and hence are useful in *green chemistry.*

The advantages and disadvantages of fluorous solvents are represented in Table 1.3. Fluorous media have a number of potentially interesting and useful properties. They are non-protic, display neither strong Lewis acidity nor basicity, are inert to radical and oxidizing conditions, and do not react with nucleophiles or electrophiles. This general lack of reactivity is one of the keys to the successful utilization of fluorous media as common replacements for more conventional solvents. A summary of the most important physical and chemical properties of fluorous media, along with their applications, is shown in Table 1.4.

Table 1.3

Advantages	Disadvantages
Low toxicity solvent	Expensive, can require functionalized ligands
Easy to recycle and reuse	Low solubility of many organic materials
Low reactivity in general	Low-boiling-point compounds are volatile
High density to create multiphases	Disposal can be problematic

Table 1.4

Property	Applications
Immiscible with organic solvents	Easy separation of organic materials from fluorous solvents by biphasic separation
High affinity of fluorous materials for fluorinated solvents	Easy separation and recycling of fluorinated materials (e.g. catalysts)
Inert in common reaction conditions	Useful medium for a variety of reaction processes
Denser than most organic solvents	Working as a liquid screen
Low viscosity	Smooth passive transport of reagents
Good solubility of gases	Advantageous in reactions involving gases

1.4 Supercritical carbon dioxide

Supercritical carbon dioxide ($scCO_2$) or dense CO_2 has the potential to be the ideal green reaction medium.[18–20] The physical and chemical properties of CO_2 are well understood, and it is non-toxic to the environment and to humans. At high concentrations, it is an asphyxiant, and a large accidental release could be problematic; however, precautions can be taken appropriately to minimize such risks. At atmospheric pressure, carbon dioxide is gaseous, which means that simple depressurization will leave no hazardous solvent effluent that requires complex or expensive waste treatment. CO_2 is a greenhouse gas; however, it can be obtained in large quantities as a byproduct of fermentation, combustion or ammonia synthesis. Its use should not lead to any net increase in CO_2 emissions – it is simply a case of exploiting the CO_2 before discharging it into the atmosphere. It is relatively cheap, particularly when compared with conventional solvents, and it is readily available on an industrial scale in a very pure form. CO_2 has a number of practical advantages.

- *It is a cheap, non-toxic and volatile solvent.* These properties make $scCO_2$ a very attractive solvent for industrial processes. Product isolation to total dryness is achieved by simple evaporation. This could prove to be particularly useful in the final steps of pharmaceutical syntheses, when even trace amounts of solvent residues are considered problematic.
- *It has a generally low reactivity.* CO_2 is non-protic, not strongly Lewis acidic or basic, and inert to radical and oxidizing conditions.[21] This general lack of reactivity is essential to the $scCO_2$ being a successful, common replacement for more conventional solvents. However, $scCO_2$ can react with nucleophiles (e.g. carbamic acid formation from amines), although this can be reversed and exploited synthetically.[22]
- *It offers an alternative to the processing of solid materials.* There are two complementary routes to particle formation with supercritical fluids in general, and $scCO_2$ in particular: rapid expansion of supercritical solutions (RESS)[23] and supercritical anti-solvent precipitation (SASP).[24] Together, these techniques allow the processing of a wide range of solid materials with high

Table 1.5

Advantages	Disadvantages
Non-toxic solvent	Specialized equipment is required (which may be considered expensive, particularly in early stages)
No solvent waste	High pressure may be dangerous if performed by unqualified personnel with unsuitable equipment
Inexpensive and readily available	Energy is required to compress CO_2 (this can be minimized on a large scale by partial decompression and recycling)
Non-flammable	Low solubility of polar substrates (but can be enhanced using co-solvents and other techniques)
Low reactivity in general	Reacts with strong nucleophiles
Reversible reactivity with weak nucleophiles (e.g. amines)	
Potential for product processing applications supercritical anti solvents, SAS; rapid expansion of supercritical solution, RESS; super fluid chromotography, SFC	

Table 1.6

Property	Applications
Variable density and solvent power	Optimization of solvent to favor a particular process/reaction and enhance selectivity; easy separation of products
High volatility	Easily removed by evaporation, easy product isolation and enhanced product purity (no solvent residues)
High solubility of light gases	Good for reactions involving H_2, O_2, CO and other gases
Inert to oxidation	Suitable for reactions involving high concentrations of O_2, which would otherwise be dangerous
Inert to radical species	Very useful medium for radical reaction processes
Low viscosity and high diffusion rates	Enhanced reaction rates
Reagent clustering	Enhanced reaction rates and substrate solubility
Reversible reactivity with weak nucleophiles (e.g. amines)	*In situ* protection of reactive functionality

control over morphology. Although these techniques are beyond the scope of this introductory chapter, they are particularly relevant in pharmaceutical formulation and materials applications, and further information can be found in recent reviews.[23–25]

A summary of the main advantages and disadvantages of the use of $scCO_2$ in synthetic chemistry is shown in Table 1.5. Important physical and chemical properties of $scCO_2$, as well as their applications in synthetic chemistry, are shown in Table 1.6.

References

[1] Nelson, W. M. *Green Solvents for Chemistry*; Oxford University Press: New York, 2003.
[2] Seddon, K. R. *J. Chem. Technol. Biotechnol.* **1997**, *68*, 351.
[3] Tanaka, K., Ed. *Solvent-Free Organic Synthesis*; Wiley-VCH: Weinheim, 2004.
[4] Eissen, M.; Metzger, J. O.; Schmidt, E.; Schneidewind, U. *Angew. Chem. Int. Ed.* **2002**, *41*, 414.
[5] Collins, T. J. In *Macmillan Encyclopedia of Chemistry*; Lagowski J. J., Ed.; Macmillan Reference USA: New York, 1997.
[6] Abraham, M. A.; Moens, L., Eds. *Clean Solvents – Alternative Media for Chemical Reactions and Processing*; ACS Symposium Series 819, American Chemical Society: Washington D.C., 2002.
[7] Blaser, H.-U.; Studer, M. *Green Chem.* **2003**, *5*, 112.
[8] Li, C.-J. *Chem. Rev.* **1993**, *93*, 2023.
[9] Grieco, P. A., Ed. *Organic Synthesis in Water*; Blacky Academic and Professional: London, 1998.
[10] Cornils, B.; Herrmann, C. B., Eds. *Aqueous-Phase Organometallic Catalysis*, 2nd Edn.; Wiley-VCH: Weinheim, 2004.
[11] Scamehorn, J. F.; Harwell, J. H. *Surfactant-Based Separation Processes*; Marcel Dekker: New York, 1989.
[12] Rogers, R.; Seddon, K., Eds. *Ionic Liquids: Industrial Applications for Green Chemistry*; ACS Symposium Series 818, American Chemical Society: Washington D.C., 2002.
[13] Wasserscheid, P.; Welton, T., Eds. *Ionic Liquids in Synthesis*; Wiley-VCH: Weinheim, 2003.
[14] Gladysz, J.; Curran, D. P.; Horváth, I. T., Eds. *Handbook of Fluorous Chemistry*; Wiley-VCH: Weinheim, 2004.
[15] Zhu, D.-W. *Synthesis* **1993**, 953.
[16] Horváth, I. T.; Rábai, J. *Science* **1994**, *266*, 72.
[17] Gladysz, J. *Science* **1994**, *266*, 55.
[18] Clifford, T. *Fundamentals of Supercritical Fluids*; Oxford University Press: New York, 1999.
[19] Jessop, P. G.; Leitner, W., Eds. *Chemical Synthesis Using Supercritical Fluids*; Wiley-VCH: Weinheim, 1999.
[20] Desimone, J. M.; Tumas, W. *Green Chemistry Using Liquid and Supercritical Carbon Dioxide*; Oxford University Press: New York, 2003.
[21] Tanko, J. M.; Blackert, J. F. *Science* **1994**, *263*, 203.
[22] Furstner, A.; Koch, D.; Langemann, K.; Leitner, W.; Six, C. *Angew. Chem., Int. Ed. Engl.* **1997**, *36*, 2466.
[23] Tom, J. W.; Debenedetti, P. G. *J. Aerosol. Sci.* **1991**, *27*, 555.
[24] Debenedetti, P. G. In *Supercritical Fluids; Fundamentals for Application*; Kluwer Academic Publishers: Dordecht, 1994, p. 719.
[25] Kendall, J. L.; Canelas, D. A.; Young, J. L.; DeSimone, J. M. *Chem. Rev.* **1999**, *99*, 543.

2 Ionic liquids

Jonathan J. Jodry and Koichi Mikami

2.1 Historical background and synthesis

2.1.1 Historical background

In the past decade, ionic liquids have developed from a curiosity to a new class of solvent with attractive properties. However, the history of these salts goes back to the early twentieth century. In 1914, Walden reported that the $[EtNH_3][NO_3]$ salt **1** (Figure 2.1), with a melting point of 8°C, was liquid at room temperature.[1,2] This interesting property did not attract a lot of interest until it was observed that mixtures of $AlCl_3$ and *N*-alkylpyridinium halide salt could be liquid at room temperature.

The first research into chloroaluminate ionic liquids was oriented toward their use in electrochemistry,[3] and in 1951 Hurley and Wier used the ethylpyridinium bromide/$AlCl_3$ ionic liquid **2** to electroplate aluminium.[4] The use of chloroaluminate ionic liquids as electrolytes attracted interest from both fundamental and applied research. Osteryoung (Colorado States Academy) and Wilkes (U.S. Air Force Academy) prepared and studied the all-chloride system butylpyridinium chloride/$AlCl_3$ **3** (Figure 2.1), which was not stable toward reduction, limiting its use as an electrolyte.[5,6] Efforts were made to develop alternative low-melting chloroaluminate ionic liquids that would be less subject to reduction, and this led to the discovery by Wilkes and Hussey in 1982 that mixtures of dialkylimidazolium chloride salts and $AlCl_3$ formed ionic liquids.[7] One of the more interesting derivatives, [1-ethyl-3-methylimidazolium][chloride] ([emim][Cl]) has a melting point below room temperature when it contains between one and two equivalents of $AlCl_3$ (corresponding to a mole fraction x between 0.5 and 0.67, structure **4**).

There followed research into pyridinium and imidazolium chloroaluminate ionic liquids by Hussey, Seddon and Welton (investigation of transition metal complexes

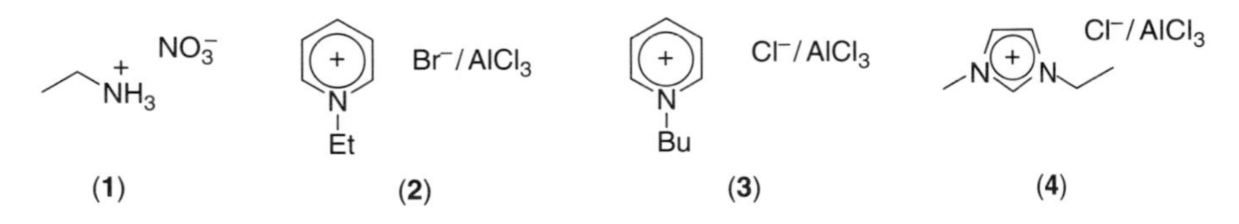

Figure 2.1

BF_4^- PF_6^-

(5) (6)

Figure 2.2

in non-aqueous but polar solvents),[8–13] Wilkes (ionic liquids as catalysts for Friedel–Crafts reactions)[14] and Chauvin (homogeneous transition metal catalysis in ionic liquids).[15] All chloroaluminate ionic liquids (as well as trichlorostannate ones)[16] have to be rigorously protected from moisture and handled in a glove box because contact with water produces corrosive HCl.

The real breakthrough in the use of ionic liquids as solvents came in 1992, when Wilkes and Zaworotko studied the formation of dialkylimidazolium salts associated with water-stable anions. This led to the discovery of the ionic liquids [emim][BF_4] **5** (Figure 2.2) and [emim][NO_3], with melting points of 15°C and 38°C, respectively.[17] These ionic liquids are hygroscopic, but they are not degraded in the presence of moisture. Shortly afterward, the synthesis of the corresponding hexafluorophosphate derivative **6** was described.[18] Today, most ionic liquids are based on imidazolium derivatives associated with non-covalent anions.

All salts have the potential to be used as ionic liquids, assuming that they are heated above their melting points. Early electrochemical applications of ionic liquids were often performed at high temperatures, and in the corresponding publications these salts are referred to as *molten salts*. This term now evokes an image of high melting-point salts, which are usually corrosive, viscous, and definitively not suitable for organic synthesis. Where is the boundary between *molten salts* and *ionic liquids*? A practical definition is to say that ionic liquids are salts with a melting point below 100°C. However, it must be emphasized that, with the development of asymmetric catalysis, ionic liquids are now expected to be liquid not only at room temperature, but also at much lower temperatures.

Ionic liquids are attracting growing academic and industrial research interest, which can be quantified in terms of the increase in the number of publications and patents concerning them.[19,20]

2.1.2 *Synthesis*

2.1.2.1 *Preparation of imidazolium halides*

Formation of 1 alkyl 3-methylimidazolium ionic salts is usually performed by condensing 1-methylimidazole, a commercially available and inexpensive substrate, and an alkyl halide. The first synthesis of [emim][Cl] **7** by Wilkes was carried out by heating 1-methylimidazole and an excess of chloroethane for 2 days, in the absence of solvent (Scheme 1).[7] This procedure was improved by using an equimolar amount of the chloride reactant.[21]

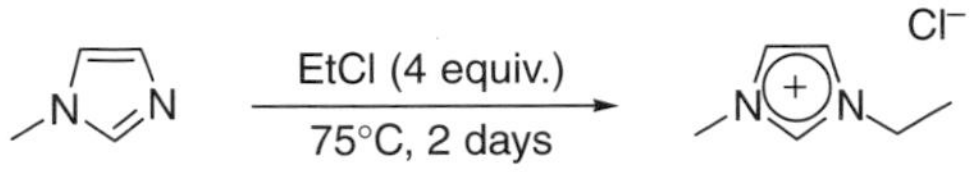

EtCl (4 equiv.)
75°C, 2 days
Cl^-
(7)

Scheme 1

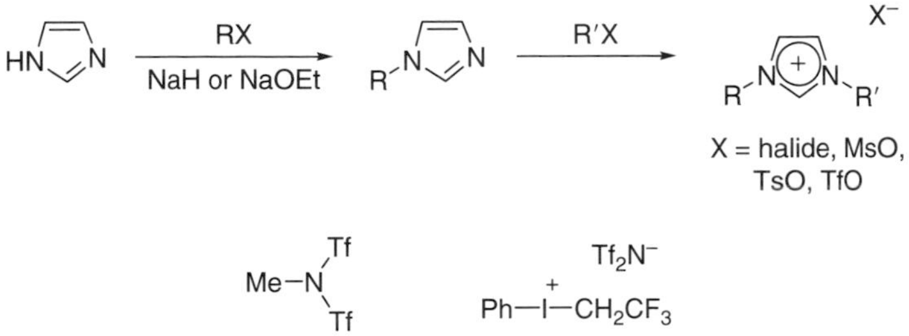

RX
NaH or NaOEt
R′X
X^-
X = halide, MsO, TsO, TfO

Me–N(Tf)(Tf)
(8)

Tf_2N^-
$Ph{-}\overset{+}{I}{-}CH_2CF_3$
(9)

Scheme 2

Efforts have been made to accelerate the reaction. Ionic liquids can be obtained in <1 min using microwave-assisted preparation,[22,23] and ultrasound-assisted synthesis leads to the rapid isolation of ionic liquids with high chemical purity.[24] If a microwave reactor with reaction condition control mechanisms is used, ionic liquids can be conveniently prepared even on a large scale (up to 2 moles), both in high yields and with short reaction times.[25]

It is also known that condensation can be accelerated by using bromide or iodide derivatives instead of chloride ones.[26] Other procedures involving the use of solvents (tetrahydrofuran (THF), chloroform, 1,1,1-trichloroethane) have been proposed, but since one of the objectives of ionic liquids is to lead to environmentally clean processes, their preparation should, as far as possible, be performed under neat conditions (absence of solvent).

Rather than starting from 1-methylimidazole, imidazole can be converted to 1-alkylimidazole following deprotonation with a strong base. Quaternization leads to the formation of imidazolium cations that can bear a variety of alkyl substituents. Instead of halide derivatives, alkyl mesylate, tosylate or triflate can also be used, resulting in the direct formation of ionic liquids with these corresponding anions (Scheme 2).[27,28] A one-step preparation of hydrophobic bis(trifluoromethanesulfonyl)amide salts has also been described using **8** and **9** as reagents for direct methylation and trifluoroethylation, respectively.[29]

Simpler procedures can be used for the preparation of symmetrical 1,3-dialkyl-imidazolium salts. Heating an excess of chloride derivative with *N*-trimethylsilylimidazole **10** in toluene at 110°C produces the desired salts in

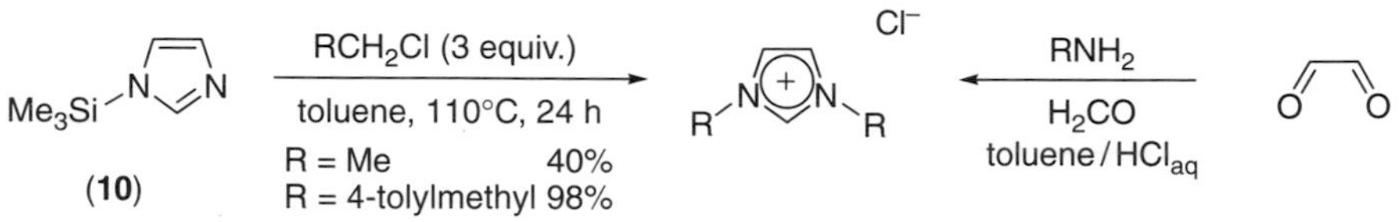

Scheme 3

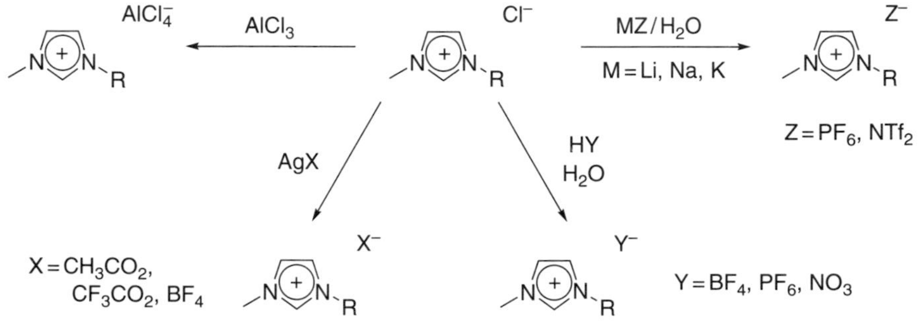

Scheme 4

moderate to excellent yields.[30] Another approach is the direct condensation of two equivalents of a primary amine with aqueous glyoxal and paraformaldehyde in an acidic medium, yielding the same salts in an one-pot procedure (Scheme 3).[31]

2.1.2.2 Anion metathesis

It has already been noted that imidazolium salts can be directly isolated with different counter-anions (Scheme 2). However, a larger selection of ionic liquids can be prepared by anion metathesis on the halide salt, leading to the replacement of the anion by a non-coordinating one such as PF_6, SbF_6 or BF_4. Different procedures for the metathesis of halide salts have been developed (Scheme 4).

- Addition of $AlCl_3$ leads to the formation of chloroaluminate derivatives with a mixture of anions ($[AlCl_4]^-$, $[Al_2Cl_7]^-$, $[Al_3Cl_{10}]^-$), depending on the ratio between the Lewis acid and the starting chloride salt. The main limitation for the use of these salts is their sensitivity to moisture.
- Water-soluble ionic liquids, typically acetate, trifluoroacetate and tetrafluoroborate salts, can be obtained by the addition of the corresponding silver salt and elimination of the precipitate. This method leads to highly pure ionic liquids containing a very low amount of the chloride anion. However, starting silver salts are expensive, and an equivalent of waste material (AgCl) is generated.
- Water-insoluble ionic liquids can be prepared by addition of a strong acid to an aqueous solution of the chloride salt. The newly formed hydrophobic phase can be directly isolated by decantation or extracted using CH_2Cl_2, and

then carefully washed with water to remove traces of the acid and halide. This procedure is often used for the synthesis of hexafluorophosphate derivatives. In the case of water-soluble ionic liquids, such as tetrafluoroborate salts, acidity is removed by the repeated addition of water followed by elimination, under reduced pressure, of the acid/water mixture. Complete removal of the halide is very difficult.

- The most common approach for water-insoluble ionic liquids is the addition of a lithium, sodium or potassium salt of the desired anion to an aqueous solution of the halide salt. The newly formed ionic liquid separates from the aqueous layer as a hydrophobic phase, allowing its clean isolation. For this procedure, an extremely thorough aqueous wash is required.[32] Other hydrophilic derivatives such as mesylate of triflate can be used as a starting material instead of halide salts.

2.1.2.3 Functionalized imidazolium ionic liquids

The interest in functionalized ionic liquids is growing because ionic liquids bearing ether, amino or alcohol functionalities have been shown to display special properties, including the ability to dissolve a larger amount of metal halide salts and to extract heavy metal ions from aqueous solutions. Imidazolium-based ionic liquids with ether and hydroxyl (see Section 2.2.1),[33] thiourea, thioether and urea (see Section 2.2.8)[34] have been prepared following the standard quaternization procedure. A straightforward approach has been described for the preparation of imidazolium (as well as pyridinium) cations with ester, ketone or cyanide functionalities: 1-methylimidazole reacts with methanesulfonic acid to provide the imidazolium salt **11**, which undergoes a Michael-type reaction with methyl vinyl ketone as α,β-unsaturated compound to produce the ionic liquid **12** (Scheme 5).[35]

2.1.2.4 Other types of ionic liquid

Imidazolium salts are the most important class of ionic liquids, but preparation of ionic liquids based on different cations has also been reported.[36] Pyridinium salts are prepared in a similar way to imidazolium, by direct alkylation of pyridine.[4,6,37] Alkylisoquinolium bis(perfluoroethylsulfonyl)imide salt **13** is similarly prepared by N-alkylation of isoquinoline followed by anion metathesis,[38] and iminium salt **14** is obtained directly associated with the triflate anion (Scheme 6).[39]

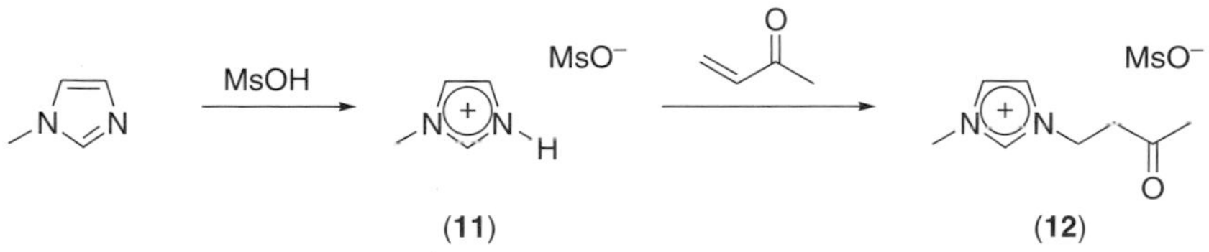

MsOH
MsO⁻
(11)
MsO⁻
(12)

Scheme 5

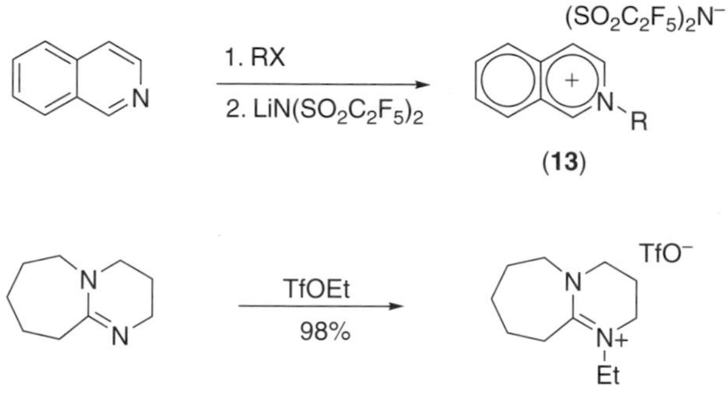

1. RX
2. $LiN(SO_2C_2F_5)_2$
$(SO_2C_2F_5)_2N^-$
(**13**)

TfOEt
98%
TfO^-
(**14**)

Scheme 6

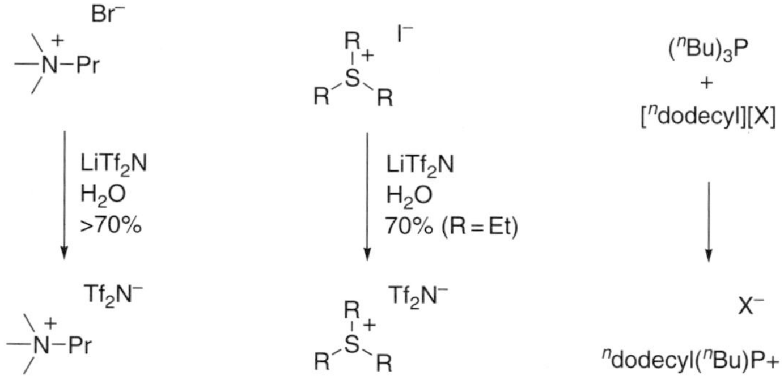

$LiTf_2N$
H_2O
>70%
(**15**)

$LiTf_2N$
H_2O
70% (R = Et)
(**16**)

$(^nBu)_3P$ + [^{n}dodecyl][X]
^{n}dodecyl(^{n}Bu)P+ X^-
(**17**)

Scheme 7

Tetraalkylammonium bis(trifluoromethanesulfonyl)amide **15** is prepared from a commercially available bromide derivative by simple anion metathesis.[40,41] A similar procedure is used for sulfonium bis(trifluoromethanesulfonyl)amide **16**.[42]

Phosphonium halides **17** are synthesized by alkylation of triphenylphosphine (Scheme 7).[43] The use of phosphonium ionic liquids has not attracted significant interest from the academic research community, although these liquids are produced by the ton industrially, including, for example, $[(n\text{-hexyl})_3P(C_{14}H_{29})][Cl]$ (CYPHOS IL 101, Cytec Canada Inc.). Indeed, a large number of industrial patents are related to both the preparation and the use as reaction solvents of phosphonium ionic liquids.[44]

In parallel, new types of anion have also been described, usually associated with the dialkylimidazolium cation. For example, dicyanamide **18** salts are water-soluble ionic liquids with low viscosity.[45,46] Other types of anion-containing metals have been reported, such as hexafluoroniobate and tantalate **19**[47] and $Co(CO)_4$ **20**, but these have found only limited applications.[48,49] Ionic liquids formed with

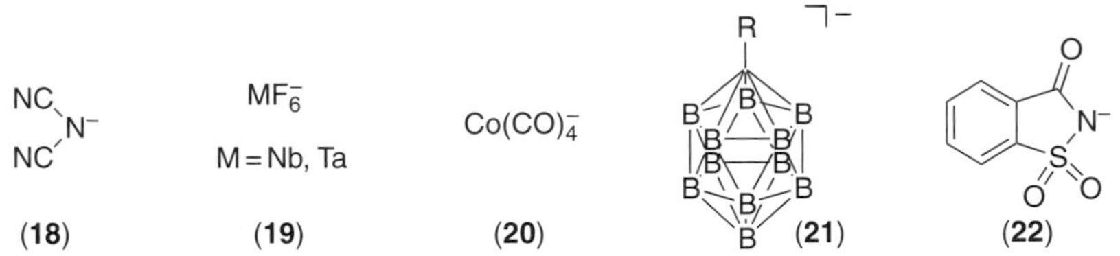

$N(CN)_2^-$ **(18)** MF_6^- M = Nb, Ta **(19)** $Co(CO)_4^-$ **(20)** **(21)** **(22)**

Figure 2.3

icosahedral carborane **21**, one of the most inert and least nucleophilic anions, are also known.[50] In 2004, the sweetener anion saccharinate **22** was associated with several imidazolium and pyrrolidinium cations to form viscous room-temperature ionic liquids (Figure 2.3).[51]

In fact, a great number of different ionic liquids can be synthesized by combining the available cations and anions: Seddon suggests that, if binary and ternary mixtures are considered, up to 10^{18} room-temperature ionic liquids can be prepared.[52]

2.1.2.5 *Purification*

Ionic liquids used as solvents have to be isolated with high chemical purity. Their non-volatility is a disadvantage in terms of their preparation because, unlike classical solvents, they cannot be purified by distillation. The starting materials are therefore purified. Typically, for the preparation of imidazolium halides, 1-methylimidazolium must be distilled over NaOH and the haloalkane should be washed with concentrated sulfuric acid (to remove coloration), neutralized with an $NaHCO_3$ solution, washed with water, dried and distilled before use.

Care must also be taken during the synthesis of the salts, which should always be colorless, free-flowing liquids. However, commercially available ionic liquids are often sold as slightly yellow liquids, and most researchers produce highly colored salts when they synthesize their first ionic liquids. The reason for the coloration is unknown, but it is associated with impurities present in the starting materials, oxidation products and overheating during the condensation reaction. It is usually considered that condensation with alkyl halide should be performed at a maximum temperature of 80°C (although lower temperatures may sometimes be necessary). When the halide salt formed is solid, multiple recrystallizations allow for the removal of color.[53] Coloration itself is not a problem (except in photochemical applications): the quantity of colored impurities is too small to be detected by nuclear magnetic resonance (NMR) imaging or other analytical methods and can only be monitored using UV/Vis spectroscopy.

After anion metathesis, halides have to be completely removed, not only because of their corrosive nature, but also because they can deactivate catalysts dissolved in the ionic liquid. Any trace of impurity (including Li^+, K^+, Na^+ and Ag^+) can have a considerable influence on the physical properties of the salts (melting point, viscosity, density). Furthermore, if PF_6 salts are not completely acid free, toxic HF can be formed in the ionic liquid (see Section 2.2.8). Careful washing with water is only a basic purification of a hydrophobic ionic liquid. For biocatalytic applications, where extremely pure media are required, a more complex purification procedure is needed. A dichloromethane solution of the ionic liquid is filtered through silica gel, washed with a saturated aqueous solution of sodium carbonate and dried over anhydrous magnesium sulfate. This ensures that no trace of chloride salt, acid or silver cation remains.[54] Stirring the ionic liquid for 24 h with activated charcoal and filtrating it through a small plug of acidic alumina will remove remaining traces of colored impurities.[53]

Finally, it is unfortunate to have to observe that the purity of ionic liquids is sometimes a neglected issue in the literature, which can lead to apparently inconsistent results.

2.2 Physical properties

The wise association of a selected cation and anion allows a chemist to form an ionic liquid with valuable physical and chemical properties. This ability to tune the medium to fit specific needs has won ionic liquids the name of *designer solvents.* Understanding the relationship between the structure and the properties of ionic liquids is therefore of central importance. Low melting points and high thermal stability characterize these salts as reaction media. Although all ionic liquids have similar polarity, their solubility can easily be changed by modification of the anion. Further properties that are less important for traditional organic solvents have to be considered: viscosity, which can be high because of the saline structure, and acidity, which is also anion dependent. As a new medium favorable for green chemistry, the environmental friendliness of ionic liquids should be studied carefully.

2.2.1 Melting point

Probably the most amazing feature of ionic liquids is that they are liquid. Intuitively, we expect salts to be solid at room temperature and to melt only at very high temperatures. Take sodium chloride (melting point 803°C): since its ions have similar size and shape, a solid, crystalline packing structure is obtained. In contrast, the ions forming ionic liquids do not pack well, which explains why they can remain liquid at low temperature.[50]

Three main factors have to be considered in the choice of the cation.

- *A good distribution of charges.*[55] Replacing an alkali metal cation by a suitable organic one lowers the melting point significantly. Since the best charge distribution is obtained for aromatic compounds, imidazolium and pyridinium cations are the most interesting derivatives, and their chloride salts melt at under 150°C (Table 2.1).

Table 2.1

	Salt	Melting point (°C)
	NaCl	803
Cl^- pyridinium N–R	R = Me ([mpy][Cl])	137
	R = Bu ([bpy][Cl])	132
Cl^- imidazolium N–R	R = Me ([mmim][Cl])	125
	R = Et ([emim][Cl])	87
	R = Bu ([bmim][Cl])	60

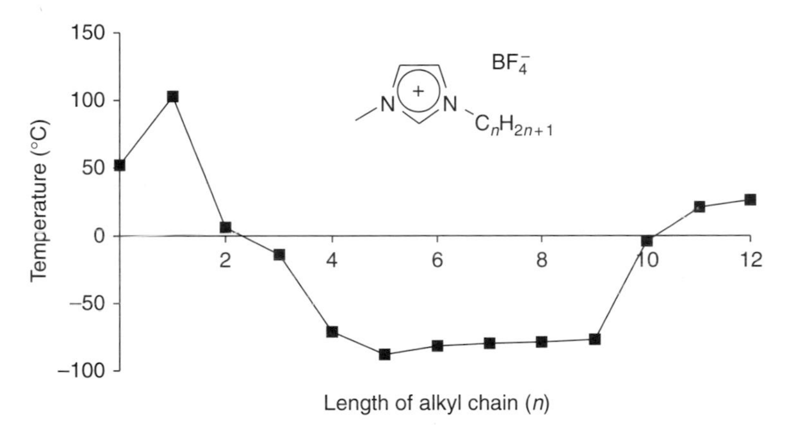

Figure 2.4

- *An unsymmetric structure*. To limit the packing between the ions, non-symmetric cations should be used. This effect is clearly observed in [1-alkyl-3-methylimidazolium][chloride] salts when the second lateral chain is extended from a methyl group ([mmim][Cl], melting point 125°C) to a butyl group ([bmim][Cl], 60°C). Similarly, studies of [1-alkyl-3-methylimidazolium][BF_4] reveal that the melting point is lowered when the alkyl chain is elongated from methyl to butyl. For $n = 4$ to $n = 9$, it remains constant at about −80°C, before again reaching room temperature for longer derivatives (Figure 2.4).[56] When especially long alkyl chains are used ($n > 12$), the salts can exhibit liquid crystalline behavior.[37,48] On the other hand, increasing the chain on *N*-alkylpyridinium chloride salts does not significantly change the melting point, since the structure remains symmetric.
- *Weak intermolecular interactions*. The presence of intermolecular interactions such as hydrogen bonding should be avoided, since the formation of more compact structures is accompanied by an increase in the melting point.

Table 2.2

Salt	Melting point (°C)
[emim][$CB_{11}H_{12}$]	122
[emim][I]	80
[emim][PF_6]	62
[emim][NO_3]	38
[emim][BF_4]	15
[emim][$AlCl_4$]	7
[emim][OTf]	−9
[emim][NTf_2]	−12
[emim][CF_3CO_2]	−14
[emim][$N(CN)_2$]	−21

Table 2.3

Me_3N^+–Pr Tf_2N^-	R_4N^+ Tf_2N^- (R = ^{n}hexyl)	Me_3S^+ Tf_2N^-	Et_3S^+ Tf_2N^-
23	**24**	**25**	**26**
m.p. = 17°C	m.p. = −7°C	m.p. = 45°C	m.p. = −36°C

m.p. = melting point.

The anion also has a strong impact on the melting point of the salt. A large and symmetric anion is more likely to be well separated from the cation. This effect is obvious from comparing different emim salts (Table 2.2). Halide derivatives have the highest melting points, larger anions such as tetrafluoroborate and hexafluorophosphate significantly lower it and the use of fluorinated organic anions leads to salts that can be liquid at under 0°C.

Among these anions, two are often used to form ionic liquids, namely, hexafluorophosphate PF_6 and bis(trifluoromethanesulfonyl)amide NTf_2. In Table 2.3, the melting points of tetraalkylammonium and trialkylsulfonium salts formed with the NTf_2 anion are indicated, clearly showing that a large variety of ionic liquids with low melting points can be prepared.

It is somewhat surprising to find that certain ionic liquids have been reported as being used as solvents at temperatures below their melting point, or that melting points are reported with important differences; for example, [emim][Br] (66 to 79°C), [emim][NTf_2] (−12 to 4°C). In fact, the melting point is not the only feature characterizing the liquidity of ionic liquids: calorimetric studies indicate that imidazolium derivatives could display important supercooling properties, since they have freezing points much below their melting ones. A typical example is [emim][PF_6]. Although it has a melting point of

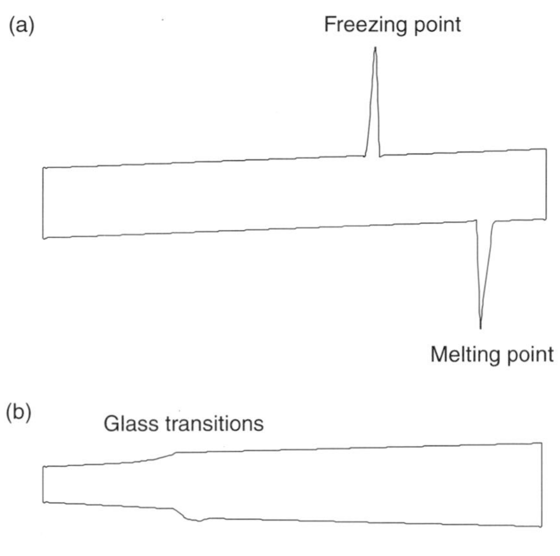

Figure 2.5

62°C, it can be liquid at room temperature as it freezes only at 5°C. Similar behavior was observed for all imidazolium salts, with differences between the melting and the freezing points as high as 200°C.[57] Once they have supercooled, ionic liquids can be kept in this state for a long period, sometimes up to weeks, but the addition of small crystals as seeds will lead to rapid crystallization.[58]

Upon supercooling, ionic liquids often become much more viscous before finally forming glasses.[56] The presence of water in ionic liquids changes the melting point, as well as the ability to supercool.[59] For all these reasons, the value of the melting point for a given ionic liquid must be considered carefully, and it should be kept in mind that it might be possible to make it liquid at lower temperatures – or else to observe that it is solid above its published melting point. A schematic representation of a differential scanning calorimetry (DSC) is given for an ionic liquid displaying supercooling ability (Figure 2.5a) and for one showing only glass transitions (Figure 2.5b).

Although it is generally stated that lower melting points are obtained by the association of asymmetric and aromatic cations with symmetric and large anions, this rule is purely empirical: it is still difficult to predict whether a given salt will be liquid at room temperature. For example, 1,3-dialkylimidazolium hexafluorophosphate salts bearing dibutyl, dipentyl, dioctyl, dinonyl or didecyl alkyl groups are expected to have rather high melting points owing to their high symmetry. Surprisingly, they are all liquid at room temperature; with the exception of the 1,3-didecylimidazolium salt (melting point −27°C), all the other salts have no observed melting points, but instead freeze at low temperatures (between −69°C and −80°C) with glass transition (Table 2.4).[28]

Table 2.4

	Salt	Melting point (°C)	Glass transition (°C)
	$R = C_4H_9$	–	−69
	$R = C_5H_{11}$	–	−72
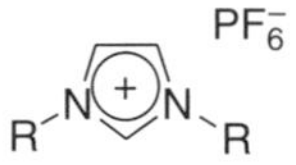 PF_6^- R-N(+)N-R			
	$R = C_8H_{17}$	–	−80
	$R = C_9H_{19}$	–	−70
	$R = C_{10}H_{21}$	−27	–

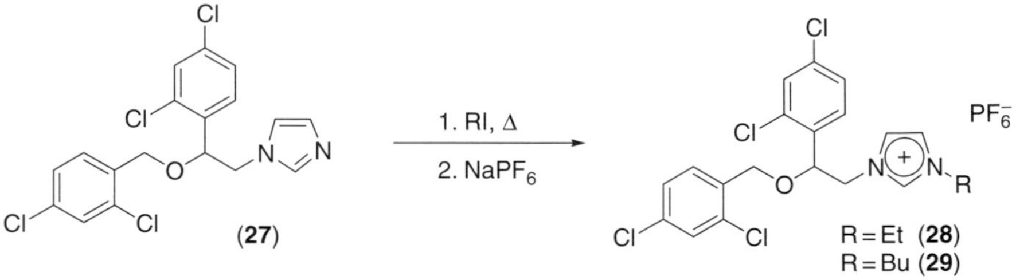

1. RI, Δ
2. $NaPF_6$

(27) → R = Et (28), R = Bu (29)

Scheme 8

The unusual supercooling property of ionic liquids can be exploited to produce functionalized room-temperature ionic liquids that have higher melting points as a result of their complex structure. Quaternarization of the imidazole ring of the antifungal drug miconazole **27** with several alkyl iodides, followed by anion metathesis, led to the formation of the first ionic liquids derived from a bioactive molecule. Because of the presence of two benzyl aromatic rings and perhaps because of the ether functionality, these ionic liquids have high melting points (86°C for **29**). However, both **28** and **29** have supercooled phases and can be kept liquid at room temperature for weeks (Scheme 8).[60]

Similarly, ionic liquids bearing a hydroxyl functionality – which leads to unfavorable intermolecular hydrogen bondings – can be used as a liquid at low temperature. Salts **30–33** are easily synthesized from 1-methylimidazole and 2-chloroethanol; they are solid at room temperature. However, because of a weak tendency to crystallize, they enter into a metastable supercooled liquid state when cooled and can therefore be used as solvents at low temperature. The glass transition temperatures are very low, between −72°C (**30**) and −111°C (**33**, Table 2.5). The influence of the anion on the glass transition temperatures, as opposed to the melting points, is not clear, as the chlorine salt **33** is a metastable liquid at lower temperature than the hexafluorophosphate and tetrafluoroborate salts **30** and **31**.[33]

Table 2.5

Synthesis of the salt	Anion X^-	Salt	Glass transition (°C)
1-methylimidazole → (1. $Cl\!\sim\!\sim OH$; 2. anion exchange) → 1-(2-hydroxyethyl)-3-methylimidazolium X^-	$[PF_6]$	**30**	−72
	$[BF_4]$	**31**	−84
	$[CF_3CO_2]$	**32**	−89
	[Cl]	**33**	−111

2.2.2 *Thermal stability*

Given that ionic liquids cannot be evaporated, the only limitation on their use at high temperatures is their thermal stability. Quaternary ammonium chloride salts have a maximum working temperature of 150°C, and the decomposition temperature of dialkylimidazolium depends on the associated anion: halide derivatives have the lowest thermal stability (150–250°C for [bmim][Cl], depending on the source of the data), but some salts, such as [bmim][BF_4], [emim][TfO] and [emim][NTf_2], have been heated up to 400°C without showing any degradation.[27,59,61] These values should be interpreted with care, as degradation of the ionic liquids will occur at lower temperatures if they are heated for a long period. For example, assuming that a maximum decomposition of 1% is acceptable, [emim][NTf_2] can be kept at 307°C for 1 h, but at only 251°C if it is heated for 10 h.[62]

Since ionic liquids can be heated to much higher temperatures than any organic solvent, they can be used at a wide range of temperatures: 300°C is usual for this medium, and a window of more than 400°C has been reported for some. Therefore, much higher kinetic control of reactions can be obtained using ionic liquids, and reactions can be carried out without solvent pressure problems arising.

2.2.3 *Polarity*

Polarity is the most common classification for solvents. There is no absolute polarity scale; in a first approximation, it can be considered that polar solvents are characterized by their ability to dissolve charged solutes. Since ionic liquids are themselves salts, they are expected to be very polar.

The polarities of different solvents can be compared using fluorescence probe molecules or solvatochromic dyes. When such molecules are dissolved in a solvent, the frequency of the maximum fluorescence or absorption, respectively, changes as a function of the polarity of the medium, and this property can then be compared with the result for standard solvents. The polarities of several ionic liquids were estimated using 5-dimethylamino-isoindole-1,3-dione **34** as a fluorescence probe and Reichardt's dye **35** for visible absorbance measurements,[63] allowing the determination of the normalized E_T^N parameter (E_T^N(tetramethylsilane) = 0, $E_T^N(H_2O) = 1$).[64–66] Another set of experiments was conducted using the solvatochromic dye Nile Red **36** (Scheme 9) to give E_{NR}.[67]

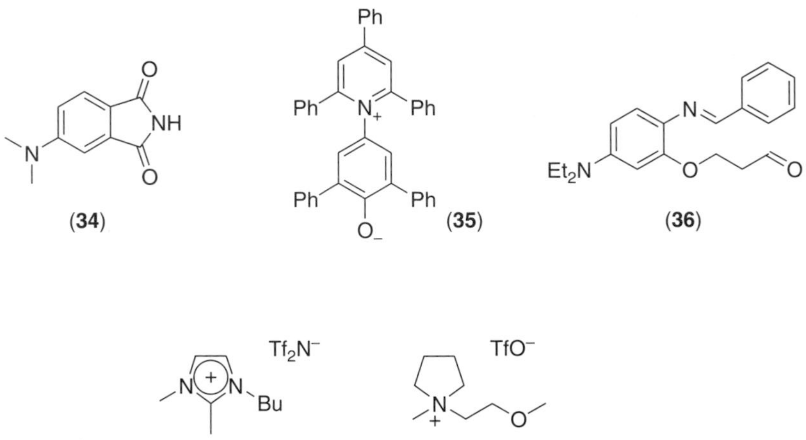

(34) (35) (36)

(37) (38)

Scheme 9

Table 2.6

Solvent	E_T^N	E_{NR}
hexane	0.009	247.0
[mmep][CF_3COO]	0.370	
acetonitrile	0.470	225.0
isopropanol	0.546	
[bmmim][NTf_2] (**37**)	0.552	–
[omim][PF_6]	0.642	–
[bmim][NTf_2]	0.642	218.0
ethanol	0.654	218.2
[bmim][PF_6]	0.667	218.5
[bmim][OTf]	0.667	–
methanol	–	217.7
[bmim][BF_4]	0.673	217.2
[mmep][OTf] (**38**)	0.910	
[$EtNH_3$][NO_3]	0.954	
water	1.000	201.7

Similar results were obtained for all studies (Table 2.6). All 1-alkyl-3-methylimidazolium salts tested had a polarity close to that of methanol and ethanol. The acidity of the C_2–H proton was confirmed, as its substitution by a methyl group led to a less polar ionic liquid ([bmmim][NTf_2], **37**). No strong influence of the associated anion was observed. Imidazolium ionic liquids belong to the class of polar solvents but, although they are salts, they are still much less polar than water. On the other hand, the polarity of pyrrolidinium ionic liquids varies strongly with the anion, as shown for 1-methyl-1-(2-methoxyethyl)pyrrolidinium derivatives: [mmep][OTf] (**38**) and $[EtNH_3][NO_3]$ have a similar polarity to that of water, whereas $[mmep][CF_3COO]$ is less polar than acetonitrile.[66]

Given that ionic liquids with similar E_T^N parameters can behave very differently when used as solvents, a different classification of the polarity has been proposed that takes into account both the hydrogen bond basicity and the dipolarity. The anion has a strong effect on the hydrogen bond basicity of the ionic liquid, whereas the contribution of the cation is negligible.[68]

2.2.4 *Solubility*

Unlike polarity, the solubility of ionic liquids depends strongly on the structure of the associated anion. In fact, the miscibility of ionic liquids with traditional solvents is one of their most interesting features and is clearly observed during the anion metathesis of imidazolium derivatives. The starting chlorine salt, for example [bmim][Cl], is completely soluble in water. After exchange of the anion, $[bmim][PF_6]$ or $[bmim][NTf_2]$ salts are obtained as hydrophobic products and therefore form a separate layer. Table 2.7 summarizes the solubility of 1-butyl-3-methylimidazolium salts as function of the anion.[61]

Almost all ionic liquids are hygroscopic, even the ones insoluble in water, and they have to be dried carefully before use. The amount of water dissolved in ionic liquids can be determined using Karl–Fischer measurements: the most hydrophobic salts, such as $[bmim][PF_6]$ and $[bmim][NTf_2]$, contain about 1.2% (w/w) of water in their saturated condition.[27,69]

Since some ionic liquids are insoluble in water, they can replace traditional organic solvents in liquid–liquid extractions. In a $[bmim][PF_6]$/water biphasic system, neutral substances dissolve preferentially in the ionic liquid phase, and charged molecules are recovered mainly in the aqueous layer.[21] This behavior can easily be seen in the case of the partitioning of thymol blue **39** in a mixture of $[bmim][PF_6]$ (lower phase) and water (upper phase, Figure 2.6), as a function of the pH. At low pH, the thymol blue exists in its neutral zwitterionic red form and is exclusively soluble in the ionic liquid phase. At pH 7, the yellow monoanion partitions into both phases, and at a basic pH, the blue dianion is found exclusively in the aqueous layer.[70]

Ionic liquids are also insoluble in several organic solvents, which is a valuable property for catalytic reactions: the synthesized products can be extracted from

Table 2.7

Salt	H_2O	MeOH	CH_2Cl_2	Et_2O
[bmim][Cl]	Soluble	Soluble	Insoluble	Insoluble
$[bmim][BF_4]$	Soluble	Soluble	Soluble	Insoluble
$[bmim][PF_6]$	Insoluble	Soluble	Soluble	Insoluble
$[bmim][NTf_2]$	Insoluble	Soluble	Soluble	Insoluble
$[bmim][AlCl_4]$	Not compatible	Not compatible	Soluble	Insoluble

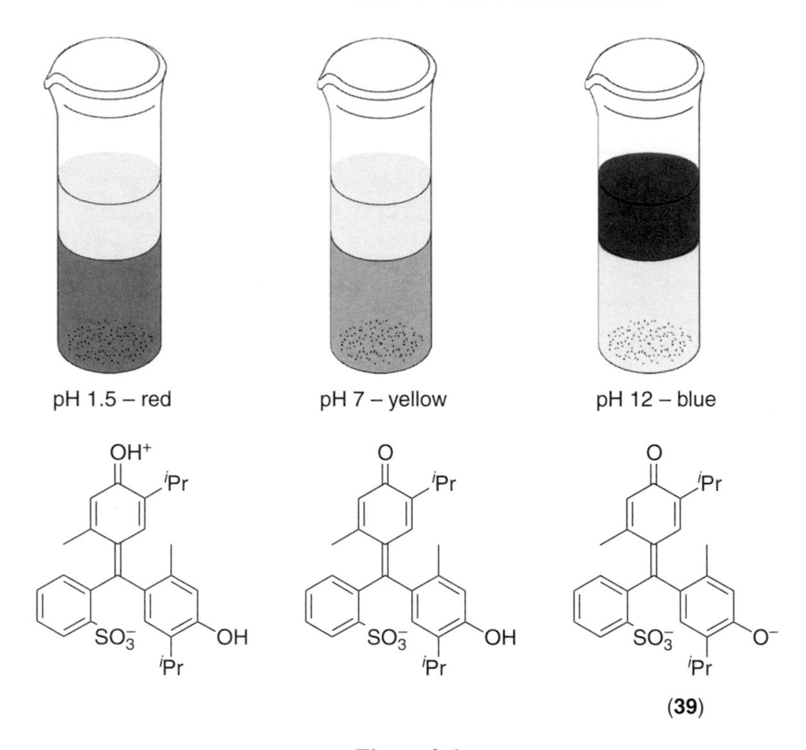

Figure 2.6

N(CN)2−

PO4 3−

4 PO4−

(40) (41) (42)

Figure 2.7

the medium using organic solvents, and the catalysts remain in the ionic liquid and can be directly recycled. This important feature is discussed below.

New anions have been used to form hydrophilic ionic liquids. The association of the dicyanamide anion with *N*,*N*′-dialkylpyrrolidinium, tetraalkylammonium and emim **40**, (Figure 2.7) leads to water-soluble ionic liquids with melting points below 0°C.[45] Similar behavior has been observed for phosphate ionic liquids, not only for the corresponding standard $[emim]_3[PO_4]$ (**41**), but also for polycationic species such as **42**. Although they are viscous, these salts are liquid at room temperature.[71,72]

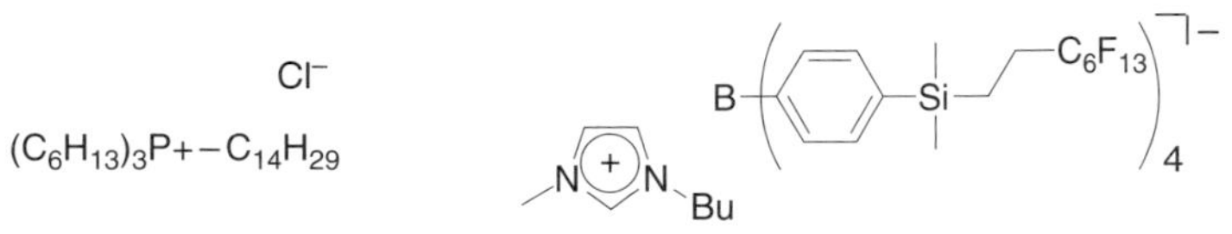

Cl^-

$(C_6H_{13})_3P^+-C_{14}H_{29}$

(43)

(44)

Figure 2.8

Adding a polar functional group on the cation will also form water-soluble ionic liquids. Salts **30–33** (Table 2.5), whose imidazolium ring has a hydroxyl functionality, are all soluble in water, even when the hydrophobic hexafluorophosphate anion is used. These salts have been developed for their ability to solubilize inorganic salts (LiCl, $HgCl_2$ and $LaCl_3$).[33]

Most of the research carried out into new types of anions leads to the formation of water-soluble ionic liquids. However, hydrophobic ones have much greater potential, in particular as a medium for metal-catalyzed cross-coupling reactions, since salts formed as byproducts during the reaction can be easily removed by washing with water (see Section 2.3.5). Hydrophobic ionic liquids are usually formed with PF_6 and NTf_2 anions and cannot easily be used on large scale because of their chemical instability or cost. A wise combination of cation and anion can lead to the formation of several room-temperature hydrophobic ionic liquids that do not require the use of PF_6 or NTf_2 anions, such as phosphonium chloride **43**[44,73] and imidazolium borate **44** (Figure 2.8).[74,75] However, these salts have not been much used in synthetic applications, and there is clearly a need for new types of hydrophobic ionic liquids.

2.2.5 *Viscosity*

The viscosity of standard organic solvents is rarely considered important. However, because of their saline structure, ionic liquids can be rather viscous; sometimes even using a magnetic stirrer is impossible. The viscosity of standard organic solvents ranges from 0.24 cP (diethylether) to 2.24 cP (DMSO). Compared with the latter, which is already considered viscous, ionic liquids have viscosities 10 to 100 times higher (Table 2.8) and more comparable to that of an oil. The viscosity of these salts appears to be governed by van der Waals interactions and hydrogen bondings.[27] Both the size and the basicity of the anion affect the viscosity. The symmetry of the anion also seems to play a role, which explains the large difference observed between symmetric ionic anions (PF_6) and organic anions of lower symmetry (NTf_2). Interestingly, although the [bmim][PF_6] salt is one the most viscous ionic liquids, it is still the one most used.[59,76]

The viscosity of ionic liquids can be lowered by the addition of a small amount of organic co-solvent. It is noteworthy that the lowering of viscosity does not depend on the nature of the solvent added, but only on its mole fraction.[77] Furthermore, a slight increase in temperature can diminish the viscosity of these solvents notably;

Table 2.8

Solvent	Viscosity (cP) (25°C)
diethylether	0.24
water	1.00
DMSO	2.24
[emim][N(CN)$_2$]	21
[emim][NTf$_2$]	28
[bmim][NTf$_2$]	52
[bmim][CH$_3$CO$_2$]	73
[bmim][OTf]	90
[bmim][PF$_6$]	207
[bmim][BF$_4$]	233

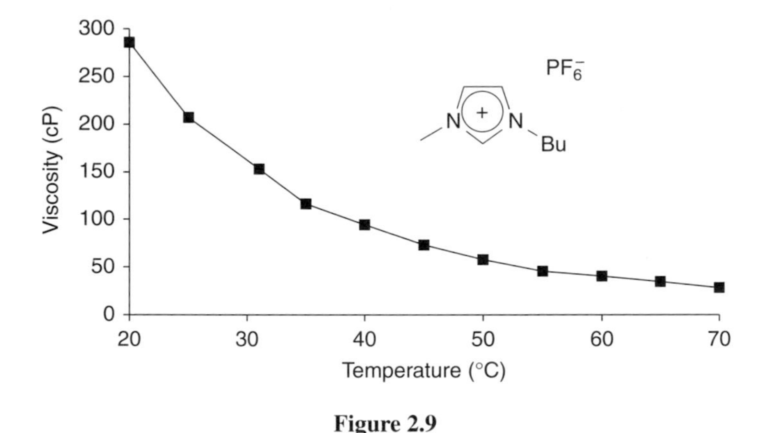

Figure 2.9

when dry [bmim][PF$_6$] is heated from 20°C to 50°C, its viscosity drops from 286 to only 58 cP (Figure 2.9).

2.2.6 *Acidity*

The acidity and coordination properties of ionic liquids are related to the nature of the anion. These properties can be easily tuned, as shown in Table 2.9.[78] From basic and strongly coordinating anions to acidic and non-coordinating ones, a large number of anions are available. The ionic liquids used most in organic synthesis are of neutral, weakly coordinating anions (BF_4, PF_6).

The Lewis acidity/basicity of imidazolium chloroaluminate can be varied significantly by changing the composition. Although imidazolium chloride salts have a strongly basic anion, the addition of 1.0 equivalent of $AlCl_3$ (50 mol%) forms the neutral [imidazolium][$AlCl_4$] salt. Further addition of Lewis acid leads to a mixture with highly acidic anions such as $Al_2Cl_7^-$ and $Al_3Cl_{10}^-$ (Scheme 10).

Table 2.9

Basic, strongly coordinating	Neutral, weakly coordinating	Acidic, non-coordinating
$Cl^-/NO_3^-/SO_4^{2-}$	$AlCl_4^-/SbF_6^-/BF_4^-/PF_6^-$	$Al_2Cl_7^-/Cu_2Cl_3^-/Al_3Cl_{10}^-$

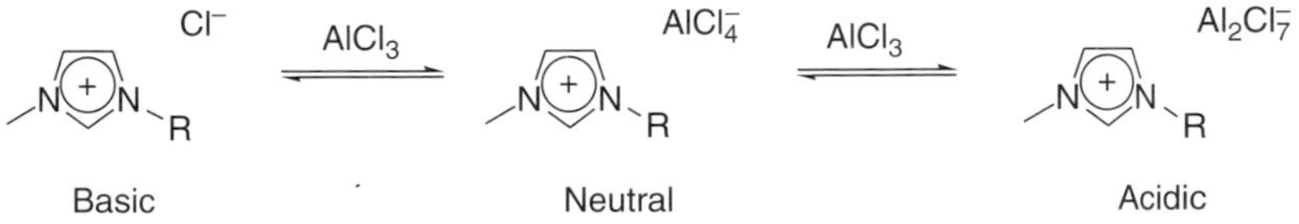

Scheme 10

Ionic liquids can also exhibit superacidity (Brønsted superacids are acids that are more acidic than pure H_2SO_4). The dissolution of gaseous HCl in acidic [emim][Cl]/$AlCl_3$ (55 mol% of Lewis acid) leads to a superacidic system[79] that has similar properties to those of liquid HF and can be used for the protonation of arenes. The Brønsted acidity of products dissolved in water-stable ionic liquids can be increased: $HNTf_2$ and TfOH display higher chemical activity in [bmim][NTf_2] and [bmim][BF_4] than in water.[80]

2.2.7 *Chirality*

Some chiral ionic liquids (CILs) have been reported. Many are based on the imidazolium cation because this structure usually leads to ionic liquids with favorable properties, namely, low melting point and viscosity. The first non-racemic chiral ionic liquid **45** was based on the standard achiral bmim cation associated with an enantiomerically pure (*S*)-lactate anion. This ionic liquid is now commercially available. Using **45**, the asymmetric Diels–Alder reaction between ethyl acrylate and cyclopentadiene was attempted, but only very low enantioselectivity was attained.[81] Since the presence of a reactive functional group is detrimental in terms of the melting point and can lead to undesired reactions, a CIL bearing planar chirality was synthesized. Salt **46** is a racemic mixture and has a melting point of −20°C. This CIL was not resolved, and the two enantiomers were observed using ^{1}H-NMR in the presence of the (1*S*)-(+)-10 camphorsulfonate anion.[82] Bromide salts **47** (R = Me, *i*-Pr, *i*-Bu) were synthesized in a 30–33% yield from amino acids.[83] Their melting points are rather low (5°C, 11°C and 15°C, respectively), but the presence of bromide might be detrimental to future catalytic applications.

Functional groups are important for the further application of CILs. Salt **48**, which bears ester functionality, was prepared in three steps from commercially available and inexpensive (*S*)-ethyl lactate, in an 82% overall yield. This CIL, as well

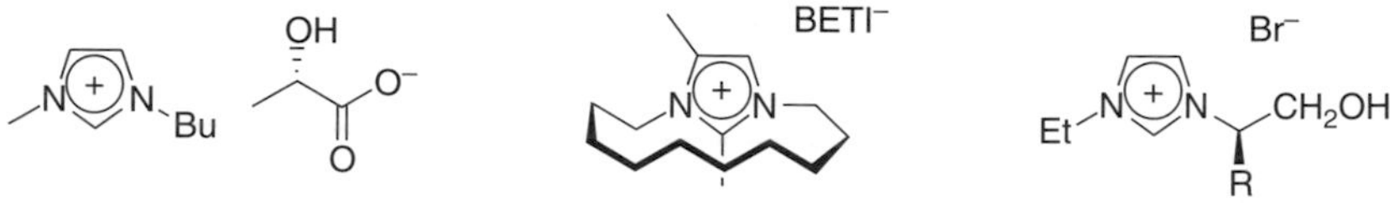

(45) (46) (47)

Figure 2.10

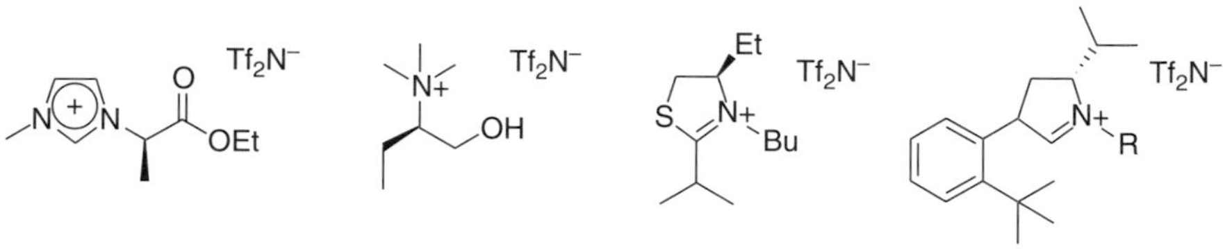

(48) (49) (50) (51)

Figure 2.11

as the $N(SO_2C_2F_5)_2$ (BETI) or $N(SO_2C_4F_9)Tf$ salts, is liquid at room temperature, exhibiting only a glass transition at around −55°C (Figure 2.10).[84] The enantiomeric purity of salt **48** was confirmed by NMR, using Lacour's TRISPHAT anion[85–87] as an NMR chiral shift reagent. Salt **49**, which was derived from (*R*)-2-aminobutan-1-ol, was prepared on a kilogram scale in two steps and a 75% overall yield. This salt has an interestingly low melting point of −18°C and can be used as an NMR chiral shift reagent with Mosher's acid sodium salt as a substrate.[88,89] Other types of CILs have also been reported, such as the thiazolium salt **50**[90] and the imidazolinium salt **51**,[91] also prepared from the chiral pool (Figure 2.11).

2.2.8 *Toxicity and environmental issues*

Ionic liquids are considered promising media for clean processes and green chemistry. They are an alternative to volatile organic solvents in particular, which are used by industry in huge amounts every year. Because ionic liquids have no effective vapor pressure, they cannot evaporate and pollute the atmosphere. This property also allows products formed in this medium to be recovered by distillation instead of by extraction with organic solvents, which further contributes to cleaner processes. Alternatively, it is possible to isolate the products by extracting them with supercritical CO_2, or even in some cases by direct decantation. Another important feature of ionic liquids is their ability to dissolve catalysts; after recovery of the product, the ionic liquid containing the catalyst can usually be recycled and reused without loss of activity, promoting cleaner processes.

Ionic liquids can be specifically designed to accomplish environmental tasks, such as the extraction of toxic heavy metal ions from specific media. Imidazolium

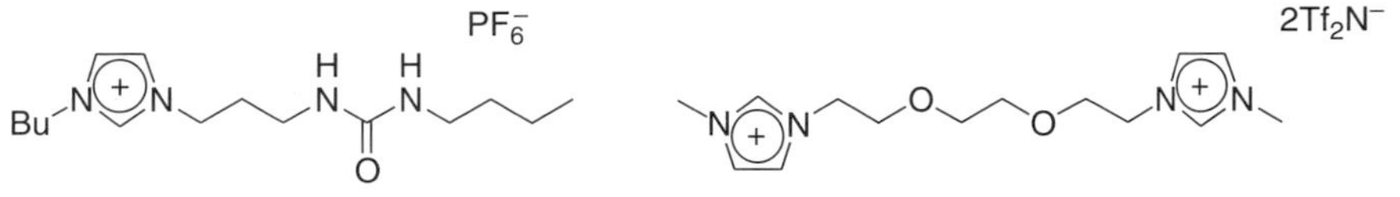

Figure 2.12

derivatives with urea functionality **52** (Figure 2.12) have been used as solvents for the extraction of mercury(II) and cadmium(II) salts from water, with a dramatically enhanced partition ratio.[34] X-ray studies with the room-temperature ionic liquid **53** incorporating an ethylene-glycol spacer suggest that mercury(II) salts are extracted via formation of a carbene. Furthermore, mercury can be stripped back from the ionic liquid to an aqueous phase by controlling the pH.[92] Standard ionic liquids such as [bmim][PF_6] and [emim][BF_4] have also found application in the selective removal of sulfur-containing compounds from gasoline.[93]

It must be stressed, however, that since only a few industrial applications using ionic liquids have been developed, the environmental fate of, and complete toxicity data for, these solvents are essentially unknown, and in most cases MSDSs (Material Safety Data Sheets) are lacking. As one of the most used ionic liquids, [bmim][PF_6] has been more carefully studied. Chemada Fine Chemicals reports LD_{50} values of 300–500 mg/kg (rat, oral) and $>$2000 mg/kg (rat, dermal) for this salt, with no dermal irritation or sensitization. The toxicological effects of 1-alkyl-3-methylimidazolium chloride salts have also been studied using *Caenorhabditis elegans*, a soil roundworm with a 3-day life cycle.[94] Both [bmim][Cl] and [omim][Cl] are classified as *not acutely toxic*; the lethality of ionic liquids increases with longer alkyl chains.

Not all ionic liquids are as stable as stated in early articles: in particular, the PF_6 anion is known to undergo hydrolysis, forming volatile products including HF and POF_3, which not only are dangerous but also damage glassware and reactors. Recently, [bmim][F]·H_2O has been identified as a crystalline decomposition product during the preparation and the drying of [bmim][PF_6] using standard procedures.[95] This represents an important drawback of this ionic liquid; although it is much used in academic research, it cannot find application in large-scale industrial processes.

2.3 Applications as reaction media

Since the use of ionic liquids has become widespread, a large number of reactions previously developed in organic solvents have been attempted in these new solvents. Virtually all known transformations can be tested in ionic liquids, leading to a plethora of publications. Some reactions are improved, but many are not. This section focuses on reactions that are improved by the use of ionic liquids as solvents, either by leading to better yield, regioselectivity or enantioselectivity or for practical reasons – the product is easier to isolate, less waste is generated or the reaction occurs more quickly.

2.3.1 Hydroformylation

The synthesis of aldehydes by hydroformylation of alkenes is an important industrial process discovered in 1938. The use of biphasic catalysis, which is a well-established method for the separation of the product and the recovery of the catalyst, was developed for this reaction in the Ruhrchemie–Rhône–Poulenc process: a water-soluble rhodium complex is used as catalyst – the reaction proceeding in water – and the organic layer is formed simply by the alkene reagent and the aldehyde products. Butanal is manufactured this way, but owing to the low solubility of longer olefins in water, this process is limited to C_2–C_5 olefins.[96] To overcome this limitation, ionic liquids have been studied as a reaction solvent for this biphasic reaction.

As early as 1972, the platinium-catalyzed hydroformylation of ethene in tetraethylammonium trichlorostannate melt (melting point 78°C) was described.[16] More recently, the same platinium metal was used for the hydroformylation of oct-1-ene in a mixture of [bmim][Cl] and $SnCl_2$ (1.04 equiv. of Lewis acid), which was liquid at room temperature (Scheme 11). High turnover frequencies (TOF $126\,h^{-1}$) were achieved; more importantly, a very good regioselectivity of 19 : 1 between *n*-**54** and *iso*-**54** was obtained. Unfortunately, the conversion was low, and a fairly large amount of undesired hydrogenated *n*-octane was isolated along with the product.[97]

Phosphonium ionic liquids have been used several times for metal-catalyzed hydroformylations. Ruthenium and cobalt metal complexes catalyze the hydroformylation of internal olefins in [^{n}Bu$_4$P][Br]; the major products are, however, the corresponding alcohols.[98] Rhodium-catalyzed hydroformylations were conducted in [Bu_3PEt][TsO] and [Ph_3PEt][TsO] melts (melting points 81°C and 94°C, respectively). The products were easily isolated by decantation of the solid medium at room temperature.[99]

Most recent research into hydroformylation has been carried out using rhodium metal complexes in imidazolium-based ionic liquids. Using a similar condition to that developed for hydrogenation, Chauvin *et al.* carried out the biphasic hydroformylation of 1-pentene in [bmim][PF_6] with neutral [$Rh(CO)_2$(acac)]/PPh_3 (acac = acetylacetonate) (Scheme 12). Similar catalytic activity (TOF $\sim 300\,h^{-1}$)

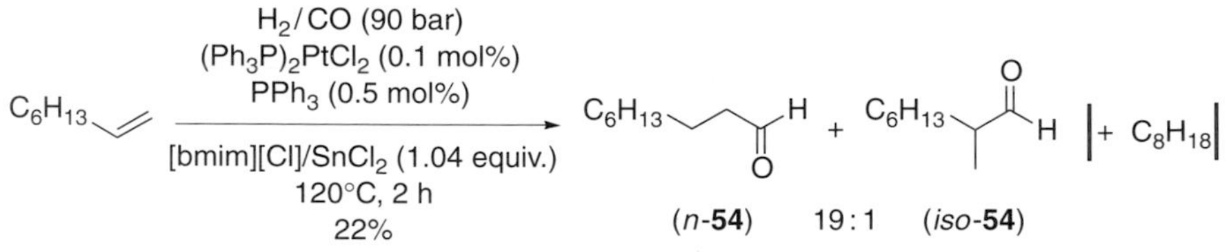

C_6H_{13}-CH=CH$_2$ → [H_2/CO (90 bar), $(Ph_3P)_2PtCl_2$ (0.1 mol%), PPh_3 (0.5 mol%); [bmim][Cl]/$SnCl_2$ (1.04 equiv.), 120°C, 2 h, 22%] → (*n*-**54**) + (*iso*-**54**) (19 : 1) + C_8H_{18}

Scheme 11

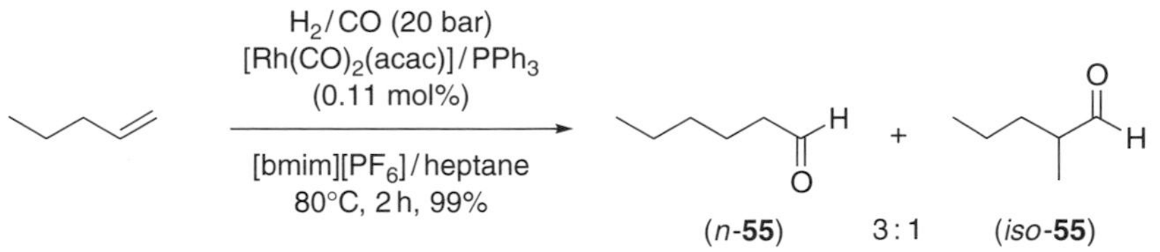

H_2/CO (20 bar)
$[Rh(CO)_2(acac)]/PPh_3$
(0.11 mol%)
$[bmim][PF_6]$/heptane
80°C, 2 h, 99%
(*n*-**55**) 3 : 1 (*iso*-**55**)

Scheme 12

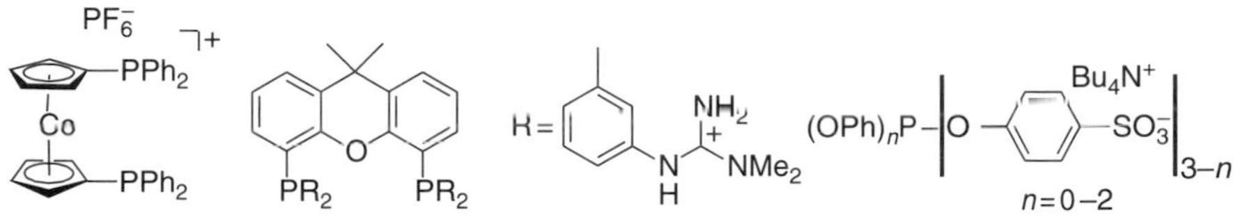

PF_6^-
PPh_2
Co
PPh_2
PR_2 PR_2
R =
NH_2
NMe_2
$(OPh)_nP$
Bu_4N^+
SO_3^-
3–*n*
n = 0–2
56 **57** **58**

Scheme 13

and selectivity (*n*-**55**/*iso*-**55** ratio 3 : 1) were observed in the ionic liquid and in toluene, but a slight leaching of the catalyst in the organic layer was detected. The catalyst has, therefore, to be immobilized using a polar ligand, namely, the sodium salt of monosulfonated triphenylphosphane. This transformation led, however, to lower catalytic activity; and the turnover frequency fell to 59 h^{-1}.[100]

Both high activity and high regioselectivity can be obtained by replacing the triphenylphosphine ligand with 1,1′-bis(diphenylphosphanyl)cobaltocenium hexafluorophosphate **56** (Scheme 13), which gives a satisfactory result for the hydroformylation of oct-1-ene (TOF 810 h^{-1}, *n*-**54**/*iso*-**54** ratio 16.2 : 1). However, a slight leaching of the catalyst was still observed (<0.2%).[101] Better results were obtained using diphosphine ligand **57** containing a xanthene backbone: not only was high selectivity achieved, but also the catalyst could be recycled ten times and almost no leaching of the catalyst occurred (<0.07%).[102]

The advantage of charged phosphine ligands was confirmed by performing the biphasic hydroformylation of hex-1-ene in a variety of ionic liquids. Optimization of the rate and the selectivity was obtained with a suitable combination of the cation (imidazolium, pyrrolidinium) and weakly coordinated anion. To minimize the leaching of the catalyst, ligands bearing cationic (guanidinium or pyridinium) or anionic (sulfonate) groups were used. Interestingly, phosphite ligand **58** (Scheme 13), which is unstable in water and therefore cannot be used for traditional biphasic (water/organic medium) hydroformylations, can yield heptanal with good regioselectivity and at a good rate in $[bmim][PF_6]$.[103]

To develop rhodium-catalyzed biphasic hydroformylation for industrial applications, a continuous flow process has been developed for the hydroformylation of oct-1-ene in $[bmim][PF_6]/scCO_2$, allowing the production of aldehydes at a constant rate for 30 h (see Section 4.4.2).[104] The main advantage of biphasic processes

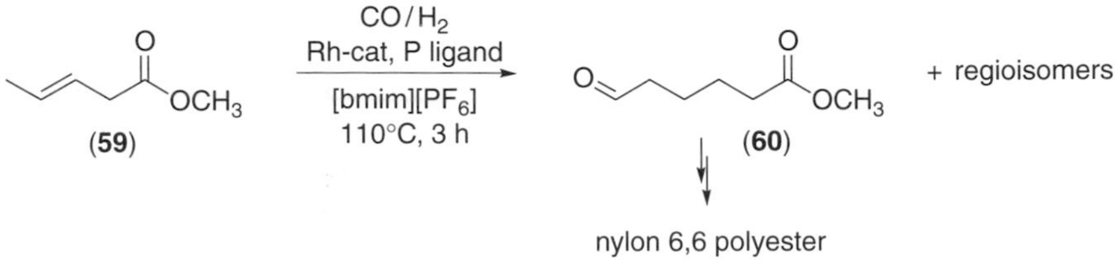

CO/H_2
Rh-cat, P ligand
[bmim][PF_6]
110°C, 3 h
(59) OCH_3 → (60) OCH_3 + regioisomers
↓
nylon 6,6 polyester

Scheme 14

is to allow easy recovery of the product and clean recycling of the catalyst. In some cases, however, monophasic hydroformylations can also be efficient, for example, the transformation of methyl-3-pentenoate **59** (obtained from 1,3-butadiene) to 5-formyl-methylpentanoate **60** by an isomerization followed by hydroformylation at the terminal position (Scheme 14). Ester **60** is an important product for industry as it is a precursor of the adipic acid used for the preparation of nylon-6,6. This product, which was obtained along with other regioisomers, was simply isolated by distillation from the reaction mixture. The ionic liquid phase, containing the rhodium catalyst and the ligand, was reused without further treatment; only a slight decrease in efficiency was observed, with the turnover frequency dropping from ~150 to 90 h^{-1} after the tenth run.[105]

2.3.2 Hydrogenation

Hydrogenation is one of the most important reactions in organic synthesis. It usually leads not only to quantitative yield but also to high enantioselectivity, even using only an infinitesimal amount of catalyst. Hydrogenation is thus a reaction of choice for industry. There was early interest in performing this reaction in ionic liquids on the basis of easier recovery of the products because of the high solubility of metal complexes in ionic liquids. Hydrogenation was one of the first reactions successfully catalyzed by transition metal complexes in this medium.

Early work focused on the rhodium-catalyzed hydrogenation of olefins. In 1995, the hydrogenation of 1-pentene in [bmim][SbF_6] was reported, using [Rh(nbd)(PPh_3)$_2$][PF_6] (nbd = norbornadiene) (Scheme 15).[100] This reaction highlights many of the advantages of ionic liquids. Since both the reagent and the product have limited solubility in the ionic liquid, hydrogenation proceeds under biphasic conditions. The product, obtained in an 83% yield (along with a small amount of pent-2-ene owing to isomerization), was directly isolated by decantation. An accelerated reaction rate was observed, five times faster than in acetone. Finally, the catalyst was immobilized in the ionic liquid, and this phase could be reused.

The efficiency of the reaction was strongly correlated with the anion of the ionic liquid. Using [bmim][PF_6], a much lower conversion to pentane was obtained, and a fairly large amount of the isomerized product pent-2-ene was isolated (41%). The [bmim][BF_4] salt, because it is water miscible and hence difficult to purify, led to

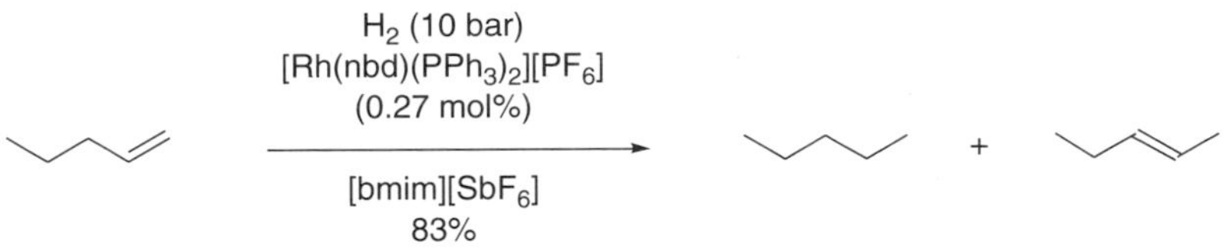

H_2 (10 bar)
$[Rh(nbd)(PPh_3)_2][PF_6]$
(0.27 mol%)
[bmim][SbF_6]
83%

Scheme 15

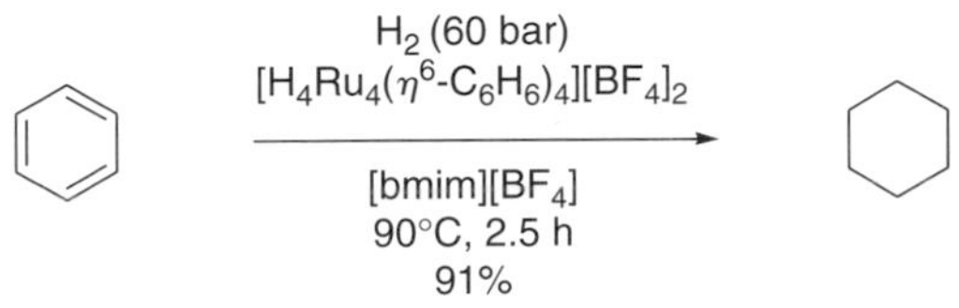

H_2 (60 bar)
$[H_4Ru_4(\eta^6\text{-}C_6H_6)_4][BF_4]_2$
[bmim][BF_4]
90°C, 2.5 h
91%

Scheme 16

only 5% of the desired product, probably because of the presence of coordinating chlorine anions. Cyclohexadiene was also hydrogenated in the same optimized biphasic conditions. Since the starting material was five times more soluble in [bmim][SbF_6] than the product, this process was very efficient (96% conversion and 98% selectivity).

A similar procedure was used for the reduction of cyclohexene in [bmim][Cl]/$AlCl_3$, [bmim][BF_4] or [bmim][PF_6].[106] Modest conversions were obtained using Wilkinson's catalyst $RhCl(PPh_3)_3$ (40%). The use of [Rh(cod)$_2$][BF_4] (cod = cyclooctadiene) led to an improvement of the yield (65%), along with a lower turnover. In both cases, catalysts were immobilized in the ionic liquids, allowing clean and easy recovery of the product by decantation.

Reduction of arenes is an important industrial process, as it is used for the generation of cleaner diesel fuels. The usual procedure of heterogeneous catalysis was successfully replaced by a biphasic [bmim][BF_4]/organic solvent system, using a ruthenium cluster catalyst. Hydrogenation of benzene was achieved with high turnovers (TOF 364 h^{-1}), the product was easily isolated and no side-products were observed (Scheme 16).[107] Stereoselective hydrogenation of aromatic compounds has also been studied in moisture-sensitive [bmim][Cl]/$AlCl_3$.[108]

Of particular importance is the use of ionic liquids in asymmetric catalytic reactions. The first example – asymmetric hydrogenation of (Z)-α-acetamidocinnamic acid to (*S*)-phenylalanine (**61**) – was reported in 1995 using [Rh(cod)(−)-diop][PF_6] (diop = 4,5-bis[(diphenylphosphanyl)-methyl]-2,2-dimethyl-1,3-dioxolan-4,5-diol) in a biphasic mixture of [bmim][SbF_6] and *i*-PrOH (Scheme 17).[100] At the end of the reaction, the product was obtained in isopropanol, and the ionic liquid containing the catalyst could be recycled. Only modest enantioselectivity was attained (64% ee), but this work opened the way to efficient asymmetric catalytic reactions in ionic liquids.

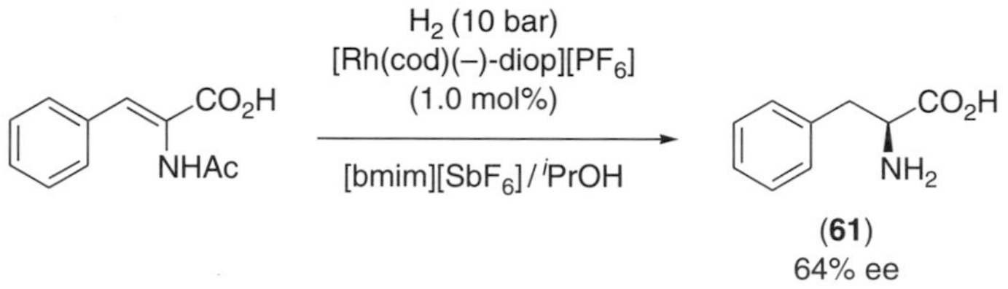

H_2 (10 bar)
$[Rh(cod)(-)\text{-}diop][PF_6]$
(1.0 mol%)
$[bmim][SbF_6]$ / ^{i}PrOH

CO_2H NHAc → CO_2H NH_2

(**61**)
64% ee

Scheme 17

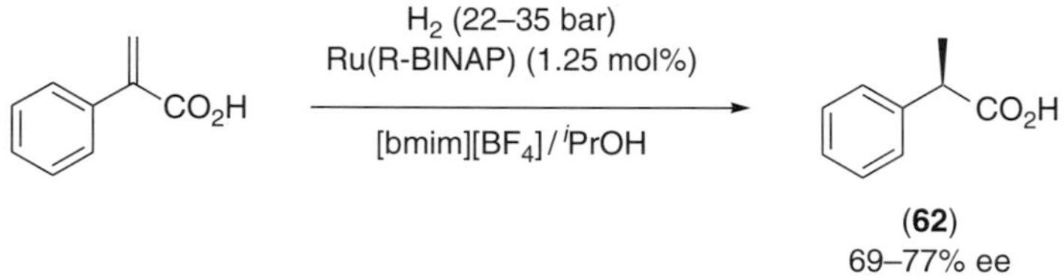

H_2 (22–35 bar)
Ru(R-BINAP) (1.25 mol%)
$[bmim][BF_4]$ / ^{i}PrOH

CO_2H → CO_2H

(**62**)
69–77% ee

Scheme 18

The asymmetric hydrogenation of 2-phenylacrylic acid was performed in a mixture of isopropanol and $[bmim][BF_4]$.[109] In a similar way to the previous example, the product was isolated in isopropanol, and the ionic liquid containing the catalyst (Ru-BINAP) was recovered and efficiently reused up to three times without significant change in catalytic activity and enantioselectivity. This method was successfully applied to the preparation of (*S*)-Naproxen **62** (quantitative yield, 69–77% ee (Scheme 18).

Another approach to the isolation of the product is to use supercritical CO_2. It was observed that organic molecules can be easily extracted from ionic liquids to $scCO_2$, with no trace of the ionic solvents being detected in the supercritical fluid.[110] Recently, procedures involving the use of ionic liquids/$scCO_2$ have been studied extensively; they will be described in Section 4.4.2.[111]

Using this strategy for the asymmetric hydrogenation of tiglic acid **63** with $Ru(O_2CMe)_2[(R)\text{-}tolBINAP]$ complex, in wet $[bmim][PF_6]$ and without any organic co-solvent (Scheme 19), a strong dependence of the enantioselectivity on the hydrogen pressure was observed,[112] which is contrary to previous results reported for $[bmim][BF_4]$.[109] This apparent contradiction was explained later by demonstrating that the important kinetic factor is the hydrogen concentration in the ionic liquid rather than the pressure. Since hydrogen is four times more soluble in $[bmim][BF_4]$ than in $[bmim][PF_6]$, a higher pressure has to be used for hydrogenation in $[bmim][PF_6]$.[113]

There are known to be some limitations to asymmetric hydrogenation of certain substrates. Asymmetric hydrogenation of tiglic acid requires a low H_2 concentration, or low H_2 pressure and mass transfer rates, which is obtained in viscous ionic liquids. On the other hand, the hydrogenation of atropic acid is more enantioselective when

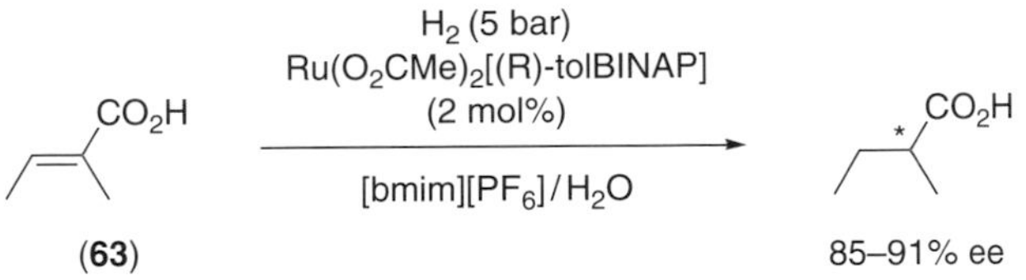

H_2 (5 bar)
$Ru(O_2CMe)_2[(R)\text{-tolBINAP}]$
(2 mol%)
$[bmim][PF_6]/H_2O$

CO_2H (**63**) → CO_2H 85–91% ee

Scheme 19

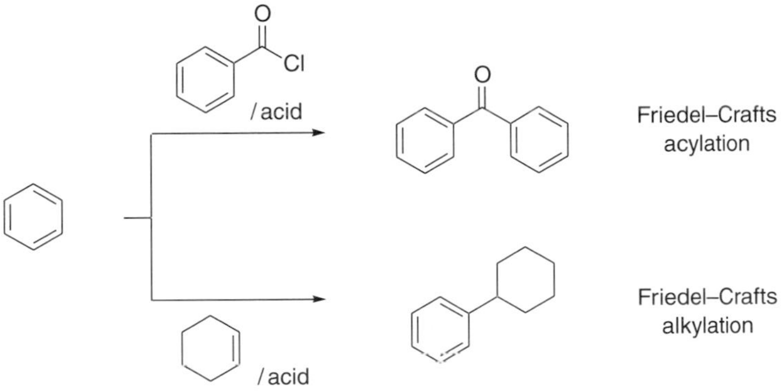

/acid — Friedel–Crafts acylation

/acid — Friedel–Crafts alkylation

Scheme 20

high concentration and high mass transfer rates of H_2 are attained, which is why this reaction is less selective in an ionic liquid than in methanol.[114]

2.3.3 The Friedel–Crafts reaction

The Friedel–Crafts reaction was one of the first to be attempted in ionic liquids (for typical examples, see Scheme 20.[115–117] Friedel–Crafts acylation, which allows easy functionalization of aromatic compounds to ketones, is of significant commercial importance. The electrophilic substitution with an acylating agent is catalyzed by an acid, typically $AlCl_3$. Since this catalyst can form a stable adduct with the carbonyl of the product, an excess of $AlCl_3$ is required, which gives rise to a copious amount of inorganic waste.

In 1986, it was discovered that both Friedel–Crafts alkylations and acylations could be performed in acidic [emim][Cl]/$AlCl_3$ (2 equiv. of $AlCl_3$). The ionic liquid behaved as both solvent and catalyst, $Al_2Cl_7^-$ being the catalytically active species.[14] Although this study is limited to the alkylation and acylation of simple benzene and the reaction is not selective (mixtures of monosubstituted and polysubstituted benzenes were obtained), the use of ionic liquids is an interesting alternative to the use of $AlCl_3$ in organic solvents. Indeed, the study demonstrates the effectiveness of acidic ionic liquids for Friedel–Crafts transformations. The [emim][Cl]/$AlCl_3$

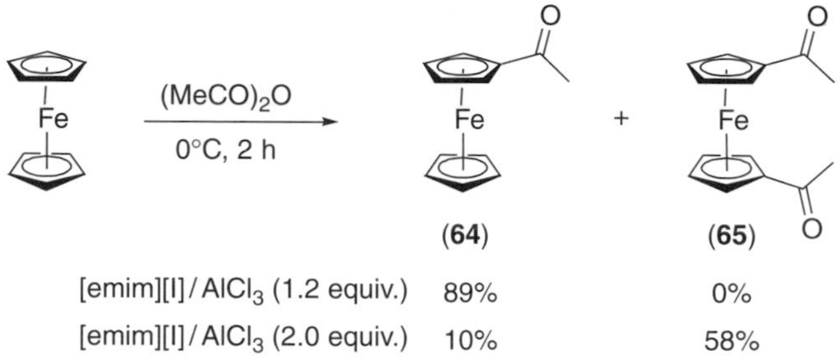

$(MeCO)_2O$, 0°C, 2 h

(64) (65)

	(64)	(65)
[emim][I]/$AlCl_3$ (1.2 equiv.)	89%	0%
[emim][I]/$AlCl_3$ (2.0 equiv.)	10%	58%

Scheme 21

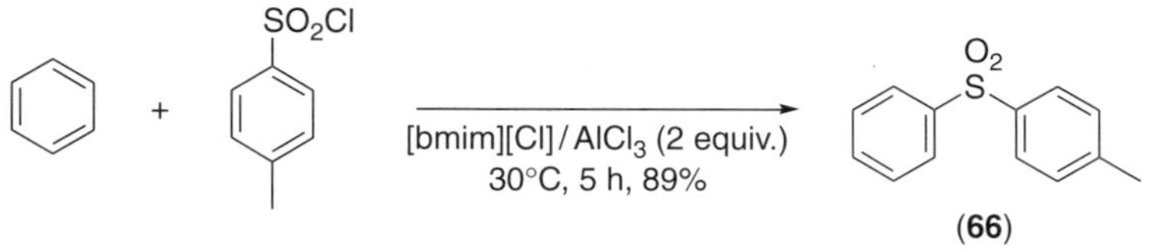

SO_2Cl [bmim][Cl]/$AlCl_3$ (2 equiv.) 30°C, 5 h, 89% (66)

Scheme 22

salt can be immobilized on a silicate or aluminate support; this solid behaves as an efficient catalyst for the alkylation of benzene.[118]

The acylation of ferrocene is usually performed using $AlCl_3$ in organic solvents under reflux. Milder conditions can be used in ionic liquids: ferrocene is acetylated by 2 equivalents of acetic anhydride in the acidic [emim][I]/$AlCl_3$ (1.2 equiv.) ionic liquid. This reaction takes place rapidly at low temperature (0°C) to give exclusively the monosubstituted ferrocene **64** in a high yield. Furthermore, using an excess of acylating agent and a more acidic medium ([emim][I]/$AlCl_3$ with 2 equiv. of $AlCl_3$), the diacylated product **65** was obtained as the major product (Scheme 21).[119]

Sulfonylation of benzene in a similar medium has been described, the sulfone **66** being obtained in a high yield (Scheme 22). 27Al-NMR studies confirmed that in a mixture of [bmim][Cl]/$AlCl_3$ with 2.0 equivalents (0.67 mol%) of Lewis acid, Al_2Cl_7 was the predominant anion. However, if the ionic liquid is present in only a slight excess (concentrated solution), the $AlCl_4$ anion predominates after the reaction, which indicates that the HCl formed during the sulfonylation is trapped by the ionic liquid, thus changing its composition.[120]

It is also possible to carry out Friedel–Crafts alkylation in water-stable [bmim][PF_6], using the $Sc(OTf)_3$ catalyst for the alkylation of arenes by simple alkenes. In organic solvents, rare earth(III) salts do not catalyze Friedel–Crafts alkylations, and no reaction is observed under these conditions. However, the reaction of benzene and cyclohexene catalyzed by $Sc(OTf)_3$ in [bmim][SbF_6] produces the substituted benzene **67** in a high yield (92%). Since benzene is used as the solvent, the product is isolated by decantation of the organic layer (Scheme 23). No reaction occurs in hydrophilic ionic liquids such as [bmim][BF_4] or [bmim][OTf], although

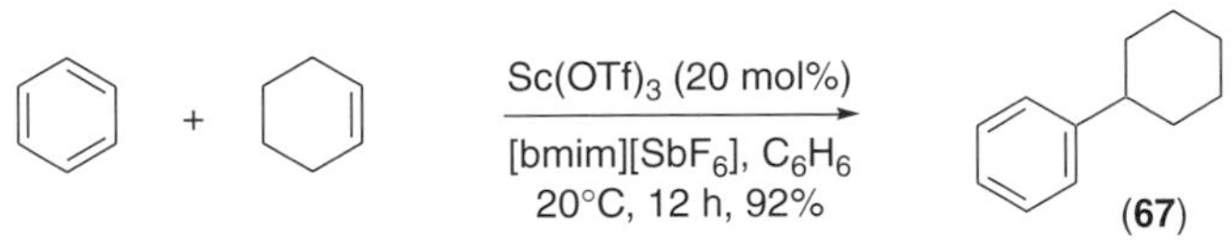

+

$Sc(OTf)_3$ (20 mol%)
$[bmim][SbF_6]$, C_6H_6
20°C, 12 h, 92%

(**67**)

Scheme 23

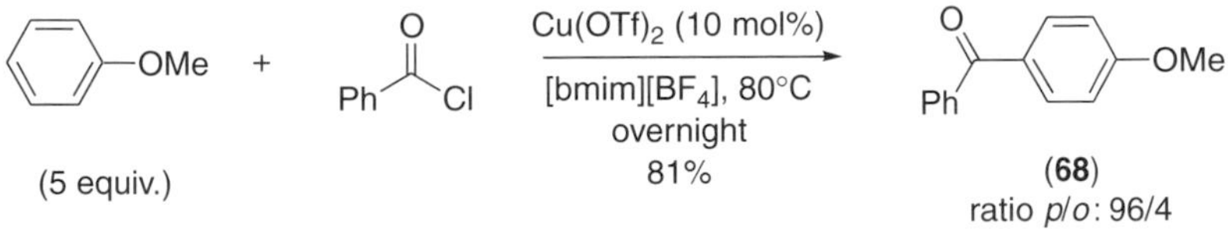

OMe + O Ph Cl

(5 equiv.)

$Cu(OTf)_2$ (10 mol%)
$[bmim][BF_4]$, 80°C
overnight
81%

O Ph OMe

(**68**)
ratio *p*/*o*: 96/4

Scheme 24

the catalyst is much more soluble in these solvents. This approach not only leads to easy-to-follow protocols, but also makes it possible to recycle and reuse the ionic liquid containing the catalyst without loss of activity.[121]

Water-stable ionic liquids were later used for Friedel–Crafts acylations, using a metal triflate catalyst. $Cu(OTf)_2$ proved to be the most efficient catalyst for this transformation, and acylation of anisole by benzoyl chloride in $[bmim][BF_4]$ gave almost exclusively the *para* adduct **68** (Scheme 24). This reaction can also be performed in organic solvents, but an accelerated rate is observed in ionic liquids.[122] Catalyst loading can be decreased (up to 1 mol%) using bismuth(III) salts as catalysts.[123]

Friedel–Crafts acylation can also be carried out in the absence of a catalyst. Seddon has described the synthesis of Pravadoline **70**, a non-steroidal anti-inflammatory drug: starting from commercially available reagents, the indole **69** was prepared by a nucleophilic displacement in $[bmim][PF_6]$. Friedel–Crafts acylation in the same solvent produced the desired product **70** (Scheme 25). The ionic liquid was reused after isolation of the product.[124]

2.3.4 *Epoxidation*

Since its discovery in 1991, methyltrioxorhenium (MTO, **80**) has attracted much interest as one of the most versatile catalysts for oxidation.[125–128] When it is associated with a stoichiometric amount of H_2O_2, the system can efficiently transform alkene to epoxide, although formation of undesired diol can occur. Alternatively, water-free conditions, using urea hydrogen peroxide (UHP), allow the formation of the desired epoxide without byproducts. A major drawback of the MTO/UHP system is its insolubility in organic solvents, leading to a kinetically slow heterogeneous system.

By using ionic liquids, in which UHP, MTO (**80**) and the intermediate peroxorhenium species are soluble, homogeneous conditions can be obtained. This approach has been used for the epoxidation of a large number of alkenes, which have been

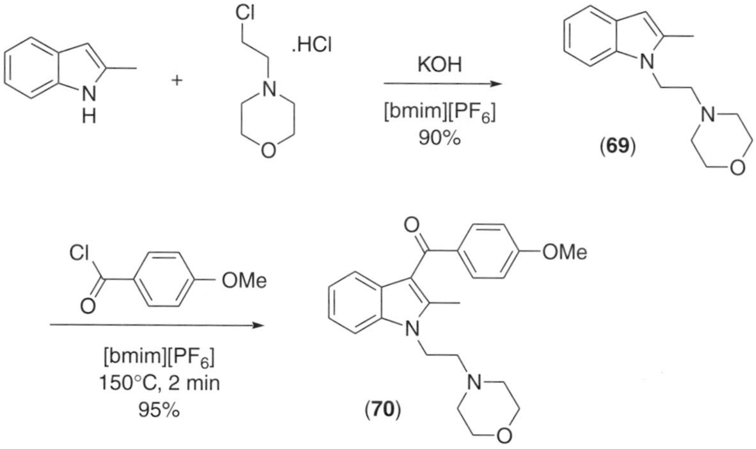

Scheme 25

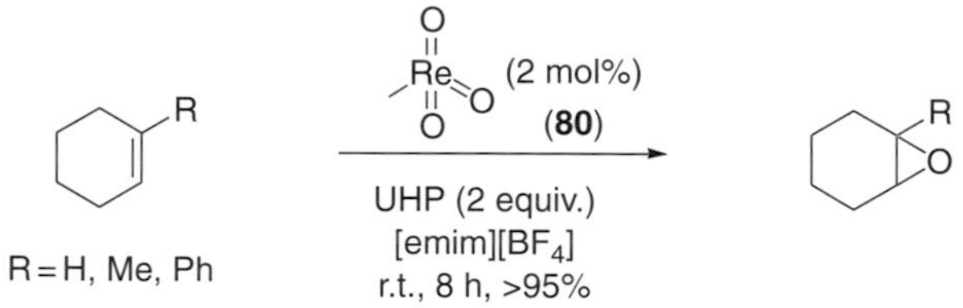

Scheme 26

transformed into their epoxide derivatives in $[emim][BF_4]$ in yields exceeding 95% (Scheme 26). With this system, no trace of diol was observed.[129,130]

Epoxidation of enone can be carried out in the absence of a metal catalyst in ionic liquid. A basic aqueous solution of hydrogen peroxide behaves as an efficient oxidant both in $[bmim][PF_6]$, which is immiscible in water, and in hydrophilic $[bmim][BF_4]$. Quantitative yields of the corresponding epoxides were obtained after very short reaction times (Scheme 27). The use of H_2O_2 in ionic liquids is a viable alternative to using water as the solvent. There is a significant rate acceleration when these salts are used as the reaction medium.[131] This strategy has been successfully applied to the epoxidation of chromones and flavonoids; in all cases, the reaction was dramatically improved in $[bmim][BF_4]$ compared with standard classical organic solvents.[132]

Ionic liquids can efficiently immobilize transition metal complexes used as catalysts for epoxidation. This phenomenon is encountered not only when specially designed, polar ligands are used,[133] but also for unmodified catalysts. For instance, several unsuccessful attempts have been made to immobilize Jacobsen's chiral Mn(III)salen epoxidation catalyst **81**. Ionic liquids are the first medium to immobilize this complex efficiently without requiring additional modification of the ligand.

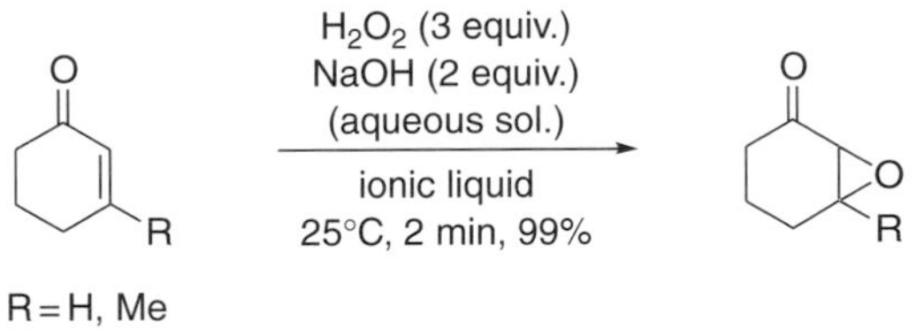

H_2O_2 (3 equiv.)
NaOH (2 equiv.)
(aqueous sol.)
ionic liquid
25°C, 2 min, 99%

R = H, Me

Scheme 27

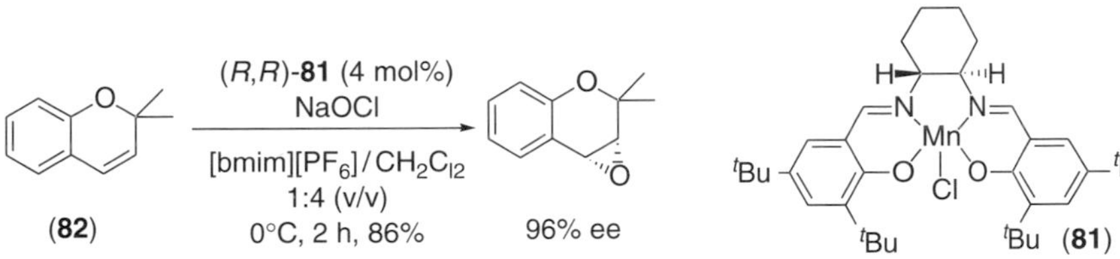

(*R*,*R*)-**81** (4 mol%)
NaOCl
[bmim][PF_6]/CH_2Cl_2
1:4 (v/v)
0°C, 2 h, 86%

(**82**) 96% ee (**81**)

Scheme 28

Asymmetric epoxidation of 2,2-dimethylchromene **82** by (*R*,*R*)-**81** (4 mol%) in a mixture of [bmim][PF_6] and CH_2Cl_2 at 0°C using NaOCl gave the epoxide in a high yield and enantioselectivity (Scheme 28). This approach was validated for different types of alkene, yielding similar results. The effect of the [bmim][PF_6] on the rate of the reaction was clearly demonstrated, since epoxidation proceeded three times more slowly in the absence of the ionic liquid. After washing with water and extraction of the product using hexane, the ionic liquid phase containing the catalyst was reused for further epoxidation. Although immobilization was made possible, a decrease was observed in the efficiency of the transformation and, after the fifth run, a much lower yield (53%) was obtained (88% ee). This problem was attributed to minor degradation of the catalyst **81** under the reaction conditions.[134]

As well as these epoxidation reactions, several oxidation reactions have also been reported in ionic liquids, including the oxidation of aromatic aldehydes to acids,[135] of alcohols to aldehydes or ketones,[53,136] of styrene to acetophenone[137] or styrene carbonate[138] and of benzyl alcohols.[139,140] Ionic liquids exhibit great stability toward oxidants, as has been demonstrated in electrochemical studies,[27] leading to an ever-growing number of publications.

2.3.5 *Palladium-catalyzed C–C bond formation*

2.3.5.1 *The Mizoroki–Heck reaction*

The Mizoroki–Heck reaction, a palladium-catalyzed coupling of olefins with aryl or vinyl halides/triflates, is a powerful method for carbon–carbon bond formation.[141–148] High efficiency is usually obtained only by starting from expensive aryl iodide (or bromide) or by using a fairly large amount of catalyst. Improvement of the catalytic activity as well as recovery and recycling of the catalyst is needed.

In the case of deactivated, electron-rich aryl chlorides in particular, the Mizoroki–Heck reaction requires a high temperature, long reaction times and a high catalyst loading. A few successful methods have been proposed to improve the catalyst or the ligand, or by adding additives.[149] Other original approaches are to form the chromium tricarbonyl chloroarene complex[150] – the $Cr(CO)_3$ moiety reducing the electronic density on the aromatic ring – or to transform *in situ* the chloroarene into the more reactive iodoarene using a nickel(II) co-catalyst and sodium iodide.[151]

Since Mizoroki–Heck reactions give the best results in polar solvents (dimethylformamide (DMF) or acetonitrile), and tetraalkylammonium salts have a beneficial effect on palladium(II) catalysts, the use of tetraalkylammonium salts as ionic liquids is favorable. The first example was reported in 1996 for the preparation of butyl *trans*-cinnamate **83** from bromobenzene in molten tetraalkylammonium and tetraalkylphosphonium bromide salts (Scheme 29). The best results were obtained using hexadecyltributylphosphonium bromide (melting point 54°C) as the ionic liquid, the reaction proceeding quantitatively even in the absence of additive ligands. An increased rate was obtained by the addition of lithium bromide or sodium acetate. The catalytic system remained stable for several reaction cycles.[152]

$[Bu_4N][Br]$ salt (melting point 103°C) is an efficient medium for the Mizoroki–Heck reaction; a screening of the standard palladium catalysts showed that, in almost all cases, a high rate acceleration was achieved using this ionic liquid instead of DMF or traditional organic solvents. The efficiency of this system was demonstrated for low-reactive chloroarene: even with the ligand-free $PdCl_2$ catalyst (5 mol%), the Mizoroki–Heck reaction of chlorobenzene and styrene produced the stilbene **84** in an excellent 89% yield (Scheme 30).[153,154]

Although the turnover number (TON) of this coupling remains low (18) and the temperature of the reaction is high (150°C), it represents an efficient use of ionic liquids because the reaction proceeds without the addition of a ligand or additive. When the same methodology was applied to the reaction between *p*-iodotoluene

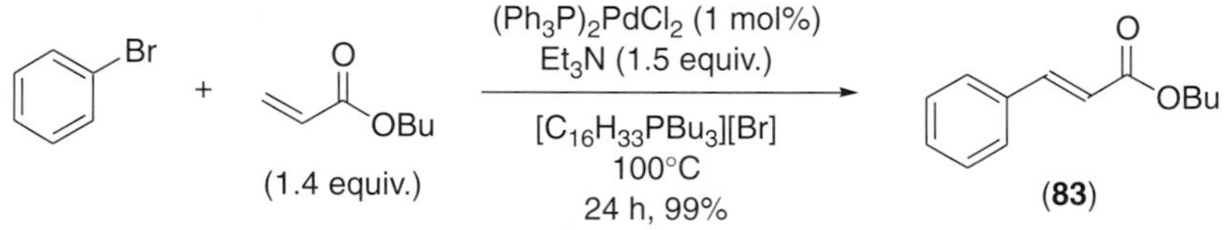

Br + OBu (1.4 equiv.)
$(Ph_3P)_2PdCl_2$ (1 mol%), Et_3N (1.5 equiv.) / $[C_{16}H_{33}PBu_3][Br]$, 100°C, 24 h, 99%
→ OBu (**83**)

Scheme 29

Cl + (1.5 equiv.)
PdCl2 (5 mol%), NaOAc (1.2 equiv.) / $[Bu_4N][Br]$, 150°C, 45 h, 89%
→ (**84**)

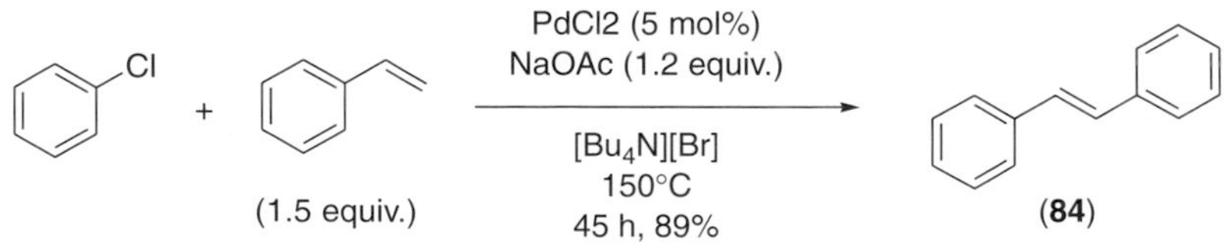

Scheme 30

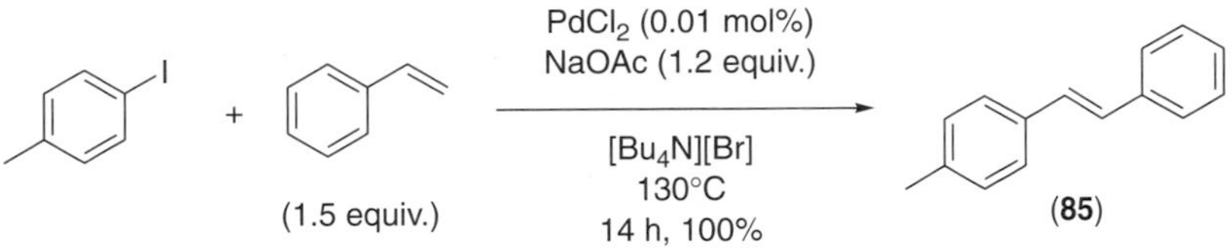

$PdCl_2$ (0.01 mol%)
NaOAc (1.2 equiv.)
(1.5 equiv.)
$[Bu_4N][Br]$
130°C
14 h, 100%
(**85**)

Scheme 31

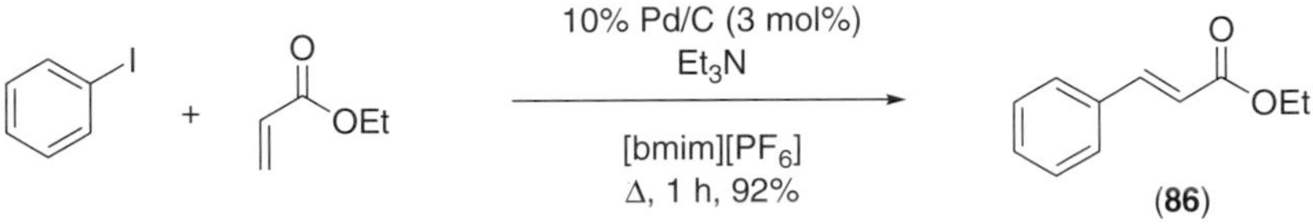

10% Pd/C (3 mol%)
Et_3N
OEt
$[bmim][PF_6]$
Δ, 1 h, 92%
O
OEt
(**86**)

Scheme 32

and styrene, alkene **85** was quantitatively obtained in a shorter reaction time, and with a catalyst loading as low as 0.01% (Scheme 31).

Furthermore, if $[NBu_4][OAc]$ is used as a base instead of sodium acetate to avoid the formation of sodium salts, the product can be distilled and the medium reused for further reactions. This recycling procedure is limited, however, by the precipitation of black palladium from the reaction mixture.

To recycle the palladium catalyst used for the Mizoroki–Heck reaction, a reaction using Pd/C as a heterogeneous catalyst was performed in $[bmim][PF_6]$. Ethyl cinnamate **86** was extracted simply from the ionic liquid using diethyl ether or hexane (Scheme 32). After the reaction, the Pd/C remained suspended in the ionic liquid, suitable for reuse. Since the $Et_3N^+I^-$ formed in the course of the Mizoroki–Heck reaction accumulates in the $[bmim][PF_6]$, slightly lower yields were obtained for successive runs. However, washing the ionic liquid with water removed any iodide salt present.[155]

The use of microwave ovens offers a solution to reducing the reaction time of organic transformations and leads to an increased yield compared with traditional methods. However, because the reaction medium is quickly heated to high temperatures, problems can occur as a result of an increase in the internal pressure in sealed vessels. Ionic liquids have proved to be an efficient aid for microwave heating. They can be heated for an extended period without signs of decomposition or an increase in pressure. Furthermore, when a small amount of ionic liquid is added to a volatile organic solvent, it can be heated to well above its boiling point.[156]

Combining ionic liquids and microwave techniques leads to an interesting tool for increasing the rate of traditionally slow Mizoroki–Heck reactions. Indeed, when performing the reaction under high temperature with microwave heating, the reaction time can be reduced impressively. The Mizoroki–Heck reaction between iodobenzene and butyl acrylate in $[bmim][PF_6]$ catalyzed by simple $PdCl_2$ in the presence of a phosphine ligand produced the adduct **87** in an excellent yield (95%). The transformation proceeded at high temperature (180°C), but for a very short

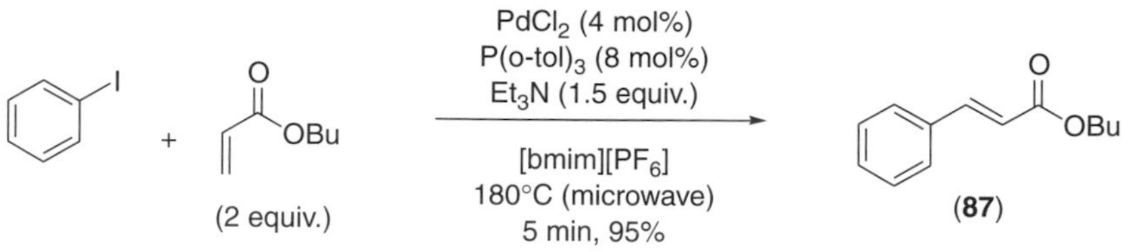

+ OBu (2 equiv.)

$PdCl_2$ (4 mol%)
P(o-tol)$_3$ (8 mol%)
Et_3N (1.5 equiv.)
[bmim][PF_6]
180°C (microwave)
5 min, 95%

OBu (**87**)

Scheme 33

OBu + Br (5 equiv.)

Pd(OAc)$_2$ (2.5 mol%)
$Ph_2P(CH_2)_3PPh_2$
(5 mol%)
Et_3N (1.2 equiv.)
[bmim][BF_4]
100°C, 24 h, 95%

BuO (α-**88**)

Scheme 34

time (Scheme 33). Almost the same conversion was obtained in the absence of the phosphine. Because of its stability, even at this high temperature, the ionic liquid containing the catalyst was successfully reused without significant loss of activity.[157]

As well as offering rate acceleration and the possibility of recovering the catalyst, ionic liquids can lead to much more regioselective Mizoroki–Heck reactions. An important feature for α-arylation and β-arylation of electron-rich enol ethers can be observed as a result of the competition between the cationic and anionic pathways. To favor the formation of an α-aryl derivative, expensive triflate starting materials can be used or else a stoichiometric amount of silver triflate or thallium acetate has to be added. This reaction should proceed via the cationic pathway in ionic liquids because of the polarity of the medium; indeed, the Mizoroki–Heck coupling between 1-bromonaphthalene and butyl vinyl ether in polar [bmim][PF_6] led exclusively to α-**88** (Scheme 34), representing a great improvement over the 75 : 25 ratio between α-**88** and β-**88** obtained in DMSO – which is the best regioselectivity attained in an organic solvent. The reaction took place slightly more slowly in the ionic liquid, however, and longer reaction times were required.[158]

Efforts have been made to explain the high rate acceleration of Mizoroki–Heck reactions in ionic liquids. The formation of the dialkylimidazol-2-ylidene palladium complex under conditions similar to those employed for the Mizoroki–Heck reaction has been studied.[159] The C_2–H proton of the imidazolium cation exhibits high acidity and can be deprotonated to form a carbene species, behaving as a good ligand for transition metals.[153,160] Therefore, in the presence of a palladium salt and a base, [bmim][Br] formed the dimeric carbene complex **89**, which further evolved to the monomeric *cis*-**90** and *trans*-**90** complexes. Each of these exists as an *anti* and a *syn* rotamer owing to the sterically demanding *N*-alkyl substituents (Scheme 35; only the *anti*-**90** rotamers are represented).

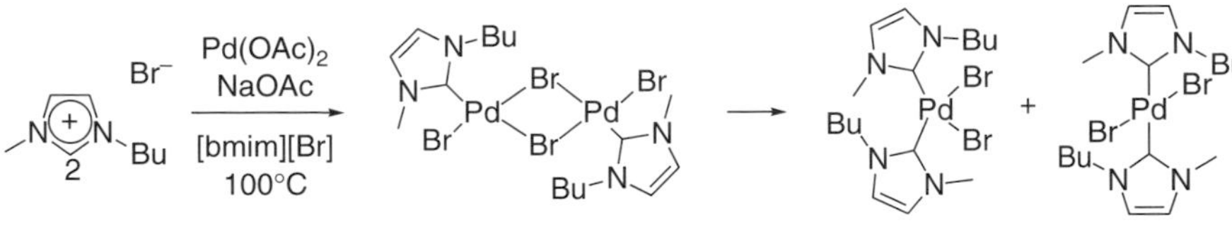

Scheme 35

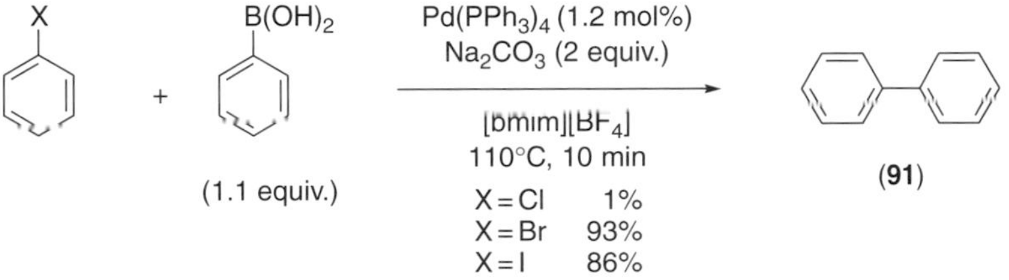

Scheme 36

The authors of this study demonstrated that similar species, along with an unidentified one, were present during the Mizoroki–Heck reaction. It was proved that isolated *trans*-**90** behaved as an efficient catalyst for this reaction. Finally, when [bmim][BF_4] was used as the starting material, no palladium carbene complex was obtained. This observation might explain why Mizoroki–Heck reactions proceed more quickly in halide ionic liquids than in non-coordinating ones such as PF_6 salts.[159] Ultrasound-promoted formation of carbene palladium complexes can also allow their *in situ* preparation at room temperature.[161]

2.3.5.2 The Suzuki–Miyaura cross-coupling reaction

The Suzuki–Miyaura cross-coupling reaction is a standard method for carbon–carbon bond formation between an aryl halide or triflate and a boronic acid derivative, catalyzed by a palladium metal complex. As with the Mizoroki–Heck reaction, this cross-coupling reaction has been developed in ionic liquids in order to recycle and reuse the catalyst. In 2000, the first cross-coupling of a halide derivative with phenylboronic acid in [bmim][BF_4] was described. As expected, the reaction proceeded much faster with bromobenzene and iodobenzene, whereas almost no biphenyl **91** was obtained using the chloride derivative (Scheme 36). The ionic liquid allowed the reactivity to be increased, with a turnover number between 72 and 78. Furthermore, the catalyst could be reused repeatedly without loss of activity, even when the reaction was performed under air.[162] Cross-coupling with chlorobenzene was later achieved – although with only a moderate yield (42%) – using ultrasound activation.[163]

These procedures require activation (heating and sonication, respectively), and yields are only moderate to good. Better results are obtained using

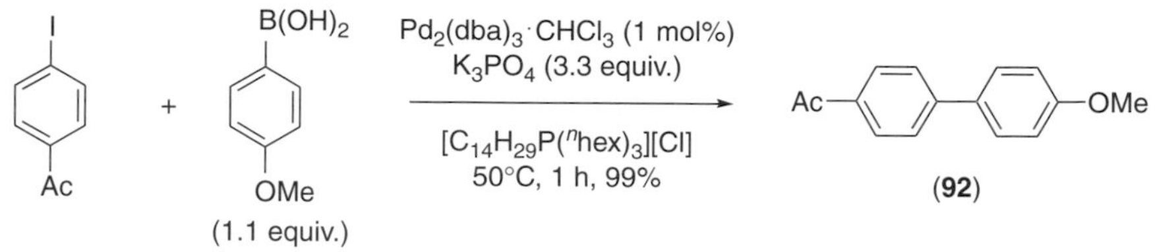

I, Ac + $B(OH)_2$, OMe (1.1 equiv.)

$Pd_2(dba)_3 \cdot CHCl_3$ (1 mol%)
K_3PO_4 (3.3 equiv.)
$[C_{14}H_{29}P(^{n}hex)_3][Cl]$
50°C, 1 h, 99%

Ac–C_6H_4–C_6H_4–OMe (**92**)

Scheme 37

I + $B(OH)_2$, Me (1.1 equiv.)

$PdCl_2$ (2 mol%)
K_2CO_3 (3 equiv.)
90°C, 8–10 h
93 : 92%
94 : 93%

C_6H_5–C_6H_4–Me

Bu, N^+, Bu, Bu, Bu BF_4^- (**93**)

N^+, Bu, Bu BF_4^- (**94**)

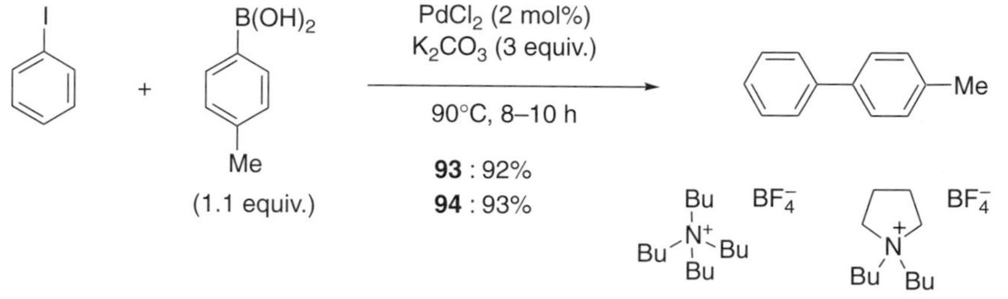

Scheme 38

hexadecyltrihexylphosphonium chloride. A typical example is the formation of adduct **92**, which proceeds quantitatively at 50°C in 1 h (Scheme 37). By adding PPh_3 as a ligand and allowing a longer reaction time (30 h), even the chloride derivative reacted at 70°C to give a good 84% yield.[73]

The Suzuki–Miyaura coupling can also be performed with a ligandless palladium catalyst, using a mixture of water and ammonium salt **93** or pyrrolidinium salt **94** (Scheme 38).[164] These BF_4 salts have high melting points (125°C and 160°C, respectively) but melt in water to form a biphasic mixture at much lower temperatures (~50°C and 80°C). The recycling was not efficient, but salts **93** and **94** can be separated and purified, unlike the case with standard ionic liquids.

2.3.5.3 *Other palladium-catalyzed cross-coupling reactions*

Ionic liquids behave not only as efficient solvents for cross-coupling reactions but also as ligands. For the Negishi cross-coupling between arylzinc halides and aryl iodides, phosphine **95**, which is derived directly from [bmim][PF_6], serves as the ligand for the palladium catalyst. The desired product **96** was obtained in a 91% yield in biphasic conditions (Scheme 39), but reuse of the ionic liquid phase led to a significant decrease in yield.[165]

A copper-free Sonogashira cross-coupling with efficient recycling and reuse of the ionic liquid containing the palladium catalyst was reported recently in 2002.[166] The possibility of recycling the palladium acetate used for the Sonogashira reaction was also demonstrated, although a loss of activity was observed from the third cycle onward as a result of leaching of the catalyst.[167] Using (bisimidazole)Pd(II) as the

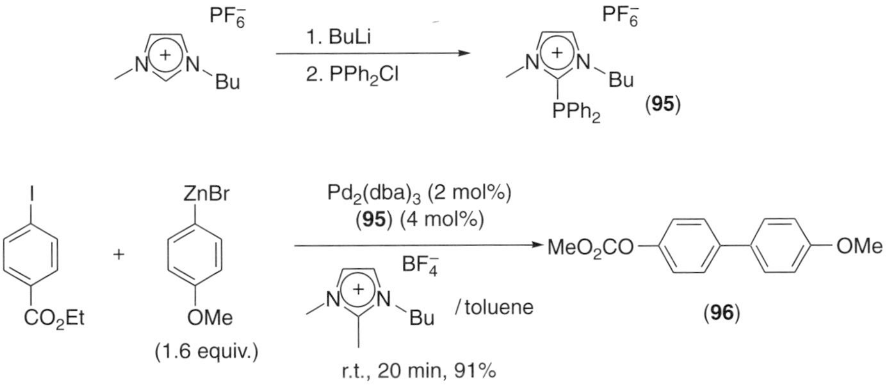

Scheme 39

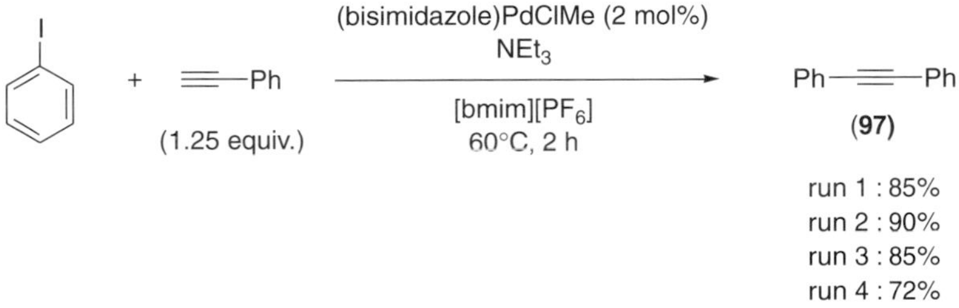

Scheme 40

catalyst, the coupling proceeds even in the absence of the copper co-catalyst and bulky phosphine ligands, and diphenylacetylene (**97**) has been obtained in yields ranging from 72% to 90% (four runs, Scheme 40).[168]

2.3.6 *The Diels–Alder reaction*

The Diels–Alder reaction is a very well known carbon–carbon bond-forming reaction. Owing to the concerted mechanism, this [4 + 2] cycloaddition can proceed with a high degree of regioselectivity and stereoselectivity. It was shown in 1997 that dialkylimidazolium salts behave as efficient Lewis acid catalysts for the Diels–Alder reaction of methacrolein and cyclopentadiene. Using a catalytic amount of diethylimidazolium salt **98** (20 mol%), the reaction proceeded at −25°C (Scheme 41). Although the yield remains very low (40%, **99** *endo* : *exo* ratio 13 : 87), it is interesting to note that no cycloaddition product is obtained in the absence of a catalyst. An asymmetric Diels–Alder reaction was attempted using the chiral imidazolium salt **100** (although it was not clearly stated in the publication, this salt is probably solid at room temperature). Although comparable yields were obtained, there was no clear evidence of asymmetric induction.[169]

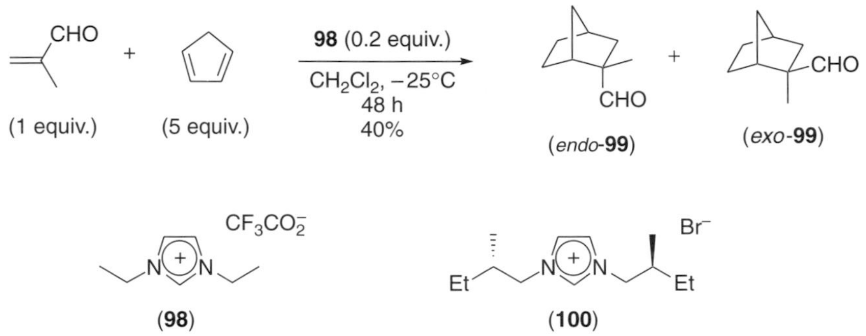

Scheme 41

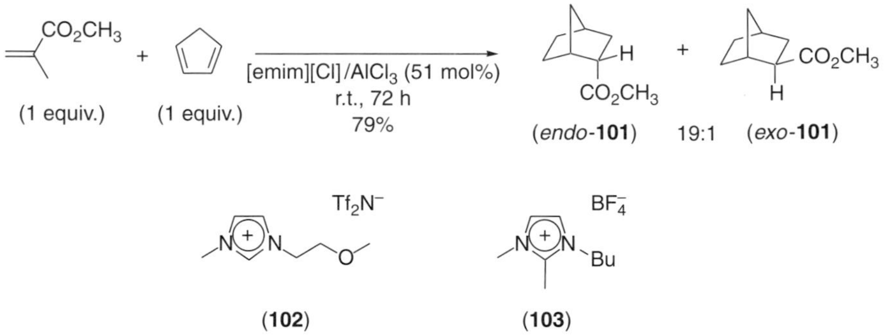

Scheme 42

These Diels–Alder reactions were performed in dichloromethane, but the efficiency of imidazolium salts as Lewis acid catalysts suggested that ionic liquids could improve the rate of such transformations. Indeed, Diels–Alder reactions result in greater stereoselectivity when performed in a 5*M* solution of $LiClO_4$ in diethyl ether.[170]

A high ratio (19 : 1) between the *endo* and the *exo* adducts **101** was obtained during the reaction of methyl acrylate and cyclopentadiene in acidic [emim][Cl]/$AlCl_3$ (51 mol%) (Scheme 42). When this reaction was performed in basic [emim][Cl]/$AlCl_3$ (48 mol%) instead, a much lower selectivity (5.25 : 1) was obtained, confirming the contribution of the acidic medium.[171] Similar transformations were also performed in moisture-stable ionic liquids, but selectivities were lower.[81,172] The enhanced *endo*-selectivity was attributed to a hydrogen bond between the ionic liquid and the methyl acrylate. The selectivity increased with the ability of the imidazolium cation to act as a hydrogen bond donor: the best results were obtained with [$EtNH_3$][NO_3] and ionic liquid **102** with ether functionality. When the acidic C_2–H proton of the imidazolium cation was replaced with a methyl group (**103**), lower selectivity was attained.[173]

+ (3 equiv.)

$Sc(OTf)_3$ (0.2 mol%)

$[bmim][PF_6]$, 20°C, 2 h

99%

Scheme 43

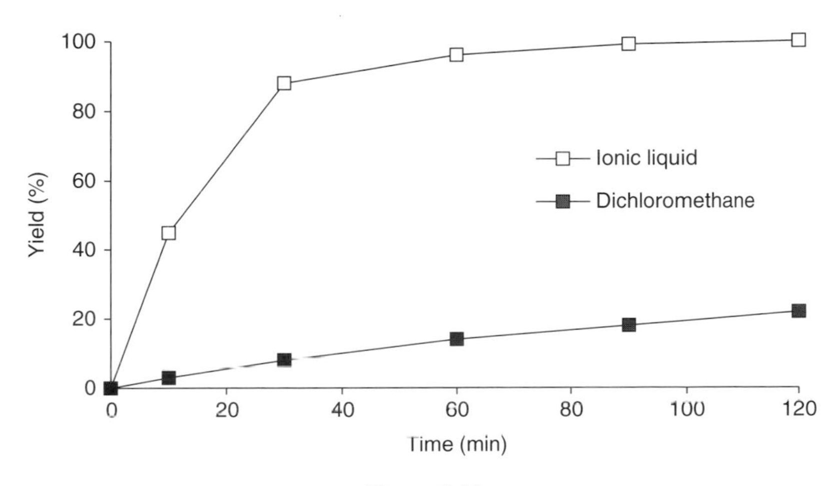

Figure 2.13

The ability of ionic liquids to dramatically improve the yields and selectivities of reactions is clearly shown by the scandium-triflate-catalyzed Diels–Alder reaction.[174] This cycloaddition usually has low turnover (TON < 10–20). Song studied this reaction in ionic liquids, which offered the possibility of recovering the catalyst. The cycloaddition between 1,4-naphthoquinone and 1,3-dimethylbutadiene catalyzed by a very small loading (0.2 mol%) of $Sc(OTf)_3$ in the standard $[bmim][PF_6]$ ionic liquid was used as a test reaction (Scheme 43).

A high rate acceleration is observed for this transformation, which is known to proceed rather slowly in organic solvents. In dichloromethane, low conversion (<25%) was observed after 2 h, whereas the reaction was almost complete in 60 min in $[bmim][PF_6]$ (Figure 2.13). Products were extracted from the ionic liquid using diethyl ether, and the ionic liquid phase was reused without loss of activity. Along with the acceleration of the reaction rate, an improved *endo*/*exo* selectivity compared with organic solvents was observed.[175]

The use of a scandium trifluoromethanesulfonate catalyst has also been reported for aza-Diels–Alder reactions. The reaction of benzaldehyde and amine **104** in [emim][OTf] as the ionic liquid led to the *in situ* formation of the corresponding imine. Cycloaddition of this imine with Danishefsky's diene gave the *N*-aryl-6-phenyl-5,6-dihydro-4-pyridone **105** in a quantitative yield (Scheme 44).

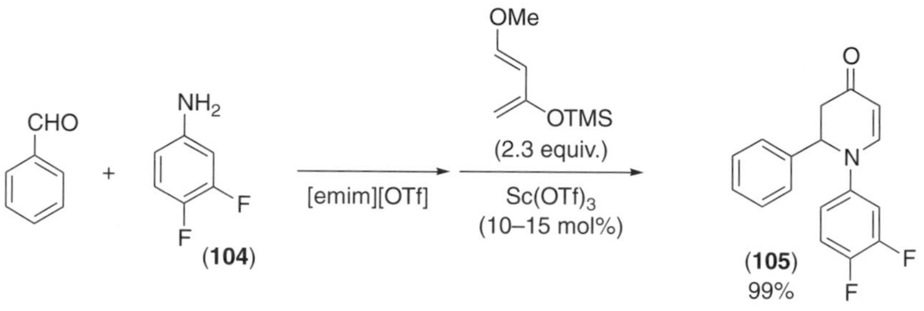

Scheme 44

Microencapsulated $Sc(OTf)_3$ was used as the catalyst, allowing clean isolation of the product from the reaction mixture by extraction with diethyl ether. The ionic liquid containing the Lewis acid was repeatedly reused.[39]

2.3.7 Biocatalysis in ionic liquids

An important number of organic reactions are now catalyzed by whole cells or isolated enzymes. However, there are still problems associated with the solubility, yield and selectivity of these biotransformations. Ever since the solubilization of alkaline phosphatase in a mixture of $[Et_3NH][NO_3]$ and water (4 : 1), it has been known that enzymes can be stable in ionic liquids.[176] Recent research shows that ionic liquids can be used efficiently as a medium for biocatalytic processes.[177]

An approach to biocatalysis that is being increasingly used combines an aqueous biomedium, in which the reaction occurs, and an organic phase that acts as a reservoir. Since substrates are usually hydrophobic, only a small amount is present in the aqueous layer, preventing inhibition of the biocatalyst. The main problem with this technique is that organic solvents can damage the cell walls of the bacteria, leading to a lowering of the catalytic efficiency. Lye demonstrated that organic solvents could be replaced by ionic liquids. The *Rhodococcus* R312 microorganism catalyzed the transformation of 1,3-dicyanobenzene to 3-cyanobenzamide and 3-cyanobenzoic acid in a biphasic mixture of water and $[bmim][PF_6]$. The cells remained in the aqueous layer, and the ionic liquid, which can dissolve a large concentration of the substrate, acted as a reservoir.[178]

The first enzyme-catalyzed reaction to be effectively performed in ionic liquids was described in 2000. The reaction of two amino acid derivatives was catalyzed by thermolysin in $[bmim][PF_6]$ (containing 5% water) to give aspartame **106** (Scheme 45). Although the rates were comparable to those observed in a mixture of ethyl acetate and water, the enzyme exhibited much higher stability in the ionic liquid and did not require immobilization.[179]

Biocatalysts can also be used in the complete absence of water. *Candida antartica* lipase B (CaLB), used either as the free enzyme or in an immobilized form, catalyzes transesterification, ammonolysis and perhydrolysis. For example, octanoic acid was quantitatively transformed to amide **107** in 4 days (Scheme 46) – a significant rate

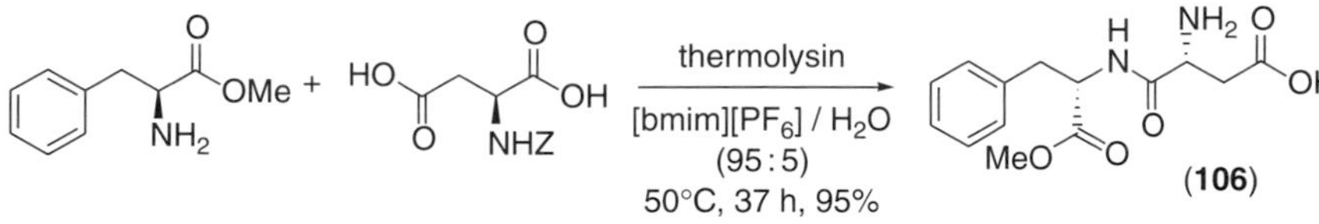

Scheme 45

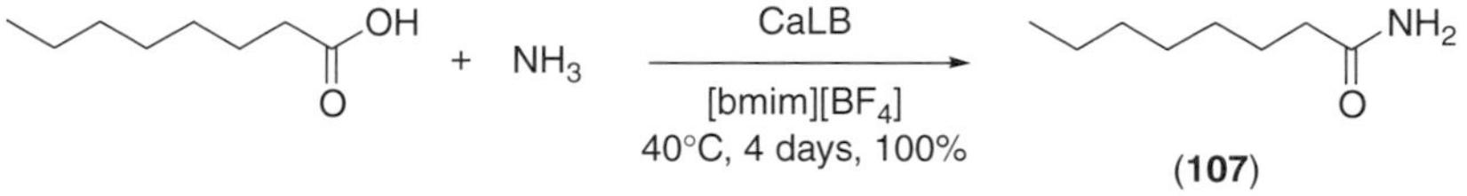

Scheme 46

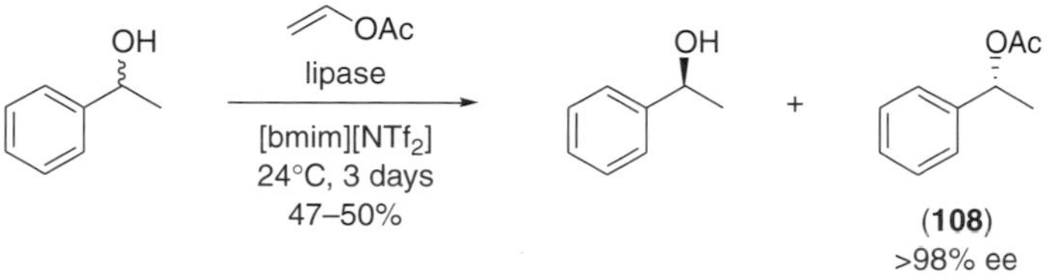

Scheme 47

improvement over the 17 days required for ammonium carbamate in methylisobutyl ketone.[180]

Lipase-catalyzed enantioselective transesterifications have been studied by different research groups. Wasserscheid studied the kinetic resolution of racemic 1-phenylethanol by enantioselective acylation with vinyl acetate in different ionic liquids. The best results were obtained using *Candida antartica* lipase B and *Pseudomonas* sp. lipase in [bmim][NTf_2] or [bmim][TfO], with up to 50% conversion and >98% enantioselectivity (Scheme 47). A slight improvement in enantioselectivity was obtained compared with the standard biotransformations in methyl *tert*-butyl ether. Product **108** can be isolated by distillation at 85°C under reduced pressure, and since the lipases are stable in the ionic liquids even upon heating, they can be reused with almost no loss of activity and similar enantioselectivity.[181] Surprisingly, the performance of this reaction was not good in structurally similar [bmim][PF_6] and [bmim][BF_4]. This was later explained by the presence of impurities in these ionic liquids.[54]

Kim performed similar transesterifications, observing that lipases were up to 25 times more enantioselective in ionic liquids ([bmim][PF_6] and [emim][BF_4]) than in conventional organic solvents.[182]

The incubation of enzymes in ionic liquids can also lead to increased reactivity. Sheldon showed that CaLB could be heated as a suspension in [bmim][PF_6] before use. Higher transesterification activity was observed than for the untreated enzyme, both when the enzyme was free (SP 525, 120% activity after 20 h incubation) and when it was immobilized (Novozym 435, 350% activity after 40 h incubation). In comparison, incubation of enzymes in *tert*-butanol led to a decrease of the biocatalytic activity.[177]

2.4 The future of ionic liquids

Although ionic liquids are usually vaunted in the literature as promising media for industrial applications, their actual use in organic processes in industry is extremely limited because they still have some important drawbacks. Intensive study has been underway for just one decade, and data on toxicology, biodegradation and environmental impact are still incomplete. Ionic liquids are believed to be harmless to both humans and the environment, but more thorough investigations need to be carried out before these solvents can be used extensively on industrial scales. It is also important to underline that, to justify the label *green solvents*, ionic liquids have to be prepared in an ecological way, without the generation of a large amount of waste material.

Today, ionic liquids are usually prepared on a small or medium scale, and their synthesis will have to be adapted so that they can be obtained both on a larger scale and, most importantly, at lower cost: although they can be recycled, ionic liquids still cost significantly more than organic solvents. On the other hand, these new solvents can be purchased from a number of companies that have started selling and producing them, sometimes in large quantities: Chemada Fine Chemicals, Covalent Associates, C-TRI (Chem Tech Research Incorporation), Cytec, Fluka, Merck, Ozark Fluorine Specialties, Sachem, Solvent Innovation and Wako. QUILL's products are also available through Acros.[183]

Intellectual property is also a concern in the future development of ionic liquids. Given that one of the main targets of these solvents is use in industrial processes, a relatively large number of patents are already protecting this field. Patents cover mainly (1) the preparation and/or new types of ionic liquids and (2) the use of ionic liquids as materials (solvent, catalyst, extraction medium, etc.).

Most tetrafluoroborate and hexafluorophosphate ionic liquids are patent free; however, owing to their cost, their insufficient stability toward hydrolysis and issues concerning their disposal, such ionic liquids are not useful for industrial applications. These problems prompted researchers to develop new types of ionic liquids with different anions, many of them being protected by state-of-the-matter patents, for example, HSO_4^- (Ref. 184), $PF_3(C_2F_5)_3^-$ (Ref. 185) and borate derivatives.[186] To solve the problem of the high cost of imidazolium-based ionic liquids, one patent claims broad coverage of cheap phosphonium ionic liquids.[187]

Scaled-up processes have already been developed using ionic liquids. The Institut Français du Pétrole (IFP) improved its nickel-catalyzed process for dimerizing butene using a chloroaluminate ionic liquid as solvent. One of the main improvements is the possibility of isolating the octenes formed by direct decantation, since they are not soluble in the ionic liquid. This process, known as Difasol, has been tested at a pilot-scale plant and is now available for licensing.[188–191]

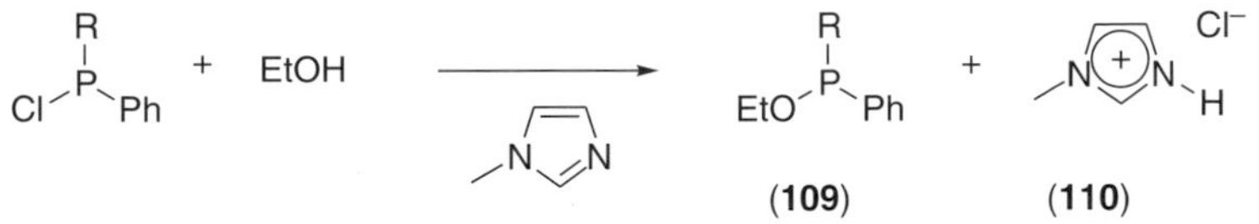

Scheme 48

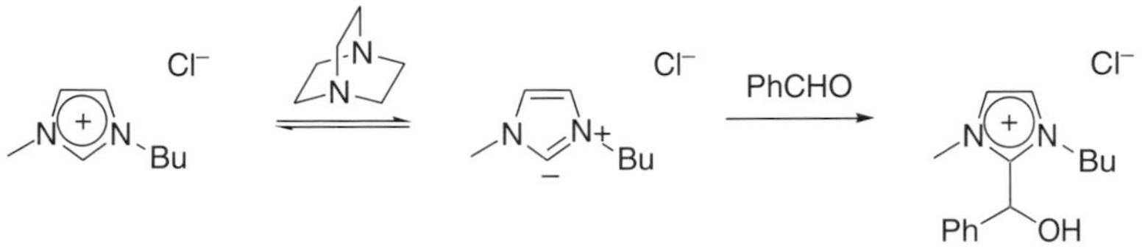

Scheme 49

In March 2003, the use was reported of ionic liquids in a commercial process by BASF. The reaction of phenylchlorophosphines and ethanol forms alkoxyphenylphosphines **109** along with HCl, which is scavenged by *N*-methylimidazole (Scheme 48). The [hmim][Cl] salt **110** formed is an ionic liquid with a melting point of 75°C; this leads to a biphasic mixture, which can be more easily stirred than a solution containing a suspension. This ionic liquid also acts as catalyst for this process of biphasic acid scavenging utilizing ionic liquids (BASIL).[192–194]

At the same time as new applications of ionic liquids are discovered on almost a daily basis, limitations of these reaction media are also uncovered. While studying the Morita–Baylis–Hillman reaction in ionic liquids, Aggarwal observed that [bmim][Cl] was deprotonated by the weak base present in the reaction mixture, leading, after reaction with benzaldehyde, to salt **111** (Scheme 49). Deprotonation of imidazolium salts with strong bases (KO^tBu or NaH) is well known, providing, for example, an easy route to Pd carbene complexes (Section 2.3.5.1). However, this observation limits the use of imidazolium-based ionic liquids even in weakly basic conditions, where they can react with electrophiles.[195] It also explains previous works reporting low yields for reactions performed in these conditions, such as the Horner–Wadsworth–Emmons reaction in [emim][PF_6] or [emim][BF_4].[196]

2.5 Experimental part

2.5.1 Preparation of [bmim][Cl]

(Ref. 53) In a 500 mL flask equipped with a reflux condenser was placed 50 mL of dry toluene under N_2, followed by 1-methylimidazole (100 mL, 1.25 mol, freshly distilled over KOH) and 1-chlorobutane (143 mL, 1.27 mmol, freshly distilled over

P_2O_5). The mixture was heated at reflux for 24 h. The biphasic solution obtained after cooling was kept at −18°C overnight. A white crystalline solid formed, and the toluene was removed under N_2. The solid was recrystallized twice from a minimum amount of acetonitrile (~75 mL) and washed with aliquots of ethyl acetate. After drying *in vacuo* for 24 h, 203 g (1.16 mol, 93%) of the [bmim][Cl] salt were isolated as a white solid.

2.5.2 *Preparation of [bmim][PF$_6$]*

(Ref. 58) Solid [bmim][Cl] (200 g, 1.15 mol) was melted in a 1000 mL flask by heating. After cooling at room temperature, it was immediately diluted with water (200 mL) to reduce the viscosity. Hexafluorophosphoric acid (60% in water, 200 mL) was added dropwise over 2 h. The mixture was stirred for 1 h, and the ionic liquid phase (bottom layer) was separated and washed with water (3 × 100 mL). After drying, 260 g (0.90 mol, 80%) of the [bmim][PF_6] salt were obtained as a very pale yellow liquid of low viscosity.

(Ref. 54) An aliquot of 0.40 mol of the ionic liquid was diluted with CH_2Cl_2 (200 mL) and filtered through silica gel (~100 g) to ensure complete removal of the chlorine salt. The solution was washed twice (40 mL) with an aqueous saturated solution of Na_2CO_3, dried over $MgSO_4$ and evaporated to give the purified ionic liquid.

2.5.3 *Preparation of a chiral imidazolium ionic liquid*

(Ref. 84) To a solution of (*S*)-2-trifluoromethanesulfonyloxy-propionic acid ethyl ester (30.14 g, 121 mmol, prepared following the standard procedure) in dry Et_2O was added dropwise at −78°C a cooled solution of 1-methylimidazole (9.910 g, 1.01 equiv.) in Et_2O. After 30 min, the mixture was warmed at room temperature, and the white solid was filtered and washed with Et_2O to give 39.28 g (119 mmol, 98%) of the triflate salt. To a solution of this salt in water (60 mL) was added a solution of *N*-lithiotrifluoromethanesulfonimide in water (37.33 g, 1.1 equiv., 40 mL). After 30 min of stirring under air, the aqueous layer was removed, and the ionic liquid was washed with water (2 × 30 mL). Drying *in vacuo* at room temperature produced 48.74 g (52.08 mmol, 89%) of the desired [1-(1-(*R*)-ethoxycarbonyl-ethyl)-3-methylimidazolium][trifluoromethanesulfonimide] (**48**).

2.5.4 *Enantioselective hydrogenation of methyl acetoacetate*

(Ref. 197) To a vial containing a mixture of [bmim][BF_4] (0.5 mL) and MeOH (0.5 mL) was added methyl acetoacetate (0.05 mL, 0.46 mmol) and the ruthenium Ru(**L**)$(DMF)_2Cl_2$ catalyst (**L** = 2,2′-bis(diphenylphosphino)-1,1′-binaphthyl-6,6′-bis(phosphonic acid), 5.0 mg, 1 mol%). The vial was placed inside a 300 mL stainless steel autoclave, which was pressurized with 100 bar of H_2, and stirred at room temperature for 22 h. The reaction was then depressurized, and the mixture extracted with hexane to yield quantitatively, after evaporation of the volatile material, the desired 3-hydroxybutyrate product with 98.3% ee (determined by chiral gas chromatogrophy (GC).

2.5.5 *Epoxidation of 2,2-dimethylchromene*

(Ref. 134) A solution of commercial household bleach was buffered to pH 11.3 with a 0.05*M* solution of Na_2HPO_4 and a few drops of NaOH 1*M*. To 110 mL of this solution cooled at 0°C was added a cooled solution (0°C) of 2,2-dimethylchromene (5.0 g, 31.3 mmol) and Jacobsen's catalyst ((*R*,*R*)-**81**, 0.79 g, 1.25 mmol, 4 mol%) in a mixture of CH_2Cl_2 (30 mL) and [bmim][PF_6] (7.5 mL). The biphasic mixture was stirred at 0°C, and the reaction monitored using thin layer chromatography. After 2 h, the organic layer was separated and washed with water. The volatile materials were removed *in vacuo*, and the ionic liquid was stirred with hexane. Concentration of the hexane layer *in vacuo* followed by column chromatography on silica gel (deactivated with 1% NEt_3 in hexane/ethyl acetate) with hexane/ethyl acetate (10 : 1) as eluent produced 4.74 g (86%) of (3*R*,4*R*)-3,4-epoxy-2,2-dimethylchromene with 96% ee. The brown ionic liquid phase containing the catalyst was reused without further treatment. Enantioselectivity was determined using chiral high performance liquid chromatography (Daicel Chiralpak AD, hexane/isopropanol (95 : 5), 0.8 mL min^{-1}).

2.5.6 *Mizoroki–Heck reaction between butyl acrylate and iodobenzene under microwave irradiation*

(Ref. 157) $PdCl_2$ (17.7 mg, 0.05 mmol), butyl acrylate (0.512 g, 4.0 mmol), triethylamine (0.303 g, 3.0 mmol), iodobenzene (0.408 g, 2.0 mmol) and [bmim][PF_6] (3.0 g, 10.5 mmol) were mixed in a vial (2–5 mL) equipped with a magnetic stirrer. The mixture was heated at 180°C for 20 min in a microwave syntheziser and distilled using a kugelrohr device. The distillate was extracted with diethyl ether to produce the desired product **87**. For recycling, butyl acrylate, iodobenzene and triethylamine were added directly to the ionic liquid phase in the same vial, and irradiation was started again.

2.5.7 *Diphenylacetylene by the Sonogashira coupling reaction*

(Ref. 168) Iodobenzene (204 mg, 1 mmol) and phenylacetylene (128 mg, 1.2 mmol) were added to the (bisimidazole)PdClMe catalyst (6.4 mg, 0.02 mmol, 2 mol%), triethylamine (1.25 mmol) and [bmim][PF_6] (2 mL). The reaction mixture was heated at 60°C for 2 h. Extraction with Et_2O yielded the desired diphenylacetylene (85% GC yield), and the ionic liquid phase was washed with water to remove the amine salt and recycled for the following runs (GC yields of 90%, 85% and 72% for runs 2–4).

References

[1] Walden, P. *Bull. Acad. Sci. St. Petersburg* **1914**, 405.
[2] Sugden, S.; Wilkins, H. *J. Chem. Soc.* **1929**, 1291.
[3] This story was detailed in an excellent review: Wilkes, J. S. *Green Chem.* **2002**, *4*, 73.
[4] Hurley, F. H.; Wier, T. P., Jr. *J. Electrochem. Soc.* **1951**, *98*, 203.
[5] Chum, H. L.; Koch, V. R.; Miller, L. L.; Osteryoung, R. A. *J. Am. Chem. Soc.* **1975**, *97*, 3264.
[6] Robinson, J.; Osteryoung, R. A. *J. Am. Chem. Soc.* **1979**, *101*, 323.
[7] Wilkes, J. S.; Levisky, J. A.; Wilson, R. A.; Hussey, C. L. *Inorg. Chem.* **1982**, *21*, 1263.
[8] Scheffler, T. B.; Hussey, C. L.; Seddon, K. R.; Kear, C. M.; Armitage, P. D. *Inorg. Chem.* **1983**, *22*, 2099.
[9] Laher, T. M.; Hussey, C. L. *Inorg. Chem.* **1983**, *22*, 3247.
[10] Scheffler, T. B.; Hussey, C. L. *Inorg. Chem.* **1984**, *23*, 1926.
[11] Hitchcock, P. B.; Mohammed, T. J.; Seddon, K. R.; Zora, J. A.; Hussey, C. L.; Ward, E. H. *Inorg. Chim. Acta* **1986**, *113*, L25.
[12] Appleby, D.; Hussey, C. L.; Seddon, K. R.; Turp, J. E. *Nature* **1986**, *323*, 614.
[13] Dent, A. J.; Seddon, K. R.; Welton, T. *J. Chem. Soc., Chem. Commun.* **1990**, 315.
[14] Boon, J. A.; Levisky, J. A.; Pflug, J. L.; Wilkes, J. S. *J. Org. Chem.* **1986**, *51*, 480.
[15] Chauvin, Y.; Gilbert, B.; Guibard, J. *J. Chem. Soc., Chem. Commun.* **1990**, 1715.
[16] Parshall, G. W. *J. Am. Chem. Soc.* **1972**, *94*, 8716.
[17] Wilkes, J. S.; Zaworotko, M. J. *J. Chem. Soc., Chem. Commun.* **1992**, 965.
[18] Fuller, J.; Carlin, R. T.; De Long, H. C.; Haworth, D. *J. Chem. Soc., Chem. Commun.* **1994**, 299.
[19] Rogers, R. D.; Seddon, K. R.; Volkov, S., Eds. *Green Industrial Applications of Ionic Liquids* (Proceedings of the NATO Advanced Research Workshop held in Heraklion, Crete, Greece 12–16 April 2000.) [In: NATO Sci. Ser., II, 2003; 92]; Kluwer: Dordrecht, 2003.
[20] Wasserscheid, P.; Welton, T., Eds. *Ionic Liquids in Synthesis*; Wiley-VCH: Weinheim, 2003.
[21] Huddleston, J. G.; Rogers, R. D. *Chem. Commun.* **1998**, 1765.
[22] Varma, R. S.; Namboodiri, V. V. *Chem. Commun.* **2001**, 643.
[23] Varma, R. S.; Namboodiri, V. V. *Pure Appl. Chem.* **2001**, *73*, 1309.
[24] Lévêque, J.-M.; Luche, J.-L.; Pétrier, C.; Roux, R.; Bonrath, W. *Green Chem.* **2002**, *4*, 357.
[25] Deetlefs, M.; Seddon, K. R. *Green Chem.* **2003**, *5*, 181.
[26] Dzyuba, S. V.; Bartsch, R. A. *J. Heterocycl. Chem.* **2001**, *38*, 265.
[27] Bonhote, P.; Dias, A.-P.; Armand, M.; Papageorgiou, N.; Kalyanasundaram, K.; Graetzel, M. *Inorg. Chem.* **1996**, *35*, 1168.
[28] Dzyuba, S. V.; Bartsch, R. A. *Chem. Commun.* **2001**, 1466.
[29] Zhang, J.; Martin, G. R.; DesMarteau, D. D. *Chem. Commun.* **2003**, 2334.
[30] Harlow, K. J.; Hill, A. F.; Welton, T. *Synthesis* **1996**, 697.
[31] Herrmann, W. A.; Koecher, C.; Goossen, L. J.; Artus, G. R. J. *Chem. Eur. J.* **1996**, *2*, 1627.
[32] A combined synthesis involving the formation of the halide salt and an *in situ* anion metathesis with a sodium salt was also developed, but requests very long reaction time (typically two weeks): Wasserscheid, P.; Hilgers, C.; Boesmann, A. Single step preparation of ionic fluids by alkylation of amines, phosphines, imidazoles, pyridines, triazoles, and pyrazoles with alkyl halides followed by ion exchange. Eur. Pat. Appl. 1182197, 2002
[33] Branco, L. C.; Rosa, J. N.; Moura Ramos, J. J.; Afonso, C. A. M. *Chem. Eur. J.* **2002**, *8*, 3671.
[34] Visser, A. E.; Swatloski, R. P.; Reichert, W. M.; Mayton, R.; Sheff, S.; Wierzbicki, A.; Davis, J. H., Jr.; Rogers, R. D. *Chem Commun.* **2001**, 135.
[35] Wasserscheid, P.; Driessen-Hoelscher, B.; van Hal, R.; Steffens, H. C.; Zimmermann, J. *Chem. Commun.* **2003**, 2038.
[36] Forsyth, S. A.; Pringle, J. M.; MacFarlane, D. R. *Aust. J. Chem.* **2004**, *57*, 113.
[37] Gordon, C. M.; Holbrey, J. D.; Kennedy, A. R.; Seddon, K. R. *J. Mater. Chem.* **1998**, *8*, 2627.
[38] Visser, A. E.; Holbrey, J. D.; Rogers, R. D. *Chem. Commun.* **2001**, 2484.
[39] Zulfiqar, F.; Kitazume, T. *Green Chem.* **2000**, *2*, 137.

[40] Matsumoto, H.; Kageyama, H.; Miyazaki, Y. *Chem. Lett.* **2001**, 182.
[41] Matsumoto, H.; Yanagida, M.; Tanimoto, K.; Nomura, M.; Kitagawa, Y.; Miyazaki, Y. *Chem. Lett.* **2000**, 922.
[42] Matsumoto, H.; Matsuda, T.; Miyazaki, Y. *Chem. Lett.* **2000**, 1430.
[43] Fry, S. E.; Pienta, N. J. *J. Am. Chem. Soc.* **1985**, *107*, 6399.
[44] Bradaric, C. J.; Downard, A.; Kennedy, C.; Robertson, A. J.; Zhou, Y. *Green Chem.* **2003**, *5*, 143.
[45] MacFarlane, D. R.; Golding, J.; Forsyth, S.; Forsyth, M.; Deacon, G. B. *Chem. Commun.* **2001**, 1430.
[46] Forsyth, S. A.; Batten, S. R.; Dai, Q.; MacFarlane, D. R. *Aust. J. Chem.* **2004**, *57*, 121.
[47] Matsumoto, K.; Hagiwara, R.; Ito, Y. *J. Fluorine Chem.* **2002**, *115*, 133.
[48] Bowlas, C. J.; Bruce, D. W.; Seddon, K. R. *Chem. Commun.* **1996**, 1625.
[49] Brown, R. J. C.; Dyson, P. J.; Ellis, D. J.; Welton, T. *Chem. Commun.* **2001**, 1862.
[50] Larsen, A. S.; Holbrey, J. D.; Tham, F. S.; Reed, C. A. *J. Am. Chem. Soc.* **2000**, *122*, 7264.
[51] Carter, E. B.; Culver, S. L.; Fox, P. A.; Goode, R. D.; Ntai, I.; Tickell, M. D.; Traylor, R. K.; Hoffman, N. W.; Davis, J. H., Jr. *Chem. Commun.* **2004**, 630.
[52] Holbrey, J. D.; Seddon, K. R. *Clean Prod. Process.* **1999**, *1*, 223.
[53] Farmer, V.; Welton, T. *Green Chem.* **2002**, *4*, 97.
[54] Park, S.; Kazlauskas, R. J. *J. Org. Chem.* **2001**, *66*, 8395.
[55] Stegemann, H.; Rohde, A.; Reiche, A.; Schnittke, A.; Füllbier, H. *Electrochim. Acta* **1992**, *37*, 379.
[56] Holbrey, J. D.; Seddon, K. R. *J. Chem. Soc., Dalton Trans.* **1999**, 2133.
[57] Ngo, H. L.; LeCompte, K.; Hargens, L.; McEwen, A. B. *Thermochim. Acta* **2000**, *357–358*, 97.
[58] Carda-Broch, S.; Berthod, A.; Armstrong, D. W. *Anal. Bioanal. Chem.* **2003**, *375*, 191.
[59] Huddleston, J. G.; Visser, A. E.; Reichert, W. M.; Willauer, H. D.; Broker, G. A.; Rogers, R. D. *Green Chem.* **2001**, *3*, 156.
[60] Davis, J. H., Jr.; Forrester, K. J.; Merrigan, T. *Tetrahedron Lett.* **1998**, *39*, 8955.
[61] Zhao, D.; Wu, M.; Kou, Y.; Min, E. *Catal. Today* **2002**, *74*, 157.
[62] Baranyai, K. J.; Deacon, G. B.; MacFarlane, D. R.; Pringle, J. M.; Scott, J. L. *Aust. J. Chem.* **2004**, *57*, 145.
[63] Reichardt, C. *Solvents and Solvent Effects in Organic Chemistry*; 2nd Edn.; Wiley-VCH: Weinheim, 1988.
[64] Aki, S. N. V. K.; Brennecke, J. F.; Samanta, A. *Chem. Commun.* **2001**, 413.
[65] Muldoon, M. J.; Gordon, C. M.; Dunkin, I. R. *J. Chem. Soc., Perkin Trans.* **2001**, *2*, 433.
[66] Kaar, J. L.; Jesionowski, A. M.; Berberich, J. A.; Moulton, R.; Russell, A. J. *J. Am. Chem. Soc.* **2003**, *125*, 4125.
[67] Carmichael, A. J.; Seddon, K. R. *J. Phys. Org. Chem.* **2000**, *13*, 591.
[68] Anderson, J. L.; Ding, J.; Welton, T.; Armstrong, D. W. *J. Am. Chem. Soc.* **2002**, *124*, 14247.
[69] Swatloski, R. P.; Visser, A. E.; Reichert, W. M.; Broker, G. A.; Farina, L. M.; Holbrey, J. D.; Rogers, R. D. *Green Chem.* **2002**, *4*, 81.
[70] Visser, A. E.; Swatloski, R. P.; Rogers, R. D. *Green Chem.* **2000**, *2*, 1.
[71] Engel, R.; Cohen, J. I.; Lall, S. I. *Phosphorus, Sulfur Silicon Relat. Elem.* **2002**, *6–7*, 1441.
[72] Lall, S. I.; Mancheno, D.; Castro, S.; Behaj, V.; Cohen, J. I.; Engel, R. *Chem. Commun.* **2000**, 2413.
[73] McNulty, J.; Capretta, A.; Wilson, J.; Dyck, J.; Adjabeng, G.; Robertson, A. *Chem. Commun.* **2002**, 1986.
[74] van den Broeke, J.; Winter, F.; Deelman, B. J.; van Koten, G. *Org. Lett.* **2002**, *4*, 3851.
[75] van den Broeke, J.; Stam, M.; Lutz, M.; Kooijman, H.; Spek, A. L.; Deelman, B.-J.; van Koten, G. *Eur. J. Inorg. Chem.* **2003**, 2798.
[76] Hagiwara, R.; Ito, Y. *J. Fluorine Chem.* **2000**, *105*, 221.
[77] Seddon, K. R.; Stark, A.; Torres, M.-J. *Pure Appl. Chem.* **2000**, *72*, 2275.
[78] Chauvin, Y.; Olivier-Bourbigou, H. *CHEMTECH* **1995**, *25*, 26.
[79] Smith, G. P.; Dworkin, A. S.; Pagni, R. M.; Zingg, S. P. *J. Am. Chem. Soc.* **1989**, *111*, 525.
[80] Thomazeau, C.; Olivier-Bourbigou, H.; Magna, L.; Luts, S.; Gilbert, B. *J. Am. Chem. Soc.* **2003**, *125*, 5264.
[81] Earle, M. J.; McCormac, P. B.; Seddon, K. R. *Green Chem.* **1999**, *1*, 23.

[82] Ishida, Y.; Miyauchi, H.; Saigo, K. *Chem. Commun.* **2002**, 2240.
[83] Bao, W.; Wang, Z.; Li, Y. *J. Org. Chem.* **2003**, *68*, 591.
[84] Jodry, J. J.; Mikami, K. *Tetrahedron Lett.* **2004**, *45*, 4429.
[85] Lacour, J.; Ginglinger, C.; Grivet, C.; Bernardinelli, G. *Angew. Chem., Int. Ed. Engl.* **1997**, *36*, 608.
[86] Lacour, J.; Ginglinger, C.; Favarger, F. *Tetrahedron Lett.* **1998**, 4825.
[87] Jodry, J. J.; Lacour, J. *Chem. Eur. J.* **2000**, *6*, 4297.
[88] Wasserscheid, P.; Keim, W.; Bolm, C.; Boesmann, A. Preparation of chiral ionic liquids. PCT Int. Appl. 0155060, 2001.
[89] Wasserscheid, P.; Boesmann, A.; Bolm, C. *Chem. Commun.* **2002**, 200.
[90] Levillain, J.; Dubant, G.; Abrunhosa, I.; Gulea, M.; Gaumont, A.-C. *Chem. Commun.* **2003**, 2914.
[91] Clavier, H.; Boulanger, L.; Audic, N.; Toupet, L.; Mauduit, M.; Guillemin, J.-C. *Chem. Commun.* **2004**, 1224.
[92] Holbrey, J. D.; Visser, A. E.; Spear, S. K.; Reichert, W. M.; Swatloski, R. P.; Broker, G. A.; Rogers, R. D. *Green Chem.* **2003**, *5*, 129.
[93] Zhang, S.; Zhang, Z. C. *Green Chem.* **2002**, *4*, 376.
[94] Swatloski, R. P.; Holbrey, J. D.; Memon, S. B.; Caldwell, G. A.; Caldwell, K. A.; Rogers, R. D. *Chem. Commun.* **2004**, 668.
[95] Swatloski, R. P.; Holbrey, J. D.; Rogers, R. D. *Green Chem.* **2003**, *5*, 361.
[96] Kuntz, E. G. *CHEMTECH* **1987**, 570.
[97] Wasserscheid, P.; Waffenschmidt, H. *J. Mol. Catal. A: Chem.* **2000**, *164*, 61.
[98] Knifton, J. F. *J. Mol. Catal.* **1987**, *43*, 65.
[99] Karodia, N.; Guise, S.; Newlands, C.; Anderson, J. A. *Chem. Commun.* **1998**, 2341.
[100] Chauvin, Y.; Mussmann, L.; Olivier, H. *Angew. Chem., Int. Ed. Engl.* **1995**, *34*, 2698.
[101] Brasse, C. C.; Englert, U.; Salzer, A.; Waffenschmidt, H.; Wasserscheid, P. *Organometallics* **2000**, *19*, 3818.
[102] Wasserscheid, P.; Waffenschmidt, H.; Machnitzki, P.; Kottsieper, K. W.; Stelzer, O. *Chem. Commun.* **2001**, 451.
[103] Favre, F.; Olivier-Bourbigou, H.; Commereuc, D.; Saussine, L. *Chem. Commun.* **2001**, 1360.
[104] Sellin, M. F.; Webb, P. B.; Cole-Hamilton, D. J. *Chem. Commun.* **2001**, 781.
[105] Keim, W.; Vogt, D.; Waffenschmidt, H.; Wasserscheid, P. *J. Catal.* **1999**, *186*, 481.
[106] Suarez, P. A. Z.; Dullius, J. E. L.; Einloft, S.; De Souza, R. F.; Dupont, J. *Polyhedron* **1996**, *15*, 1217.
[107] Dyson, P. J.; Ellis, D. J.; Welton, T.; Parker, D. G. *Chem. Commun.* **1999**, 25.
[108] Adams, C. J.; Earle, M. J.; Seddon, K. R. *Chem. Commun.* **1999**, 1043.
[109] Monteiro, A. L.; Zinn, F. K.; De Souza, R. F.; Dupont, J. *Tetrahedron Asymmetry* **1997**, *8*, 177.
[110] Blanchard, L. A.; Hancu, D.; Beckman, E. J.; Brennecke, J. F. *Nature* **1999**, *399*, 28.
[111] Jessop, P. G. *Yuki Gosei Kagaku Kyokaishi* **2003**, *61*, 484.
[112] Brown, R. A.; Pollet, P.; McKoon, E.; Eckert, C. A.; Liotta, C. L.; Jessop, P. G. *J. Am. Chem. Soc.* **2001**, *123*, 1254.
[113] Berger, A.; de Souza, R. F.; Delgado, M. R.; Dupont, J. *Tetrahedron Asymmetry* **2001**, *12*, 1825.
[114] Jessop, P. G.; Stanley, R. R.; Brown, R. A.; Eckert, C. A.; Liotta, C. L.; Ngo, T. T.; Pollet, P. *Green Chem.* **2003**, *5*, 123.
[115] Roberts, R. M.; Khalaf, A. A. *Friedel-Crafts Alkylation Chemistry A, Century of Discovery*; Dekker: New York, 1984.
[116] Olah, G. A.; Krishnamurit, R.; Prakash, G. K. S. In *Comprehensive Organic Synthesis*; Trost, B. M.; Fleming, I., Eds.; Pergamon Press: Oxford, 1991.
[117] Olah, G. A. *Friedel-Crafts and Related Reactions* Vol. II; Wiley-Interscience: New York, 1964.
[118] DeCastro, C.; Sauvage, E.; Valkenberg, M. H.; Holderich, W. F. *J. Catal.* **2000**, *196*, 86.
[119] Stark, A.; MacLean, B. L.; Singer, R. D. *J. Chem. Soc., Dalton Trans.* **1999**, 63.
[120] Nara, S. J.; Harjani, J. R.; Salunkhe, M. M. *J. Org. Chem.* **2001**, *66*, 8616.
[121] Song, C. E.; Roh, E. J.; Shim, W. H.; Choi, J. H. *Chem. Commun.* **2000**, 1695.
[122] Ross, J.; Xiao, J. *Green Chem.* **2002**, *4*, 129.

[123] Gmouh, S.; Yang, H.; Vaultier, M. *Org. Lett.* **2003**, *5*, 2219.
[124] Earle, M. J.; Seddon, K. R.; McCormac, P. B. *Green Chem.* **2000**, *2*, 261.
[125] Herrmann, W. A.; Wagner, W.; Flessner, U. N.; Vokhardt, U.; Komber, H. *Angew. Chem., Int. Ed. Engl.* **1991**, *30*, 1636.
[126] Romao, C. C.; Kuehn, F. E.; Herrmann, W. A. *Chem. Rev.* **1997**, *97*, 3197.
[127] Espenson, J. H. *Chem. Commun.* **1999**, 479.
[128] Owens, G. S.; Arias, J.; Abu-Omar, M. M. *Catal. Today* **2000**, *55*, 317.
[129] Owens, G. S.; Durazo, A.; Abu-Omar, M. M. *Chem. Eur. J.* **2002**, *8*, 3053.
[130] Owens, G. S.; Abu-Omar, M. M. *Chem. Commun.* **2000**, 1165.
[131] Bortolini, O.; Conte, V.; Chiappe, C.; Fantin, G.; Fogagnolo, M.; Maietti, S. *Green Chem.* **2002**, *4*, 94.
[132] Bernini, R.; Mincione, E.; Coratti, A.; Fabrizi, G.; Battistuzzi, G. *Tetrahedron* **2004**, *60*, 967.
[133] Srinivas, K. A.; Kumar, A.; Chauhan, S. M. S. *Chem. Commun.* **2002**, 2456.
[134] Song, C. E.; Roh, E. J. *Chem. Commun.* **2000**, 837.
[135] Howarth, J. *Tetrahedron Lett.* **2000**, *41*, 6627.
[136] Yadav, J. S.; Reddy, B. V. S.; Basak, A. K.; Narsaiah, A. V. *Tetrahedron* **2004**, *60*, 2131.
[137] Namboodiri, V. V.; Varma, R. S.; Sahle-Demessie, E.; Pillai, U. R. *Green Chem.* **2002**, *4*, 170.
[138] Sun, J.; Fujita, S.-i.; Bhanage, B. M.; Arai, M. *Catal. Commun.* **2004**, *5*, 83.
[139] Seddon, K. R.; Stark, A. *Green Chem.* **2002**, *4*, 119.
[140] Xie, H.; Zhang, S.; Duan, H. *Tetrahedron Lett.* **2004**, *45*, 2013.
[141] Heck, R. F. *J. Am. Chem. Soc.* **1968**, *90*, 5518.
[142] Mizoroki, T.; Mori, K.; Ozaki, A. *Bull. Chem. Soc. Jap.* **1971**, *44*, 581.
[143] Heck, R. F.; Nolley J. P., Jr. *J. Org. Chem.* **1972**, *37*, 2320.
[144] Bräse, S.; de Meijere, A. In *Metal Catalyzed Cross Coupling Reactions*; Stang, P. J.; Diederich, F., Eds.; Wiley-VCH: Weinheim, 1997, pp. 99–166.
[145] de Meijere, A.; Meyer, F. E. *Angew. Chem., Int. Ed. Engl.* **1994**, *33*, 2379.
[146] Heck, R. F. *Palladium Reagents in Organic Synthesis*; Academic Press: London, 1985.
[147] Herrmann, W. A. In *Applied Homogeneous Catalysis with Organometallic Compounds*; Cornils, B., Herrmann, W. A., Eds.; Wiley-VCH: Weinheim, 1996, pp. 712–726.
[148] Tsuji, J. *Palladium Reagents and Catalysts*; Wiley: Chichester, 1995.
[149] Reetz, M. T.; Lohmer, G.; Schwickardi, R. *Angew. Chem., Int. Ed. Engl.* **1998**, *37*, 481.
[150] Kündig, E. P.; Ratni, H.; Crousse, B.; Bernardinelli, G. *J. Org. Chem.* **2001**, *66*, 1852.
[151] Bozell, J. J.; Vogt, C. E. *J. Am. Chem. Soc.* **1988**, *110*, 2655.
[152] Kaufmann, D. E.; Nouroozian, M.; Henze, H. *Synlett* **1996**, *8*, 1091.
[153] Herrmann, W. A.; Bohm, V. P. W. *J. Organomet. Chem.* **1999**, *572*, 141.
[154] Bohm, V. P. W.; Herrmann, W. A. *Chem. Eur. J.* **2000**, *6*, 1017.
[155] Hagiwara, H.; Shimizu, Y.; Hoshi, T.; Suzuki, T.; Ando, M.; Ohkubo, K.; Yokoyama, C. *Tetrahedron Lett.* **2001**, *42*, 4349.
[156] Leadbeater, N. E.; Torenius, H. M. *J. Org. Chem.* **2002**, *67*, 3145.
[157] Vallin, K. S. A.; Emilsson, P.; Larhed, M.; Hallberg, A. *J. Org. Chem.* **2002**, *67*, 6243.
[158] Xu, L.; Chen, W.; Ross, J.; Xiao, J. *Org. Lett.* **2001**, *3*, 295.
[159] Xu, L.; Chen, W.; Xiao, J. *Organometallics* **2000**, *19*, 1123.
[160] Zhang, C.; Huang, J.; Trudell, M. L.; Nolan, S. P. *J. Org. Chem.* **1999**, *64*, 3804.
[161] Deshmukh, R. R.; Rajagopal, R.; Srinivasan, K. V. *Chem. Commun.* **2001**, 1544.
[162] Mathews, C. J.; Smith, P. J.; Welton, T. *Chem. Commun.* **2000**, 1249.
[163] Rajagopal, R.; Jarikote, D. V.; Srinivasan, K. V. *Chem. Commun.* **2002**, 616.
[164] Zou, G.; Wang, Z.; Zhu, J.; Tang, J.; He, M. Y. *J. Mol. Catal. A: Chem.* **2003**, *206*, 193.
[165] Sirieix, J.; Ossberger, M.; Betzemeier, B.; Knochel, P. *Synlett* **2000**, 1613.
[166] Fukuyama, T.; Shinmen, M.; Nishitani, S.; Sato, M.; Ryu, I. *Org. Lett.* **2002**, *4*, 1691.
[167] Kmentova, I.; Gotov, B.; Gajda, V.; Toma, S. *Monatshefte für Chemie* **2003**, *134*, 545.
[168] Park, S. B.; Alper, H. *Chem. Commun.* **2004**, 1306.
[169] Howarth, J.; Hanlon, K.; Fayne, D.; McCormac, P. *Tetrahedron Lett.* **1997**, *38*, 3097.
[170] Waldmann, H. *Angew. Chem., Int. Ed. Engl.* **1991**, *30*, 1306.
[171] Lee, C. W. *Tetrahedron Lett.* **1999**, *40*, 2461.
[172] Fischer, T.; Sethi, A.; Welton, T.; Woolf, J. *Tetrahedron Lett.* **1999**, *40*, 793.
[173] Aggarwal, A.; Lancaster, N. L.; Sethi, A. R.; Welton, T. *Green Chem.* **2002**, *4*, 517.
[174] Kobayashi, S. *Synlett* **1994**, *9*, 689.

[175] Song, C. E.; Roh, E. J.; Lee, S.-g.; Shim, W. H.; Choi, J. H. *Chem. Commun.* **2001**, 1122.
[176] Magnusson, D. K.; Bodley, J. W.; Adams, D. F. *J. Sol. Chem.* **1984**, *13*, 583.
[177] For a review, see: Sheldon, R. A.; Lau, R. M.; Sorgedrager, M. J.; van Rantwijk, F.; Seddon Kenneth, R. *Green Chem.* **2002**, *4*, 147.
[178] Cull, S. G.; Holbrey, J. D.; Vargas-Mora, V.; Seddon, K. R.; Lye, G. J. *Biotechnol. Bioeng.* **2000**, *69*, 227.
[179] Erbeldinger, M.; Mesiano, A. J.; Russell, A. J. *Biotechnol. Prog.* **2000**, *16*, 1131.
[180] Lau, R. M.; van Rantwijk, F.; Seddon, K. R.; Sheldon, R. A. *Org. Lett.* **2000**, *2*, 4189.
[181] Schöfer, S. H.; Kaftzik, N.; Kragl, U.; Wasserscheid, P. *Chem. Commun.* **2001**, 425.
[182] Kim, K.-W.; Song, B.; Choi, M.-Y.; Kim, M.-J. *Org. Lett.* **2001**, *3*, 1507.
[183] Davis James, H., Jr.; Fox, P. A. *Chem. Commun.* **2003**, 1209.
[184] Keim, W.; Korth, W.; Wasserscheid, P. Preparation of ionic liquids for catalysts. PCT Int. Appl. 0016902, 2000.
[185] Schmidt, M.; Heider, U.; Geissler, W.; Ignatyev, N.; Hilarius, V. Ionic Liquids II. PCT Int. Appl. 0015884, 2002.
[186] Hilarius, V.; Heider, U.; Schmidt, M. Ionic liquids. Eur. Pat. Appl. 1160249, 2001.
[187] Robertson, A. J. Preparation of phosphonium salts as ionic liquids. PCT Int. Appl. 0187900, 2001.
[188] Chauvin, Y.; Einloft, S.; Olivier, H. *Ind. Eng. Chem. Res.* **1995**, *34*, 1149.
[189] Chauvin, Y.; Olivier, H.; Wyrvalski, C. N.; Simon, L. C.; De Souza, R. F. *J. Catal.* **1997**, *165*, 275.
[190] Olivier, H.; Commereuc, D.; Forestiere, A.; Hugues, F. Process and apparatus for reaction such as dimerization, oligomerization or metathesis of an organic feedstock such as an olefin in the presence of a nonaqueous polar phase containing a metal catalyst. Eur. Pat. Appl. 882691, 1998.
[191] Freemantle, M. *Chem. Eng. News* **1998**, *76*, 32.
[192] Volland, M.; Seitz, V.; Maase, M.; Flores, M.; Papp, R.; Massonne, K.; Stegmann, V.; Halbritter, K.; Noe, R.; Bartsch, M.; Siegel, W.; Becker, M.; Huttenloch, O. Method for the separation of acids from chemical reaction mixtures by means of ionic fluids. PCT Int. Appl. 03062251, 2003.
[193] Freemantle, M. *Chem. Eng. News* **2003**, *81*, 9.
[194] Seddon, K. R. *Nat. Mater.* **2003**, *2*, 363.
[195] Aggarwal, V. K.; Emme, I.; Mereu, A. *Chem. Commun.* **2002**, 1612.
[196] Kitazume, T.; Tanaka, G. *J. Fluorine Chem.* **2000**, *106*, 211.
[197] Ngo, H. L.; Hu, A.; Lin, W. *Chem. Commun.* **2003**, 1912.

3 Fluorous solvents

Ilhyong Ryu and Hiroshi Matsubara (Sections 3.1 and 3.3.1),
Charlotte Emnet and John A. Gladysz (Section 3.2),
Seiji Takeuchi and Yutaka Nakamura (Sections 3.3.2–3.3.4) and
Dennis P. Curran (Sections 3.4 and 3.5)

3.1 Historical background

Organic synthesis relies largely on solution phase chemistry. Owing to the safety problems inherent in the use of volatile organic materials, as well as environmental concerns, it has become important to reduce dependency on organic solvents in favor of a reaction medium that can be considered to be *green* and where the following requirements are met:

- low volatility, to decrease environmental contamination
- low toxicity for all species living on earth
- easy recyclability, to lower disposal concerns and minimize the volume of solvents that must be manufactured and stored
- sufficient or superior abilities as a reaction medium compared with existing organic solvents

Current regulations require that chemical laboratories strictly control the use of solvents that are harmful to human health, such as benzene or dichloromethane, and avoid their release into the atmosphere. In response to this trend toward green chemistry, chemists, whose raison d'être is the manipulation of molecules, have been encouraged to embark on new challenges to invent and develop green reaction media that will eliminate the use of harmful solvents in organic synthesis. Fluorous chemistry offers some attractive features in terms of green solvents. Although the road to the complete replacement of organic solvents will be long and arduous, the journey has started. Needless to say, new developments in fluorous reaction media stand together with some other novel and traditional reaction media, such as ionic liquids, supercritical liquids and water. There are many candidates competing for superiority.

Several traditional organofluorine solvents, such as trifluoroacetic acid, hexafluoroisopropanol and trifluoroethanol, have been applied in organic synthesis, where they serve as polar solvents. However, as a result of the evolution of fluorous chemistry in the 1990s,[1,2,40] unforeseen opportunities now exist to employ fluorous

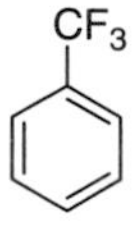

BTF
(benzotrifluoride or
α,α,α-trifluorotoluene)

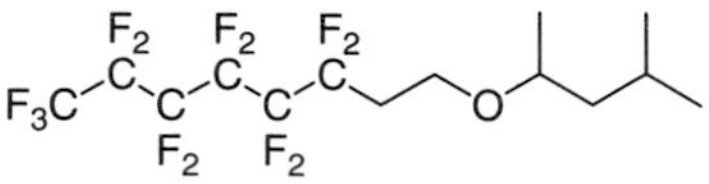

F-626
(1H,1H,2H,2H-perfluorooctyl
1,3-dimethylbutyl ether)

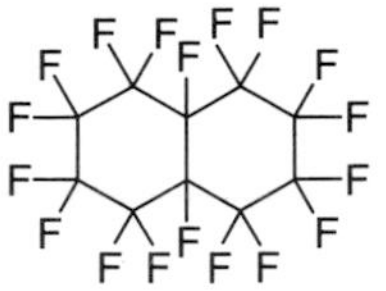

Perfluorodecalin

Figure 3.1

technologies and reaction media. This chapter is a guide to state-of-the-art fluorous reactions including fluorous/organic biphasic reaction systems, and also fluorous reactions that are carried out exclusively in a fluorous solvent or even in an organic solvent alone.

It is noteworthy that fluorous chemistry has evolved to include "hybrid" reaction media such as BTF (benzotrifluoride)[3] and F-626 (Figure 3.1).[4] BTF (boiling point 102°C, melting point −29°C) is a colorless, free-flowing liquid with a relatively low toxicity; it is slightly less polar than chloroform and dichloromethane. BTF has been successfully utilized as an amphiphilic solvent that dissolves both fluorous reagents and non-fluorous substrates in reactions involving the use of fluorous tin hydride[5] and allyltin reagents.[6] BTF has been used as a substitute for toxic dichloromethane in many common synthetic reactions, including Swern oxidation, Dess–Martin oxidation and Hosomi–Sakurai allylation reactions (Scheme 1).[3] There are some limitations to its use – for example, at lower temperatures or in the presence of aluminium trichloride, which reacts with it – however, BTF represents an interesting fluorous reaction medium for numerous synthetic reactions.

Perfluoropolyethers are used as ingredients in cosmetics and are active in the prevention and treatment of irritant dermatitis.[7] One of these, the high-boiling "light" fluorous ether F-626 (boiling point 214°C, glass transition −110°C), developed by chemists at the Kao Corporation,[8] is also a useful organic reaction medium.[4] The partition coefficients measured in a biphasic system composed of acetonitrile and perfluorohexane (FC-72) suggest that F-626 is more fluorous than BTF. F-626 can be used as a medium for reactions involving the use of fluorous reagents, fluorous substrates or both. It can also be used as a substitute for traditional high-boiling organic solvents such as diethylene glycol, DMF (dimethyl formamide) and *o*-dichlorobenzene in Vilsmeier formylations, Wolff–Kischner reductions and Diels–Alder reactions (Scheme 2).[4] The yields obtained using F-626 are comparable to those achieved using organic solvents. The fluorous nature of F-626 allows its easy recovery from the reaction mixture via a fluorous/organic biphasic treatment.

Three decades have passed since the high solubility of gases in perfluorocarbons was reported with the striking demonstration of a mouse breathing while immersed in perfluorocarbons.[9] Perfluorodecalin (Figure 3.1), which has a low surface tension

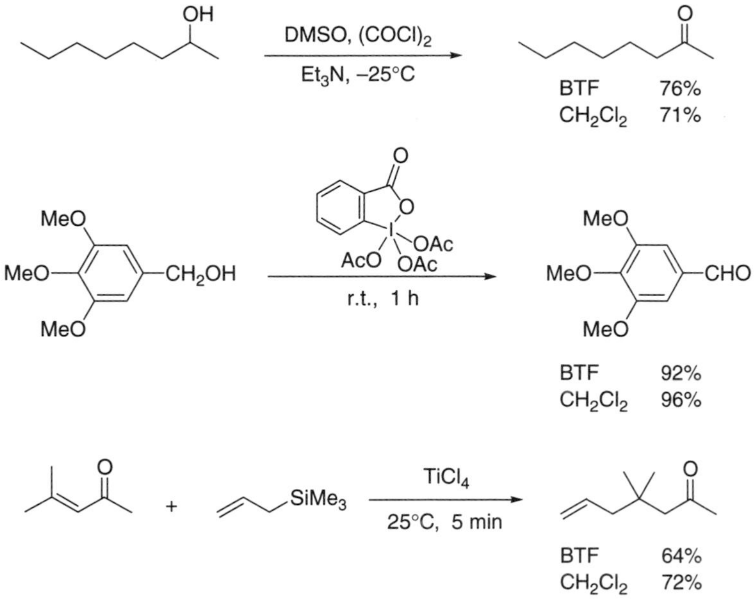

OH

DMSO, $(COCl)_2$

Et_3N, –25°C

O

BTF 76%
CH_2Cl_2 71%

MeO
MeO—CH$_2$OH
MeO

O
O
I
AcO OAc OAc

r.t., 1 h

MeO
MeO—CHO
MeO

BTF 92%
CH_2Cl_2 96%

O + SiMe$_3$

$TiCl_4$

25°C, 5 min

O

BTF 64%
CH_2Cl_2 72%

Scheme 1

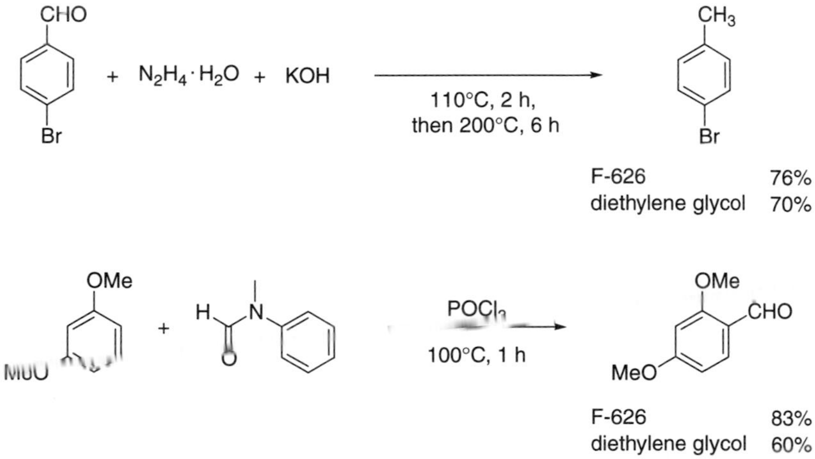

CHO
Br
+ $N_2H_4 \cdot H_2O$ + KOH

110°C, 2 h,
then 200°C, 6 h

CH_3
Br

F-626 76%
diethylene glycol 70%

OMe
MeO
+ H N O

$POCl_3$

100°C, 1 h

OMe
CHO
MeO

F-626 83%
diethylene glycol 60%

Scheme 2

(16.9 dyn/cm^2)[10] relative to alkanes (18–25 dyn/cm^2), is an excellent medium for dissolving such gases as oxygen. This compound gained attention as a potential substitute for blood. Its remarkable utility as a reaction medium for gas/liquid reactions is now receiving renewed attention.

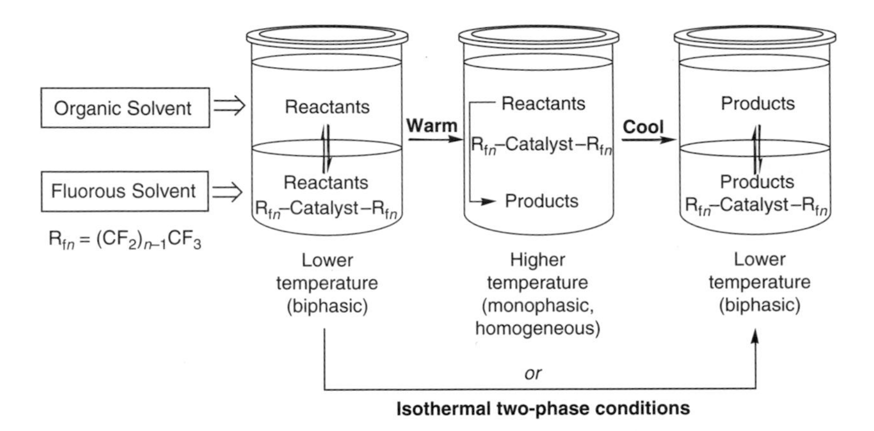

Figure 3.2

3.2 Physical properties

3.2.1 Key design elements in fluorous/organic liquid biphasic reactions

The miscibilities of fluorous and non-fluorous solvents are strongly temperature dependent, and bilayers are commonly obtained at room temperature (Figure 3.2). Moderate heating usually gives a single phase. Lighter organic solvents, such as pentane or ether, are more likely to give one phase at room temperature with, for example, perfluoromethylcyclohexane (PFMC). These phenomena are sensitive functions of any dissolved solutes, as well as the solvent mole ratios. In any event, chemistry can be performed under homogeneous, one-liquid-phase conditions or heterogeneous, two-liquid-phases conditions. The non-fluorous and fluorous products are then separated at the low-temperature two-phase limit.

Molecules can be engineered to have various affinities for fluorous phases by attaching fluoroalkyl groups of the formula $(CH_2)_m(CF_2)_{n-1}CF_3$. These are termed *ponytails* when they are attached in a more permanent manner (e.g. as an intrinsic part of the carbon skeleton) and *tags* when they are designed to be easily introduced and removed (e.g. as part of a protecting group) [2a,11–14]. They are often abbreviated $(CH_2)_mR_{fn}$ and serve a "like dissolves like" function. When such phase labels are present in sufficient length or quantity, the fluorous phase affinity (fluorophilicity) of the compound can be extremely high. This characteristic is used in the design of dyes that adhere strongly to Teflon. Molecules that have large numbers of such fluorine atoms – typically from 39 to >100 – are called *heavy fluorous* compounds. However, for certain applications moderate fluorous phase affinities suffice, and for the compounds used in such applications, the term *light fluorous* is sometimes utilized (Table 3.1).

The $(CH_2)_m$ spacers provide tuning elements that can be adjusted to insulate the active site from the electron-withdrawing perfluoroalkyl or R_{fn} segments (higher m values) or enhance Lewis acidity (lower m values). These electronic effects have

Table 3.1

	Heavy fluorous molecules	Light fluorous molecules
Ponytails with integral (permanent) fluorinated domains	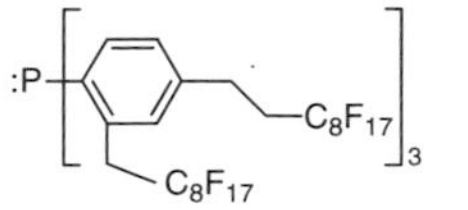 :P, C_8F_{17}, C_8F_{17}, 3	$C_8F_{17}CH_2CH_2Sn(Me)_2H$
Protecting groups/tags with removable (temporary) fluorinated domains	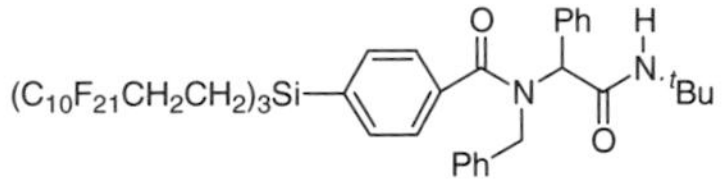 $(C_{10}F_{21}CH_2CH_2)_3Si$, O, Ph, H, N, N, ^tBu, O, Ph	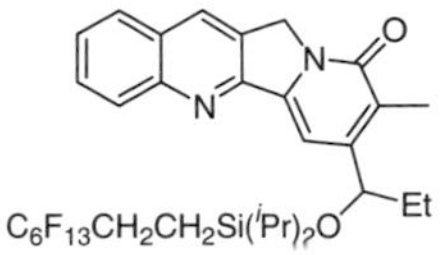 N, N, O, Et, $C_6F_{13}CH_2CH_2Si(^iPr)_2O$

been studied in several series of donor ligands.[15,16] The electron-withdrawing influence of the R_{fn} groups can be detected through a surprisingly large number of methylene groups.

As reflected by the fluorous reactions described in Section 3.3, and quantified by data below, most organic compounds have very low affinities for fluorous phases compared with organic phases. Thus, products can often be separated from heavy fluorous catalysts or spent reagents using a simple liquid/liquid phase separation, as shown in Figure 3.2. If necessary, the fluorous phase can be extracted using additional organic solvent.

In contrast, light fluorous molecules may be distributed between both phases, and fluorous chromatography is the standard method of separating these molecules. This technique is introduced in Section 3.4.

3.2.2 *Commercial availability*

Fluorous solvents are by far the most frequently used fluorous reaction media;[17] however, fluorous greases and low-melting-point solids hold promise. As summarized in Table 3.2, numerous fluorous solvents are commercially available. Perfluorinated alkanes are the most common, followed by perfluorinated dialkyl polyethers and then perfluorinated trialkyl amines. The lone pairs in such ethers and amines are extremely low in energy and are not a source of appreciable intermolecular interactions. Solvents with a wide selection of boiling points are available. Nearly all the major vendors sell fluorous solvents. Oakwood Products, ABCR, Fluorochem, Lancaster, ACROS and Apollo currently offer the largest selections. Note that the densities are always much greater than those of common organic solvents, including CCl_4 (1.589 g/mL).

The most common solvent for fluorous chemistry is perfluorohexane or FC-72 (entry 5 in Table 3.2). However, like "hexanes," perfluorohexane is a mixture of isomers, and other minor fluorous impurities can be present.[18] For preparative chemistry, this is definitely not of consequence.[18] However, for physical measurements or

Table 3.2

Entry	Solvent	Trade name	Formula [CAS #]	Boiling point (°C)	Melting point (°C)	Density (g/mL)	2003 Vendors[a]
1	perflucromethylcyclohexane	PFMC	$CF_3C_6F_{11}$ [355-02-2]	76.1	−37	1.787	*a–e*, *g–i*, *m*
2	perfluoro-1,2-dimethylcyclohexane	Flutec (PP3), FlutecR (PP3)	C_8F_{16} [306-98-9]	101.5	−56	1.867	*a–d*, *g*, *m*
3	perfluoro-1,3-dimethylcyclohexane	–	C_8F_{16} [335-27-3]	101–102	−55	1.828	*a–e*, *g*, *h*, *m*
4	perfluoro-1,3,5-trimethylcyclohexane	–	C_9F_{18} [374-76-5]	125–128	−68	1.888	*a–d*, *m*
5	perfluorohexane[b,c]	Fluorinert (FC-72)	C_6F_{14} [355-42-0]	57.1	−90	1.669	*a–e*, *g*, *h*, *m*
6	perfluoroheptane[b,d]	–	C_7F_{16} [335-57-9]	82.4	−78	1.745	*a–d*, *g*, *h*, *m*
7	perfluorooctane(s)[b,e]	–	C_8F_{18} [307-34-6]	103–104	−25	1.766	*a–e*, *g*, *m*
8	1-bromoperfluorooctane	–	$C_8F_{17}Br$ [423-55-2]	142	6	1.930	*a–e*, *g*, *h*
9	perfluorodecalin	–	$C_{10}F_{18}$ [306-94-5]	142	−10	1.908	*a–c*, *e*, *g–i*, *m*
10	α,α,α-trifluorotoluene	BTF	$CF_3C_6H_5$ [98-08-8]	102	−29	1.199	*a–e*, *g–j*, *m*

11	perfluorotributylamine[b,f]	Fluorinert (FC-43)	$C_{12}F_{27}N$ [311-89-7]	178	–	1.883	*a–e, g, h, k, m*
12	perfluorotripentylamine[b,g]	Fluorinert (FC-70)	$C_{15}F_{33}N$ [338-84-1]	212–218	–	1.93	*a–e, k, m*
13	perfluorotrihexylamine	Fluorinert (FC-71), FluorinertR (FC-71)	$C_{18}F_{39}N$ [432-08-6]	250–260	33	1.90	*a–c, m*
14	perfluoro-2-butyltetrahydrofuran	Fluorinert (FC-75), FluorinertR (FC-75)	$C_8F_{16}O$ [335-36-4]	99–107	−88	1.77	*a–e, h, m*
15	perfluoropolyether[h]	Galden HT55	Mw ≈ 340	57	–	1.65	*l*
16	perfluoropolyether[h]	Galden HT70, GaldenR HT70	Mw ≈ 410 [69991-67-9]	70	–	1.68	*c, l, m*
17	perfluoropolyether[h]	Galden HT90, GaldenR HT90	Mw ≈ 460 –	90	–	1.69	*c, l, m*
18	perfluoropolyether[h]	Galden HT110, GaldenR HT110	Mw ≈ 580	110	–	1.72	*c, l, m*

[a] Codes for vendors are as follows: *a* = Oakwood Products; *b* = ABCR; *c* = Fluorochem; *d* = Lancaster; *e* = Acros Organics; *f* = 3M; *g* = Aldrich; *h* = Fluka; *i* = Merck; *j* = Oxychem; *k* = Sigma; *l* = Solvay Solexis; *m* = Apollo Scientific Ltd.

[b] Several fluorous solvents are available in technical grades that have separate CAS numbers and common names.

[c] [86508-42-1], Fluorinert (FC-72), FluorinertR (FC-72), Flutec (PP1), FlutecR (PP1), *a, c, f, m*.

[d] [86508-42-1], Fluorinert (FC-84), FluorinertR (FC-84), *a, c, f, m*.

[e] [86508-42-1], [52623-00-4], [52923-00-4], Fluorinert (FC-77), FluorinertR (FC-77), *a, c, e, f, h, k, m*.

[f] [86508-42-1], Fluorinert (FC-43), FluorinertR (FC-43), *a, c, f, m*.

[g] [86508-42-1], Fluorinert (FC-70), FluorinertR (FC-70), *f, m*.

[h] General formula $CF_3[(OCF(CF_3)CF_2)_m(OCF_2)_n]OCF_3$.

mechanistic studies, $CF_3C_6F_{11}$ (PFMC), a more expensive but homogeneous solvent, is often favored (entry 1). Note that most commercial perfluoropolyethers (entries 15–18)[19] contain multiple stereocenters and are therefore inhomogeneous mixtures of diastereomers.

Brominated organic solvents are seldom employed, but 1-bromoperfluorooctane (entry 8) is quite commonly applied in fluorous chemistry. The availability of this compound derives from its use in artificial blood (e.g. *Oxygent*™, a product of the Alliance Pharmaceutical Company).[20] There are also ongoing efforts to develop new fluorous solvents, including fluorous ionic liquids.[21]

A few solvents exhibit intermediate properties. For example, both BTF (Ref. 22) and F-626 (Ref. 23; see Section 3.1) are able to dissolve appreciable quantities of both fluorous and non-fluorous solutes. Other solvents are often mistakenly assumed to be fluorous. Of these, the most important are perfluoroarenes such as hexafluorobenzene. The arene π clouds and sp^2 carbon–fluorine bonds lead to significant intermolecular bond dipoles, induced dipoles and quadrupolar interactions with non-fluorous molecules.[24,25]

3.2.3 *Polarity*

Perfluorinated solvents exhibit extremely low polarities, which can be quantified in many ways. As analyzed elsewhere,[26,27] one of the best scales in terms of modeling the ability of a solvent to solvate or complex a solute or transition state involves the shift of the absorption maximum of a perfluoroheptyl-substituted dye. This dye was optimized to be soluble in both fluorocarbons and very polar solvents such as DMSO (dimethyl sulfoxide). Over 100 solvents have been assayed,[26] and some of the resulting P_S or "Spectral Polarity Index" values are given in Table 3.3.

The data in Table 3.3 show that perfluoroalkanes are much less polar than the corresponding alkanes (first five entries in each column; P_S 0.00–0.99 vs. 2.56–4.07). They also show that perfluorinated trialkylamines exhibit similarly low polarities, but that those of fluorinated arenes are higher. Nonetheless, hexafluorobenzene is less polar than benzene (P_S 4.53 vs. 6.95). Interestingly, highly fluorinated alcohols such as $CF_3CF_2CF_2CH_2OH$ and $(CF_3)_2CHOH$ exhibit P_S values higher than those of similar non-fluorinated alcohols, which has been attributed to strong hydrogen bonding.

3.2.4 *Solute solubilities*

It is important to distinguish between *absolute* solubility, which is defined by a K_{sp} value or similar parameter, and *relative* solubility, which is defined by a partition coefficient. The former is treated in this section, and the latter analyzed below. Quantitative data for the types of solutes of greatest utility for fluorous chemistry are unfortunately scarce. Tables of solubilities of representative monofunctional fluorous and non-fluorous molecules, for example, would be exceedingly useful.

Table 3.3

Solvent	Formula	P_s	Solvent	Formula	P_s
perfluoromethylcyclohexane	$CF_3C_6F_{11}$	0.46	methylcyclohexane	$CH_3C_6H_{11}$	3.34
perfluoro-1,3-dimethylcyclohexane	C_8F_{16}	0.58	1,3-dimethyl-cyclohexane	C_8H_{16}	3.31
perfluorohexane	C_6F_{14}	0.00	*n*-hexane	C_6H_{14}	2.56
perfluorooctane	C_8F_{18}	0.55	*n*-octane	C_8H_{18}	2.86
perfluorodecalin	$C_{10}F_{18}$	0.99	decalin	$C_{10}H_{18}$	4.07
perfluorotributylamine	$(C_4F_9)_3N$	0.68	tri-*n*-butylamine	$(C_4H_9)_3N$	3.93
hexafluorobenzene	C_6F_6	4.53	benzene	C_6H_6	6.95
BTF	$CF_3C_6H_5$	7.03	toluene	$CH_3C_6H_5$	6.58
2,2,2-trifluoroethyl trifluoroacetate	$CF_3CO_2CH_2CF_3$	7.74	ethyl acetate	$CH_3CO_2CH_2CH_3$	6.96
1*H*,1*H*-perfluoro-1-butanol	$CF_3(CF_2)_2CH_2OH$	9.76	1-butanol	$CH_3(CH_2)_2CH_2OH$	7.62
1,1,1,3,3,3-hexafluoroisopropanol	$(CF_3)_2CHOH$	11.08	2-propanol	$(CH_3)_2CHOH$	7.85

Some fluorous molecules are poorly soluble in fluorous solvents at room temperature. This is most often observed with, but not limited to, molecules with longer R_{fn} segments ($n > 8$). Such species are even less soluble in organic solvents. One representative example is the fluorous sulfoxides $O{=}S(CH_2)_mR_{f8}$ ($m = 2, 3$).[28] Also, solubilities in all solvents dramatically decrease in the series of fluorous phosphines $P((CH_2)_2R_{f6})_3$, $P((CH_2)_2R_{f8})_3$ and $P((CH_2)_2R_{f10})_3$.[29,30] Some fluorous arenes, such as the boronic acid 3,5-$(R_{f10})_2C_6H_3B(OH)_2$, also exhibit very poor solubilities.[31]

One way to conceptualize this phenomenon is to view the ponytails as short pieces of Teflon, which does not dissolve in any common solvent. As the ponytails become longer, some physical properties of the molecule approach those of Teflon. However, just as the miscibilities of fluorous liquid phases and organic liquid phases are highly temperature dependent, so are the solubilities of fluorous solids in fluorous or non-fluorous liquid phases.[30] Hence, much higher solubilities can be achieved at elevated temperatures. This phenomenon can be used to conduct homogeneous reactions at elevated temperatures, with catalyst or reagent recovery by solid/liquid phase separation at lower temperatures.[30]

The solubilities of many small molecules in fluorous solvents have been measured.[32] For example, a solution of perfluoroheptane that is saturated with octane contains 11.2 mol% octane at 27.5°C, 31.8 mol% octane at 60.0°C and 45.1 mol% octane at 65.8°C.[32e] The smaller hydrocarbon heptane is approximately twice as soluble (21.4 mol% at 27.3°C) and similar to chloroform (22.4 mol% at 24.6°C). The data for octane demonstrate the strong temperature dependence of solubilities in fluorous phases.

The solubilities of small molecules in fluorous solvents are determined to a large extent by two parameters: solute polarity and size. The first is an extension of the "like dissolves like" paradigm. The second is uniquely important to perfluorinated solvents; because of low intermolecular forces they have large cavities (free volumes) that can accommodate small molecules. The solubilities of gases in fluorocarbons are also well established and show a correlation with the isothermal compressibility of the solvent,[33] supporting the cavity-based solubility model.

Some additional perspectives on gas solubilities are provided in Table 3.4. Literature data are normally reported as mol fractions, a unit not commonly employed by kineticists or preparative chemists. These constitute the origin of the widespread statement that "gases are much more soluble in fluorocarbons." This generalization is indeed appropriate with reference to water, where a strong hydrogen bonding network must be disrupted, but is much less so for organic solvents.[34]

The solubilities of O_2 in the fluorous solvent $CF_3C_6F_{11}$ and the organic solvent THF differ by a factor of five when expressed as mol fractions or nearly equivalent mol ratios (Table 3.4). However, the molecular weights of fluorocarbons tend to be higher than those of organic solvents, and with $CF_3C_6F_{11}$ and THF differ by a factor of approximately five. This gives (0.00453 mol O_2)/(350.05 g $CF_3C_6F_{11}$) and (0.000815 mol O_2)/(72.11 g THF), resulting in nearly equal *molal* concentrations (mol/kg; Table 3.4).

Table 3.4

Solute [concentration unit]	Solubility $CF_3C_6F_{11}$ (25°C, 1 bar)	Solubility THF (25°C, 1 bar)	Difference
O_2 [mol fraction]	0.00456[a]	0.000816[a]	factor of ~5
O_2 [mol ratio]	0.00453	0.000815	factor of ~5
Mol wt [g/mol]	350.05	72.11	[factor of ~5]
O_2 [mol/kg]	0.0129	0.0112	~equal
Density [g/mL]	1.787	0.889	[factor of ~2]
O_2 solubility [mol/L]	0.0232	0.0100	factor of ~2
H_2 [mol fraction]	0.0012[b]	0.000274[c]	factor of ~4.5
H_2 [mol ratio]	0.00119	0.000274	factor of ~4.5
H_2 [mol/kg]	0.0034	0.0038	~equal
H_2 [mol/L]	0.0061	0.0034	factor of ~2

[a] *IUPAC Solubility Data Series*, Battino, R., Ed.; Pergamon: New York, 1981, Volume 7, pp 301 and 320.
[b] Patrick, C. R. In *Preparation, Properties, and Industrial Applications of Organofluorine Compounds*, Banks, R. E., Ed.; Ellis Horwood: New York, 1982, p 333.
[c] *IUPAC Solubility Data Series*, Young, C. L., Ed.; Pergamon: New York, 1981, Volume 5/6, p 219.

The densities of fluorocarbons also tend to be higher than those of non-halogenated organic solvents. In the case of $CF_3C_6F_{11}$ and THF, they differ by a factor of two. This tranforms the preceding values to (0.00453 mol O_2)/(195.9 mL $CF_3C_6F_{11}$) and (0.000815 mol O_2)/(81.1 mL THF), resulting in *molar* concentrations that differ by a factor of slightly more than two (Table 3.4). The solubility of H_2 in $CF_3C_6F_{11}$ and THF is slightly lower. However, the relationships between molal and molar concentrations turn out to be nearly the same as with O_2 (Table 3.4).

For understandable reasons, fluorous solvents are often proposed as good media for reactions involving gases. However, the data given above clearly show that compared with organic solvents, any solubility-based rate accelerations must by necessity be modest. Furthermore, the lower polarity of fluorous media has been shown to retard additions of O_2 to iridium complexes.[35]

3.2.5 *Fluorous solvent miscibilities*

Although fluorous reactions can be conducted under heterogeneous liquid/liquid biphasic conditions, from a rate standpoint it will normally be advantageous to operate under homogeneous monophasic conditions, as in the top sequence in Figure 3.2. For this reason it is important to know at what temperatures various fluorous and non-fluorous solvents become miscible (Table 3.5).[11,32c,36]

With binary solvent systems, it is common to determine a "consolute" or "upper critical solution" temperature above which phase separation cannot occur, whatever the composition.[37] Consolute temperatures are usually found for ~50 : 50 mixtures; however, the full phase diagrams show that solvents can become miscible in other

Table 3.5

Solvent system[a]	Two phases at (°C)	One phase at (°C)	Ref.
$CF_3C_6F_{11}/CCl_4$	r.t.	$\geq$ 26.7[b]	32c
$CF_3C_6F_{11}/CHCl_3$	r.t.	$\geq$ 50.1[b]	32c
$CF_3C_6F_{11}/C_6H_6$	r.t.	$\geq$ 84.9[b]	32c
$CF_3C_6F_{11}/CH_3C_6H_5$	r.t.	$\geq$ 88.6[b]	32c
$CF_3C_6F_{11}/ClC_6H_5$	r.t.	$\geq$126.7[b]	32c
$C_8F_{17}Br/CH_3C_6H_5$	r.t.	50–60[c]	36a
perfluorodecalin/$CH_3C_6H_5$	r.t.	64[c]	36a
$CF_3C_6F_{11}$/hexane/$CH_3C_6H_5$[d]	r.t.	36.5[c]	11
$CF_3C_6F_{11}$/hexane	0	r.t.[c]	36b
$CF_3C_6F_{11}$/pentane	−16	r.t.[c]	36b
$CF_3C_6F_{11}$/ether	0	r.t.[c]	36b

[a] All data for a 1 : 1 volume ratio unless otherwise stated.
[b] Consolute temperature.
[c] Experimental observation; not a consolute temperature.
[d] Volume ratio 3 : 3 : 1.
r.t. = room temperature.

proportions at much lower temperatures. Furthermore, consolute temperatures can be strongly affected by solutes or dissolved species. Thus, Table 3.5 provides only a rough guide to a property that is a function of many parameters.

The mixing of two phases is, of course, favorable from an entropic standpoint. From an enthalpic standpoint, intermolecular attractive interactions will always be greater within the pure non-fluorous phase (which has a much greater polarity) than within the pure fluorous phase (which has a much lower polarity). Upon mixing the two phases, the stronger intermolecular interactions in the former will be markedly diluted, and the intermolecular interactions felt by the fluorous molecules will increase only slightly. Hence, no enthalpic gain is to be expected.

Molecules from both phases of a liquid/liquid biphasic system are commonly found in each phase. One familiar example is ether/water, where drying agents are needed to render the ether layer anhydrous after separating the aqueous layer. With toluene/$CF_3C_6F_{11}$ at 25°C, ratios are 98.4 : 1.6 (molar), 94.2 : 5.8 (mass) and 97.1 : 2.9 (volume) in the upper organic layer.[38] The corresponding ratios are 3.8 : 96.2, 1.0 : 99.0 and 2.0 : 98.0 in the lower fluorous layer. This results in some leaching of the fluorous solvent into the non-fluorous solvent under the conditions of Figure 3.2.

3.2.6 *Partition coefficients and fluorophilicities*

In order to extract non-fluorous products from reactions involving fluorous solvents in a rational way, partition coefficients must be known. The design and optimization of fluorous catalysts and reagents require analogous data. In 1999, only a few partition coefficients involving fluorous and organic phases had been measured.[2c] Now there is a wealth of data, to which sophisticated analysis techniques have been applied.[39] Selected values have been measured in the Gladysz laboratory using the solvent system toluene/$CF_3C_6F_{11}$ and are summarized in Table 3.6. Representative data from other research groups involving hydrocarbon/fluorous solvent mixtures are also included. A more comprehensive list is provided elsewhere.[40]

In rigorous work, partition coefficients are determined over a range of concentrations and extrapolated to infinite dilution. However, the concentrations used in Table 3.6 are close to those encountered in "real-life" fluorous/organic liquid/liquid biphasic separations, and the values for various classes of molecules are believed to have excellent cross-comparability. The measurement techniques have been critiqued elsewhere.[2c] It is preferable to express partition coefficients as ratios that have been normalized to 100, but some researchers express them as fractions or log quantities.

Entries I-1 to I-6 in Table 3.6 give partition coefficients (toluene/$CF_3C_6F_{11}$) for *n*-alkanes, and entries II-1 to II-6 give data for the corresponding terminal alkenes. The alkanes, although highly non-polar, show high affinities for the toluene phase. The coefficients increase monotonically with alkane size (from 94.6 : 5.4 for decane to 98.9 : 1.1 for hexadecane). The *n*-alkenes have slightly higher toluene phase affinities, consistent with their slightly greater polarities. A comparable monotonic size trend is found (95.2 : 4.8 to 99.1 : 0.9).

Entries III-1 and III-2 show that the much more polar ketones cyclohexanone and 2-cyclohexen-1-one also possess very high toluene phase affinities. The alcohol cyclohexanol (entry IV-1) exhibits a higher partition coefficient than cyclohexanone, and a phenyl-containing silyl ether derivative (entry XII-2) shows a value even higher (99.2 : 0.8).

Section IV of Table 3.6 lists several simple fluorous alcohols. The short-chain or R_{f1}/R_{f2} alcohols in entries IV-2 and IV-3 show very poor fluorous phase affinities. As the perfluoroalkyl segment lengthens in the series $R_{f6}(CH_2)_3OH$, $R_{f8}(CH_2)_3OH$ and $R_{f10}(CH_2)_3OH$ (entries IV-5, IV-7, IV-8), the fluorophilicities increase from 55.9 : 44.1 to 19.5 : 80.5. As expected, when a methylene group is removed from the first two compounds, the fluorophilicities also increase (47.4 : 52.6 and 26.5 : 73.5, entries IV-4, IV-6). Similar trends are found with all other functional groups in Table 3.6.

The thiol $R_{f8}(CH_2)_3SH$ (entry XIII-1), iodide $R_{f8}(CH_2)_3I$ (entry VIII-4) and primary amine $R_{f8}(CH_2)_3NH_2$ (entry IX-1) exhibit partition coefficients broadly comparable to that of the corresponding alcohol (35.7 : 64.3). Thus, more than one R_{f8} ponytail is needed to effectively immobilize simple monofunctional organic compounds in fluorous solvents. The effect of the number of ponytails is clearly seen in amines of the formula $[R_{f8}(CH_2)_3]_xNH_{3-x}$ (entries IX-1, IX-5, IX-10). As x increases from 1 to 3, the fluorous phase affinities increase monotonically from 30.0 : 70.0 to 3.5 : 96.5, to the point where no concentration in toluene remains that can be detected using gas liquid chromatography (GLC). Thus, $[R_{f8}(CH_2)_3]_3N$ represents a highly immobilized fluorous base. When the number of the methylene

Table 3.6

Entry[a]	Substance[b]	Solvent system	Partitioning % (organic : fluorous)	Ref.
I – Alkanes				
1	$CH_3(CH_2)_8CH_3$	$CH_3C_6H_5 : CF_3C_6F_{11}$	94.6 : 5.4	1
2	$CH_3(CH_2)_9CH_3$	$CH_3C_6H_5 : CF_3C_6F_{11}$	95.8 : 4.2	1
3	$CH_3(CH_2)_{10}CH_3$	$CH_3C_6H_5 : CF_3C_6F_{11}$	96.6 : 3.4	1,2
4	$CH_3(CH_2)_{11}CH_3$	$CH_3C_6H_5 : CF_3C_6F_{11}$	97.6 : 2.4	1
5	$CH_3(CH_2)_{12}CH_3$	$CH_3C_6H_5 : CF_3C_6F_{11}$	98.1 : 1.9	1
6	$CH_3(CH_2)_{14}CH_3$	$CH_3C_6H_5 : CF_3C_6F_{11}$	98.9 : 1.1	1
II – Alkenes				
1	$CH_3(CH_2)_7CH{=}CH_2$	$CH_3C_6H_5 : CF_3C_6F_{11}$	95.2 : 4.8	3
2	$CH_3(CH_2)_8CH{=}CH_2$	$CH_3C_6H_5 : CF_3C_6F_{11}$	96.3 : 3.7	3
3	$CH_3(CH_2)_9CH{=}CH_2$	$CH_3C_6H_5 : CF_3C_6F_{11}$	97.5 : 2.5	2
4	$CH_3(CH_2)_{10}CH{=}CH_2$	$CH_3C_6H_5 : CF_3C_6F_{11}$	98.1 : 1.9	3
5	$CH_3(CH_2)_{11}CH{=}CH_2$	$CH_3C_6H_5 : CF_3C_6F_{11}$	98.4 : 1.6	3
6	$CH_3(CH_2)_{13}CH{=}CH_2$	$CH_3C_6H_5 : CF_3C_6F_{11}$	99.1 : 0.9	3
7[c]	$R_{f8}CH{=}CH_2$	$CH_3C_6H_5 : CF_3C_6F_{11}$	6.5 : 93.5	4a
For further alkenes see III, XI and XII				
III – Ketones and aldehydes				
1	cyclohexanone	$CH_3C_6H_5 : CF_3C_6F_{11}$	97.8 : 2.2	2
2	2-cyclohexen-1-one	$CH_3C_6H_5 : CF_3C_6F_{11}$	98.3 : 1.7	2
3	$R_{f8}(CH_2)_3$–C_6H_4–$C(O)(CH_2)_2R_{f8}$	$CH_3C_6H_5 : CF_3C_6F_{11}$	15.4 : 84.6	5
4[c]	3, 5-$(R_{f8})_2C_6H_3(CHO)$	$CH_3C_6H_5 : CF_3C_6F_{11}$	1.4 : 98.6	4a
IV – Alcohols				
1	cyclohexanol	$CH_3C_6H_5 : CF_3C_6F_{11}$	98.4 : 1.6	2
2[c]	CF_3CH_2OH	$CH_3C_6H_5 : CF_3C_6F_{11}$	85.5 : 14.5	6
3[c]	$(CF_3)_2CHOH$	$CH_3C_6H_5 : CF_3C_6F_{11}$	73.5 : 26.5	6
4[c]	$R_{f6}(CH_2)_2OH$	$CH_3C_6H_5 : CF_3C_6F_{11}$	47.4 : 52.6	6
5[c]	$R_{f6}(CH_2)_3OH$	$CH_3C_6H_5 : CF_3C_6F_{11}$	55.9 : 44.1	6
6[c]	$R_{f8}(CH_2)_2OH$	$CH_3C_6H_5 : CF_3C_6F_{11}$	26.5 : 73.5	6
7[c]	$R_{f8}(CH_2)_3OH$	$CH_3C_6H_5 : CF_3C_6F_{11}$	35.7 : 64.3	6
8[c]	$R_{f10}(CH_2)_3OH$	$CH_3C_6H_5 : CF_3C_6F_{11}$	19.5 : 80.5	6
9[c]	$3,5\text{-}(R_{f8})_2C_6H_3$–$CH_2OH$	$CH_3C_6H_5 : CF_3C_6F_{11}$	2.6 : 97.4	4a
V – Carboxylic acids and derivatives				
1[c]	$R_{f7}C(O)O$–$CH_2C_6H_4$–OCF_3	$CH_3C_6H_5 : CF_3C_6F_{11}$	4.1 : 95.9	4a
2[c]	$R_{f8}C_6H_4$–CO_2CH_3	$CH_3C_6H_5 : CF_3C_6F_{11}$	46.9 : 53.1	4a
3[c]	$3,5\text{-}(R_{f8})_2C_6H_3$–$CO_2CH_3$	$CH_3C_6H_5 : CF_3C_6F_{11}$	1.2 : 98.8	4a
4[c]	$[R_{f8}(CH_2)_3]_2CHCO_2H$	$CH_3C_6H_5 : CF_3C_6F_{11}$	2.9 : 97.1	7

Table 3.6 *Continued.*

Entry[a]	Substance[b]	Solvent system	Partitioning % (organic : fluorous)	Ref.
VI – Aromatic compounds				
1[c]	C_6H_6	$CH_3C_6H_5 : CF_3C_6F_{11}$	94.1 : 5.9	4a
2	C_6HF_5	$CH_3C_6H_5 : CF_3C_6F_{11}$	77.6 : 22.4	1
3	C_6F_6	$CH_3C_6H_5 : CF_3C_6F_{11}$	72.0 : 28.0	1
4	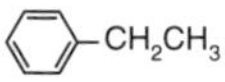 $-CH_2CH_3$	$CH_3C_6H_5 : CF_3C_6F_{11}$	98.8 : 1.2	1
5[c]	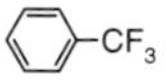 $-CF_3$	$CH_3C_6H_5 : CF_3C_6F_{11}$	87.6 : 12.4	4a
6[c]	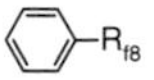 $-R_{f8}$	$CH_3C_6H_5 : CF_3C_6F_{11}$	22.5 : 77.5	4a
7[c]	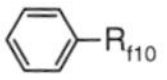 $-R_{f10}$	$CH_3C_6H_5 : CF_3C_6F_{11}$	14.6 : 85.4	4a
8[c]	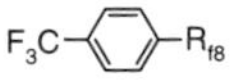 F_3C- $-R_{f8}$	$CH_3C_6H_5 : CF_3C_6F_{11}$	10.6 : 89.4	4a
9[c]	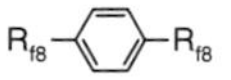 $R_{f8}-$ $-R_{f8}$	$CH_3C_6H_5 : CF_3C_6F_{11}$	0.7 : 99.3	4a
10[c]	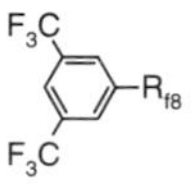 F_3C, $-R_{f8}$, F_3C	$CH_3C_6H_5 : CF_3C_6F_{11}$	1.7 : 98.3	4a
11	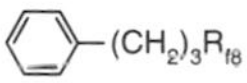 $-(CH_2)_3R_{f8}$	$CH_3C_6H_5 : CF_3C_6F_{11}$	50.5 : 49.5	1
12[c]	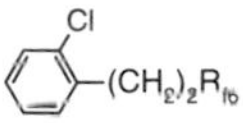 Cl, $-(CH_2)_2R_{f6}$	$CH_3C_6H_5 : CF_3C_6F_{11}$	65.5 : 34.5	4a
13[c]	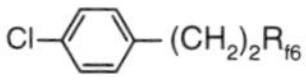 $Cl-$ $-(CH_2)_2R_{f6}$	$CH_3C_6H_5 : CF_3C_6F_{11}$	73.5 : 26.5	4a
14[c]	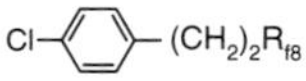 $Cl-$ $-(CH_2)_2R_{f8}$	$CH_3C_6H_5 : CF_3C_6F_{11}$	59.2 : 40.8	4a
15	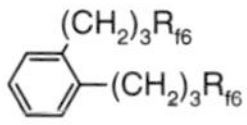 $(CH_2)_3R_{f6}$, $-(CH_2)_3R_{f6}$	$CH_3C_6H_5 : CF_3C_6F_{11}$	26.3 : 73.7	1
16	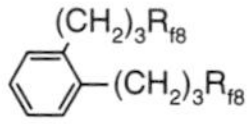 $(CH_2)_3R_{f8}$, $-(CH_2)_3R_{f8}$	$CH_3C_6H_5 : CF_3C_6F_{11}$	8.8 : 91.2	1
17	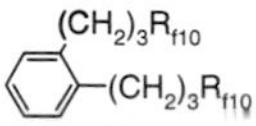 $(CH_2)_3R_{f10}$, $-(CH_2)_3R_{f10}$	$CH_3C_6H_5 : CF_3C_6F_{11}$	2.6 : 97.4	1
18	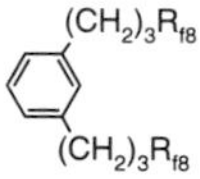 $(CH_2)_3R_{f8}$, $(CH_2)_3R_{f8}$	$CH_3C_6H_5 : CF_3C_6F_{11}$	9.3 : 90.7	1
19	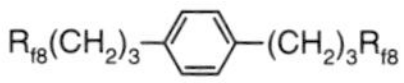 $R_{f8}(CH_2)_3-$ $-(CH_2)_3R_{f8}$	$CH_3C_6H_5 : CF_3C_6F_{11}$	8.9 : 91.1	1
20	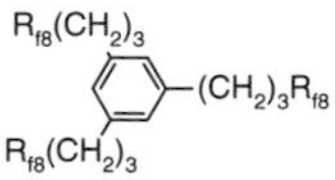 $R_{f8}(CH_2)_3$, $-(CH_2)_3R_{f8}$, $R_{f8}(CH_2)_3$	$CH_3C_6H_5 : CF_3C_6F_{11}$	<0.3 : >99.7	1
21[c]	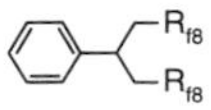 $-R_{f8}$, $-R_{f8}$	$CH_3C_6H_5 : CF_3C_6F_{11}$	2.1 : 97.9	8

Table 3.6 *Continued.*

Entry[a]	Substance[b]	Solvent system	Partitioning % (organic : fluorous)	Ref.
22	I–C_6H_3–$(CH_2)_3R_{f8}$, $(CH_2)_3R_{f8}$	$CH_3C_6H_5 : CF_3C_6F_{11}$	30.5 : 69.5	9
23	$R_{f8}(CH_2)_3$, I–C_6H_3–$(CH_2)_3R_{f8}$	$CH_3C_6H_5 : CF_3C_6F_{11}$	25.3 : 74.7	9
24	I, $R_{f8}(CH_2)_3$–C_6H_3–$(CH_2)_3R_{f8}$	$CH_3C_6H_5 : CF_3C_6F_{11}$	26.1 : 73.9	9
25	$R_{f8}(CH_2)_3$, I–C_6H_2–$(CH_2)_3R_{f8}$, $R_{f8}(CH_2)_3$	$CH_3C_6H_5 : CF_3C_6F_{11}$	2.0 : 98.0	9
26[d]	$(R_{f8}(CH_2)_2)_3Si$, OH, OH, $(R_{f8}(CH_2)_2)_3Si$ (binaphthol)	$CH_3C_6H_5 : C_6F_{14}$	1 : 99	10
For further aromatic compounds see III–V, VII, IX, X, XII–XIV				
VII – Nitrogen heterocycles				
1[c]	N, R_{f8} (2-substituted pyridine)	$CH_3C_6H_5 : CF_3C_6F_{11}$	36.8 : 63.2	4a
2[c]	N, R_{f8} (3-substituted pyridine)	$CH_3C_6H_5 : CF_3C_6F_{11}$	29.3 : 70.7	4a
3[c]	R_{f8}–C_5H_4N (4-substituted pyridine)	$CH_3C_6H_5 : CF_3C_6F_{11}$	31.1 : 68.9	4a
4	$R_{f8}(CH_2)_2$, N, $(CH_2)_2R_{f8}$	$CH_3C_6H_5 : CF_3C_6F_{11}$	6.2 : 93.8	11
5	$R_{f8}(CH_2)_3$, N, $(CH_2)_3R_{f8}$	$CH_3C_6H_5 : CF_3C_6F_{11}$	9.6 : 90.4	11
6	$R_{f8}(CH_2)_2$, N, $(CH_2)_2R_{f8}$, $(CH_2)_2R_{f8}$	$CH_3C_6H_5 : CF_3C_6F_{11}$	$<0.3 : >99.7$	11
7	$R_{f8}(CH_2)_3$, H, N, $(CH_2)R_{f8}$ (piperidine)	$CH_3C_6H_5 : CF_3C_6F_{11}$	6.4 : 93.6	11
VIII – Halides				
1[c]	$R_{f6}\mathbf{I}$	$CH_3C_6H_5 : CF_3C_6F_{11}$	21.2 : 78.8	4a
2[c]	$R_{f8}\mathbf{I}$	$CH_3C_6H_5 : CF_3C_6F_{11}$	11.5 : 88.5	4a
3[c]	$R_{f10}\,\mathbf{I}$	$CH_3C_6H_5 : CF_3C_6F_{11}$	5.5 : 94.5	4a
4	$R_{f8}(CH_2)_3\mathbf{I}$	$CH_3C_6H_5 : CF_3C_6F_{11}$	49.3 : 50.7	1
For further halides see VI				

Table 3.6 *Continued.*

Entry[a]	Substance[b]	Solvent system	Partitioning % (organic : fluorous)	Ref.
IX – Amines and imines				
1	$R_{f8}(CH_2)_3\mathbf{NH_2}$	$CH_3C_6H_5 : CF_3C_6F_{11}$	30.0 : 70.0	4b,12
2	$R_{f8}(CH_2)_4\mathbf{NH_2}$	$CH_3C_6H_5 : CF_3C_6F_{11}$	36.8 : 63.2	12
3	$R_{f8}(CH_2)_5\mathbf{NH_2}$	$CH_3C_6H_5 : CF_3C_6F_{11}$	43.1 : 56.9	12
4[c]	$R_{f8}(CH_2)_3\mathbf{NH}(CH_3)$	$CH_3C_6H_5 : CF_3C_6F_{11}$	29.1 : 70.9	4b
5	$[R_{f8}(CH_2)_3]_2\mathbf{NH}$	$CH_3C_6H_5 : CF_3C_6F_{11}$	3.5 : 96.5	4b,12
6	$[R_{f8}(CH_2)_4]_2\mathbf{NH}$	$CH_3C_6H_5 : CF_3C_6F_{11}$	4.9 : 95.1	12
7	$[R_{f8}(CH_2)_5]_2\mathbf{NH}$	$CH_3C_6H_5 : CF_3C_6F_{11}$	7.0 : 93.0	12
8[c]	$R_{f8}(CH_2)_3\mathbf{N}(CH_3)_2$	$CH_3C_6H_5 : CF_3C_6F_{11}$	20.2 : 79.8	4b
9[c]	$[R_{f8}(CH_2)_3]_2\mathbf{N}(CH_3)$	$CH_3C_6H_5 : CF_3C_6F_{11}$	2.6 : 97.4	4b
10	$[R_{f8}(CH_2)_3]_3\mathbf{N}$	$CH_3C_6H_5 : CF_3C_6F_{11}$	<0.3 : >99.7	4b,12
11	$[R_{f8}(CH_2)_4]_3\mathbf{N}$	$CH_3C_6H_5 : CF_3C_6F_{11}$	<0.3 : >99.7	12
12	$[R_{f8}(CH_2)_5]_3\mathbf{N}$	$CH_3C_6H_5 : CF_3C_6F_{11}$	0.5 : 99.5	12
13	$R_{f6}(CH_2)_2C(=N(CH_2)_2R_{f8})C_6H_4(CH_2)_3R_{f8}$	$CH_3C_6H_5 : CF_3C_6F_{11}$	1.3 : 98.7	5,13
X – Phosphorus compounds				
1[e]	$[R_{f6}(CH_2)_2]_3\mathbf{P}$	$CH_3C_6H_5 : CF_3C_6F_{11}$	1.2 : 98.8	14
2[e]	$[R_{f8}(CH_2)_2]_3\mathbf{P}$	$CH_3C_6H_5 : CF_3C_6F_{11}$	<0.3 : >99.7	14
3[e]	$[R_{f10}(CH_2)_2]_3\mathbf{P}$	$CH_3C_6H_5 : CF_3C_6F_{11}$	<0.3 : >99.7	14
4[e]	$[R_{f8}(CH_2)_3]_3\mathbf{P}$	$CH_3C_6H_5 : CF_3C_6F_{11}$	1.2 : 98.8	14
5[e]	$[R_{f8}(CH_2)_4]_3\mathbf{P}$	$CH_3C_6H_5 : CF_3C_6F_{11}$	1.1 : 98.9	14
6[e]	$[R_{f8}(CH_2)_5]_3\mathbf{P}$	$CH_3C_6H_5 : CF_3C_6F_{11}$	1.1 : 98.9	15
7[e]	$[R_{f6}(CH_2)_2]_3\mathbf{P}{=}O$	$CH_3C_6H_5 : CF_3C_6F_{11}$	<0.3 : >99.7	14
8[d]	$[R_{f6}(CH_2)_2-C_6H_4]_3P$	$CH_3C_6H_5 : C_6F_{14}$	57.1 : 42.9	16
9[e]	$[R_{f6}(CH_2)_3-C_6H_4]_3P$	$CH_3C_6H_5 : CF_3C_6F_{11}$	80.5 : 19.5	17
10[e]	$[R_{f8}(CH_2)_3-C_6H_4]_3P$	$CH_3C_6H_5 : CF_3C_6F_{11}$	33.4 : 66.6	17
11[c]	$[R_{f8}(CH_2)_3-C_6H_4]_3P \rightarrow BH_3$	$CH_3C_6H_5 : CF_3C_6F_{11}$	62.7 : 37.3	8
	$[R_{fn}(CH_2CH_2)_xSi(CH_3)_{3-x}-C_6H_4]_3P$			18,19
12[f]	$x = 1, n = 6$	$n\text{-}C_8H_{18} : CF_3C_6F_{11}$	47.6 : 52.4	18
13[f]	$x = 1, n = 8$	$n\text{-}C_8H_{18} : CF_3C_6F_{11}$	17.9 : 82.1	18
14[f]	$x = 2, n = 6$	$n\text{-}C_8H_{18} : CF_3C_6F_{11}$	5.6 : 94.4	18
15[f]	$x = 2, n = 8$	$n\text{-}C_8H_{18} : CF_3C_6F_{11}$	3.4 : 96.6	18
16[f]	$x = 3, n = 6$	$CH_3C_6H_5 : CF_3C_6F_{11}$	18.9 : 81.1	18
17[f]	$x = 3, n = 6$	$n\text{-}C_8H_{18} : CF_3C_6F_{11}$	9.6 : 90.4	18

Table 3.6 *Continued.*

Entry[a]	Substance[b]	Solvent system	Partitioning % (organic : fluorous)	Ref.
18[f]	$x = 3, n = 6$	n-$C_5H_{12} : CF_3C_6F_{11}$	6.2 : 93.8	18
19[f]	$x = 3, n = 8$	$CH_3C_6H_5 : CF_3C_6F_{11}$	32.3 : 67.7	18
20[f]	$x = 3, n = 8$	n-$C_8H_{18} : CF_3C_6F_{11}$	7.7 : 92.3	18
21[f]	$x = 3, n = 8$	n-$C_5H_{12} : CF_3C_6F_{11}$	4.8 : 95.2	18
XI – Stannanes				
1[d]	$[CH_3(CH_2)_3]_3$**Sn**H	$C_6H_6 : C_6F_{14}$	>99.7 : <0.3	20
2[d]	$[R_{f4}(CH_2)_2]_3$**Sn**H	$C_6H_6 : C_6F_{14}$	8.7 : 91.3	20
3[d]	$[R_{f6}(CH_2)_2]_3$**Sn**H	$C_6H_6 : CF_3C_6F_{11}$	2.2 : 97.8	21
4[d]		$C_6H_6 : C_6F_{14}$	2.2 : 97.8	20
5[d]	$[R_{f10}(CH_2)_2]_3$**Sn**H	$C_6H_6 : C_6F_{14}$	< 0.3 : >99.7	20
6[d]	$[R_{f4}(CH_2)_3]_3$**Sn**H	C_6H_6	45.5 : 54.5	20
7[d]	$[R_{f6}(CH_2)_3]_3$**Sn**H	$C_6H_6 : C_6F_{14}$	9.1 : 90.9	20
8[d]	$[R_{f6}(CH_2)_2]_3$ **Sn**$CH_2CH{=}CH_2$	$C_6H_6 : C_6F_{14}$	2.0 : 98.0	22
XII – Silicon compounds				
1	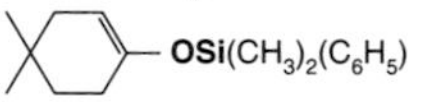 (4-*tert*-butylcyclohexenyl)–**OSi**$(CH_3)_2(C_6H_5)$	$CH_3C_6H_5 : CF_3C_6F_{11}$	99.4 : 0.6	23a
2	 (cyclohexyl)–**OSi**$(CH_3)_2(C_6H_5)$	$CH_3C_6H_5 : CF_3C_6F_{11}$	99.2 : 0.8	23b
For further compounds see VI, X and XIV				
XIII – Sulfur Compounds				
1[c]	$R_{f8}(CH_2)_3$**S**H	$CH_3C_6H_5 : CF_3C_6F_{11}$	44.1 : 55.9	4a
2[c]	CF_3**S**C_6H_5	$CH_3C_6H_5 : CF_3C_6F_{11}$	92.1 : 7.9	4a
3[c]	R_{f8}**S**C_6H_5	$CH_3C_6H_5 : CF_3C_6F_{11}$	35.7 : 64.3	4a
4[c]	$R_{f7}C(O)$**S**CH_3	$CH_3C_6H_5 : CF_3C_6F_{11}$	23.9 : 76.1	4c
5	$[R_{f8}(CH_2)_2]_2$**S**	$CH_3C_6H_5 : CF_3C_6F_{11}$	1.3 : 98.7	24
6	$[R_{f8}(CH_2)_3]_2$**S**	$CH_3C_6H_5 : CF_3C_6F_{11}$	3.4 : 96.6	24
7	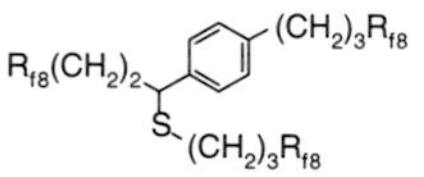 $R_{f8}(CH_2)_2$–CH(S–$(CH_2)_3R_{f8}$)–C_6H_4–$(CH_2)_3R_{f8}$	$CH_3C_6H_5 : CF_3C_6F_{11}$	0.5 : 99.5	5
XIV – Transition metal compounds				
1[e]	$[\{R_{f6}(CH_2)_2\}_3P]_3$**Rh**Cl	$CH_3C_6H_5 : CF_3C_6F_{11}$	0.14 : 99.86	25
2[e]	$[\{R_{f8}(CH_2)_2\}_3P]_3$**Rh**Cl	$CH_3C_6H_5 : CF_3C_6F_{11}$	0.12 : 99.88	25
3[e]	$[\{R_{f6}(CH_2)_2\}_3P]_3$ **Rh**(H)(CO)	$CH_3C_6H_5 : CF_3C_6F_{11}$	0.5 : 99.5	4a,26
4[g]	$[R_{f10}(CH_2)_2C_5H_4]$ **Rh**$(CO)_2$	$CH_3C_6H_5 : CF_3C_6F_{11}$	55.5 : 44.5	26
5[g]	$[R_{f10}(CH_2)_2C_5H_4]$ **Rh**(CO) $[P\{(CH_2)_2R_{f6}\}_3]$	$CH_3C_6H_5 : CF_3C_6F_{11}$	3.3 : 96.7	26
6	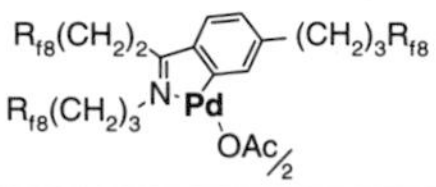 $R_{f8}(CH_2)_2$, $(CH_2)_3R_{f8}$, $R_{f8}(CH_2)_3$–N·**Pd** OAc/2 (cyclopalladated imine)	$CH_3C_6H_5 : CF_3C_6F_{11}$	4.5 : 95.5	5,13

Table 3.6 *Continued.*

Entry[a]	Substance[b]	Solvent system	Partitioning % (organic : fluorous)	Ref.
7	$R_{f8}(CH_2)_2$, $(CH_2)_3R_{f8}$, S, $R_{f8}(CH_2)_3$, **Pd**, OAc, /2	$CH_3C_6H_5 : CF_3C_6F_{11}$	9.3 : 90.7	5
	$(R_3P)_3$**Rh**Cl			19
8[f]	$R = C_6H_4$-*p*-$Si(CH_3)_2$ $(CH_2)_2R_{f6}$	*n*-$C_8H_{18} : CF_3C_6F_{11}$	0.3 : 99.7	19
9[c]		*n*-$C_8H_{18} : CF_3C_6F_{11}$	1.3 : 98.7	19
10[f]	$R = C_6H_4$-*p*-$Si(CH_3)_2$ $(CH_2)_2R_{f8}$	*n*-$C_8H_{18} : CF_3C_6F_{11}$	0.1 : 99.9	19

[a] All measurements obtained at 24°C unless otherwise stated.
[b] $R_{fn} = (CF_2)_{n-1}CF_3$.
[c] 25°C.
[d] Ambient temperature implied.
[e] 27°C.
[f] 0°C.
[g] 20°C.

References: (1) Rocaboy, C.; Rutherford, D.; Bennett, B. L.; Gladysz, J. A. *J. Phys. Org. Chem.* **2000**, *13*, 596. (2) Rutherford, D.; Juliette, J. J. J.; Rocaboy, C.; Horváth, I. T.; Gladysz, J. A. *Catal. Today* **1998**, *42*, 381. (3) Rutherford, D. unpublished results. (4) (a) Kiss, L. E.; Kövesdi, I.; Rábai, J. *J. Fluorine Chem.* **2001**, *108*, 95. (b) Szlávik, Z.; Tárkányi, G.; Gömöry, Á.; Tarczay, G.; Rábai, J. *J. Fluorine Chem.* **2001**, *108*, 7. (c) Kiss, L. E.; Rábai, J.; Varga, L.; Kövesd, I. *Synlett* **1998**, 1243. (5) Rocaboy, C.; Gladysz, J. A. *New J. Chem.* **2003**, *27*, 39. (6) Szlávik, Z.; Tárkányi, G.; Tarczay, G.; Gömöry, Á.; Rábai, J. *J. Fluorine Chem.* **1999**, *98*, 83. (7) Loiseau, J.; Fouquet, E.; Fish, R. H.; Vincent, J.-M.; Verlhac, J.-B. *J. Fluorine Chem.* **2001**, 108, 195. (8) Wende, M.; Seidel, F.; Gladysz, J. A. *J. Fluorine Chem.* **2003**, *124*, 45. (9) Rocaboy, C.; Gladysz, J. A. *Chem. Eur. J.* **2003**, *9*, 88. (10) Nakamura, Y.; Takeuchi, S.; Ohgo, Y. *J. Fluorine Chem.* **2003**, *120*, 121. (11) Rocaboy, C.; Hampel, F.; Gladysz, J. A. *J. Org. Chem.* **2002**, *67*, 6863. (12) Rocaboy, C.; Bauer, W.; Gladysz, J. A. *Eur. J. Org. Chem.* **2000**, 2621. (13) Rocaboy, C.; Gladysz, J. A. *Org. Lett.* **2002**, *4*, 1993. (14) Alvey, L. J.; Rutherford, D.; Juliette, J. J. J.; Gladysz, J. A. *J. Org. Chem.* **1998**, *63*, 6302. (15) Alvey, L. J.; Meier, R.; Soós, T.; Bernatis, P.; Gladysz J. A. *Eur. J. Inorg. Chem.* **2000**, 1975. (16) Zhang, Q.; Luo, Z.; Curran, D. P. *J. Org. Chem.* **2000**, *65*, 8866. (17) Soós, T.; Bennett, B. L.; Rutherford, D.; Barthel-Rosa, L. P.; Gladysz, J. A. *Organometallics* **2001**, *20*, 3079. (18) Richter, B.; de Wolf, E.; van Koten, G.; Deelman, B.-J. *J. Org. Chem.* **2000**, *65*, 3885. (19) Richter, B.; Spek, A. L.; van Koten, G.; Deelman, B.-J. *J. Am. Chem. Soc.* **2000**, *122*, 3945. (20) Curran, D. P.; Hadida, S.; Kim, S.-Y.; Luo, Z. *J. Am. Chem. Soc.* **1999**, *121*, 6607. (21) Curran, D. P.; Hadida, S. *J. Am. Chem. Soc.* **1996**, *118*, 2531. (22) Curran, D. P.; Luo, Z.; Degenkolb, P. *Bioorg. Med. Chem. Lett.* **1998**, *8*, 2403. (23) (a) Dinh, L. V.; Gladysz, J. A. *Tetrahedron Lett.* **1999**, *40*, 8995. (b) Dinh, L. V. unpublished results. (24) Rocaboy, C.; Gladysz, J. A. *Tetrahedron* **2002**, *58*, 4007. (25) Juliette, J. J. J.; Rutherford, D.; Horváth, I. T.; Gladysz, J. A. *J. Am. Chem. Soc.* **1999**, *121*, 2696. (26) Herrera, V.; de Rege, P. J. F.; Horváth, I. T.; Husebo, T. L.; Hughes, R. P. *Inorg. Chem. Commun.* **1998**, *1*, 197.

groups in each ponytail is increased to five, a small amount of the amine can again be detected in the toluene phase (0.5 : 99.5; entry IX-12).

Similar trends are observed with tertiary fluorous phosphines. Here, researchers have varied the lengths of the R_{fn} as well as the $(CH_2)_m$ segments (entries X-1 to X-6), and very high fluorous phase affinities can be achieved (<0.3 : >99.7 for

$[R_{f8}(CH_2)_2]_3P$ and $[R_{f10}(CH_2)_2]_3P$). Where comparisons are possible, values are slightly lower than for analogous amines. Also, oxidation to a phosphine oxide slightly increases the fluorophilicity (entry X-7 vs. entry X-1).

Thioethers, which can accommodate only two ponytails around the central heteroatom, possess fluorous phase affinities slightly lower than those of comparable amines and phosphines (entries XIII-5, XIII-6). As shown in section XI of Table 3.6, appropriately designed trialkyltin hydrides can also be highly fluorophilic. Although the solvent systems used are slightly different, values for $[R_{f6}(CH_2)_2]_3SnH$ (entries XI-3, XI-4) are similar to those for the corresponding phosphines.

Data for simple arenes are collected in section VI of Table 3.6. Both pentafluorobenzene and hexafluorobenzene preferentially partition into toluene (77.6 : 22.4 and 72.0 : 28.0; entries VI-2, VI-3), consistent with their non-fluorous nature, as described in Section 3.2.2. Benzene exhibits an even greater toluene affinity (94.1 : 5.9; entry VI-1). However, the introduction of a single ponytail of formula $R_{f8}(CH_2)_3$ levels this out, giving a partition coefficient of 50.5 : 49.5 (entry VI-11). This value is similar to those obtained when $R_{f8}(CH_2)_3$ is capped with an iodide or thiol. The compound $R_{f8}C_6H_5$ (entry VI-6), which lacks methylene spacers, is more fluorophilic still, but the electronic properties of the arene ring are strongly perturbed.

As shown in entries VI-16, VI-18 and VI-19, benzenes with two ponytails of formula $R_{f8}(CH_2)_3$ exhibit appreciable fluorophilicities, with partition coefficients of 8.8 : 91.2 to 9.3 : 90.7. The substitution pattern has little influence. As with other compounds above, when the perfluoroalkyl segment of the ponytail is shortened, the fluorous phase affinity decreases (entry VI-15), and when it is lengthened, the fluorous phase affinity increases (entry VI-17). Importantly, benzenes with three ponytails of formula $R_{f8}(CH_2)_3$ partition (within detection limits) completely into $CF_3C_6F_{11}$ (entry VI-20).

Entries VI-22 through VI-25 feature monoiodide derivatives of some of the preceding fluorous benzenes. In all cases, the fluorophilicities decrease. Only for entry VI-25, a triply ponytailed compound, is a highly biased partition coefficient maintained (2.0 : 98.0). When a more polar non-fluorous solvent, such as methanol, is employed, fluorous phase affinities increase. Nevertheless, it is clear that with monofunctional arenes, at least three ponytails of formula $R_{f8}(CH_2)_3$ are required for high degrees of fluorous phase immobilization. Entries III-4, IV-9 and V-3 show that two ponytails of formula R_{f8} (i.e. without spacers) are essentially as effective (1.4 : 98.6, 2.6 : 97.4 and 1.2 : 98.8).

These points are further illustrated by some of the fluorous triarylphosphines in Section X of Table 3.6. Entry X-9 shows that with one ponytail of formula $R_{f6}(CH_2)_3$ per ring, the toluene phase affinity is higher (80.5 : 19.5). With one ponytail of formula $R_{f8}(CH_2)_3$ per ring, the fluorous phase affinity becomes higher (33.4 : 66.6; entry X-10). It has proved problematic to replace additional aryl hydrogen atoms with $R_{f8}(CH_2)_3$ ponytails. However, a clever way around this problem has been developed;[41,42] namely, silicon-based ponytails of the formula $(R_{fn}(CH_2)_2)_xSi(CH_3)_{3-x}$ have been used as anchors for as many as three R_{f6}

or R_{f8} groups per ring. As summarized in entries X-16 through X-21, this gives phosphines with much higher fluorous phase affinities (up to 4.8 : 95.2 for $x = 3$ and $n = 8$).[42a,b]

The pyridines in entries VII-1 through VII-3, which are R_{f8}-monosubstituted *ortho, meta* and *para* isomers, show modest fluorophilicities in the narrow range of 36.8 : 63.2 to 29.3 : 70.7. The pyridine in entry VII-5 is rigorously comparable to the benzenoid analog in entry VI-18 (N:/CH exchange) and gives an essentially identical partition coefficient. Hence, the polar pyridine nitrogen has little influence. Entry VII-6 shows that three ponytails of the formula $R_{f8}(CH_2)_2$ provide essentially complete fluorous phase immobilization. The hydrogenation of the pyridine in entry VII-5 to the piperidine in entry VII-7 increases the fluorophilicity (9.6 : 90.4 vs. 6.4 : 93.6).

Section XIV of Table 3.6 features some rhodium complexes and palladacycles that are good catalyst precursors. Other metal-containing species are tabulated elsewhere.[40] The most interesting trend involves entry XIV-8. The central rhodium is surrounded by three fluorous triarylphosphines that have only one ponytail per ring and an n-$C_8H_{18}/CF_3C_6F_{11}$ partition coefficient of 47.6 : 52.4 at 0°C (entry X-12). Nonetheless, the rhodium complex is highly fluorophilic, with a partition coefficient of 0.3 : 99.7 under analogous conditions. Similar phenomena, in which the sum is greater than the parts, have been observed with other compounds that are aggregates of fluorous building blocks. It is presumed that the ponytails are deployed in a maximally efficient way around the periphery of the molecule. With the fluorous palladacycles (entries XIV-6 and XIV-7), note that there are three ponytails of formula $R_{f8}(CH_2)_m$ per arene ring. The resulting partition coefficients (4.5 : 95.5 and 9.3 : 90.7) are somewhat lower than those of the free non-palladated ligands in entries IX-13 and XIII-7 (1.3 : 98.7 and 0.5 : 99.5).

There have been intensive efforts to parameterize the above data so that fluorophilicities can be predicted.[39,43] Indeed, this effort appears to have been successful within some series of compounds. An empirical "60% rule" is often stated – that is, 60% of the molecular weight should be fluorine derived to ensure high fluorophilicity.[2a] However, exceptions are known, such as for the rhodium complexes in entries XIV-8 and XIV-10. In some compounds, but not others, a short perfluoroalkyl segment (i.e. an integrated CF_3 group or a pig-tail) seems to impart a fluorophilicity far beyond what might be expected. One example is the trisubstituted benzene in entry VI-10.

3.2.7 *Toxicity and environmental issues*

Fluorous solvents are recognized as holding an important place among the green solvents.[44] Saturated perfluorocarbons present few if any toxicity problems, and extensive use is made of them in household cookware.[45] However, all classes of chemicals are subject to increasing scrutiny, and it would not be surprising if some type of functionalized fluorous compounds were found to have adverse physiological consequence in certain organisms in the future.[46] The same holds for ionic liquids and probably all other green solvent candidates except supercritical CO_2.

No saturated perfluorocarbon has been found to contribute in any way to ozone depletion.[47] However, these compounds do raise another environmental issue; namely, they are extremely persistent. Environmental half-lives have been estimated as 4.1×10^3 years for C_5F_{12} and 3.1×10^3 years for C_6F_{12}.[48] Thus, even though there are no known adverse effects, accumulation can be detected in certain flora and fauna.[49] Attention has to be paid to containment, particularly for any large-scale application, and the priority will necessarily be the development of biodegradable fluorous solvents. Indeed, there are currently efforts in industry directed at a broad spectrum of biodegradable fluorinated consumer products, and new commercial offerings can be expected in the near future.

3.3 Applications as reaction media

This section describes applications of fluorous compounds in fluorous solvents. Most of the compounds presented contain a large number of fluorine atoms – *heavy fluorous* molecules – and display a high affinity for fluorous solvents. This property allows the easy separation of the fluorinated material, either by extraction or by filtration, and in many cases its recycling.

The use of fluorous catalysts is presented for general reactions (Section 3.3.1) and for enantioselective transformations (Section 3.3.2); fluorous analogues of traditional reagents are detailed in Section 3.3.3, followed in Section 3.3.4 by an account of the use of fluorous protecting groups.

3.3.1 Fluorous catalysts for fluorous biphasic systems

3.3.1.1 Hydroformylation

Hydroformylation is an industrially important carbonylation process that is used in the synthesis of aldehydes from olefins, CO and molecular hydrogen by homogeneous catalysis. Efficient catalyst recovery is crucial for the process to be economically viable. Pioneering work in fluorous chemistry was published in 1994 by Horváth and Rabái, who studied the fluorous hydroformylation of olefins catalyzed by an Rh complex with a fluorous phosphine as a ligand.[11,50]

The fluorous phosphine ligand $P(CH_2CH_2C_6F_{13})_3$ is designed to be soluble in a fluorous solvent, and the ethylene spacers serve as an insulator from the electron-withdrawing perfluorohexyl group. Reactions are carried out using a binary mixture of toluene and perfluoromethylcyclohexane (PFMC) as the reaction medium. These solvents are immiscible at room temperature, but a single phase is formed on heating to a reaction temperature of 100°C. After the reaction, upon cooling, phase separation occurs. The products are isolated from the upper organic phase, and the lower fluorous layer contains the catalyst, which can be recycled repeatedly (Scheme 3).

Later work modified the conditions for the Rh-catalyzed fluorous hydroformylation of 1-octene by omitting toluene.[51] Thus, the fluorous hydroformylation was carried out with PFMC as the sole solvent and a fluorous Rh catalyst derived from

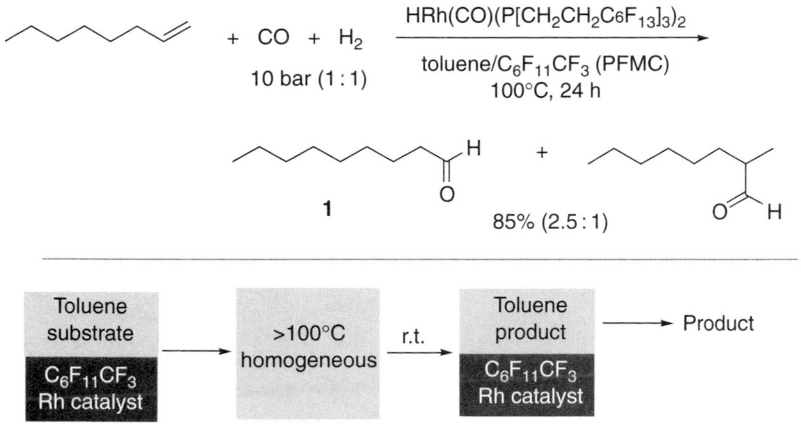

+ CO + H2 10 bar (1 : 1) HRh(CO)(P[CH2CH2C6F13]3)2 toluene/C6F11CF3 (PFMC) 100°C, 24 h

1 85% (2.5 : 1)

Toluene substrate | C6F11CF3 Rh catalyst → >100°C homogeneous → r.t. → Toluene product | C6F11CF3 Rh catalyst → Product

Scheme 3

$[Rh(acac)(CO)]_2$ and $P(C_6H_4\text{-}p\text{-}C_6F_{13})_3$. These conditions resulted in faster rates and better linear : branched ratios of the products. At room temperature, the substrate 1-octene is miscible with perfluoromethylcyclohexane but nonanal (**1**) is not, resulting in the phase separation of the product and the catalyst layer.

3.3.1.2 Hydrogenation

Wilkinson's complex, tris(triphenylphosphine)rhodium(I) chloride, is routinely used in the hydrogenation of alkenes. The fluorous tris(triarylphosphine)rhodium(I) chloride complex **2** ($RhCl[\{C_6H_4\text{-}p\text{-}SiMe_2(CH_2)_2C_8F_{17}\}_3]_3$) shows high activity in the hydrogenation of 1-alkenes under single-phase fluorous conditions using perfluoromethylcyclohexane at 80°C.[52] After the reaction, the hydrogenated products can be easily separated from the catalyst phase by cooling the reaction mixture to 0°C. The recycling efficiencies of the catalyst are >98%. The weakly electron-donating silicon atom may compensate for the electron-withdrawing perfluoroalkyl ponytail.

The fluorous version of 1,2-bis(diphenylphosphino)ethane (dppe) was used as a ligand in the Rh-catalyzed hydrogenation of 4-octyne to *cis*-4-octene **3** (Scheme 4).[53] The catalyst was recycled with 97.5% efficiency in a binary system of PFMC and acetone (1 : 1 v/v).

3.3.1.3 Catalytic hydroboration and hydrosilylation

Fluorous Wilkinson's complexes, such as $ClRh[P(C_2H_4C_6F_{13})_3]_3$ and $ClRh[P(C_2H_4C_8F_{17})_3]_3$, exhibited good catalytic activity for the hydroboration of alkenes and alkynes using catecholborane, and the hydrosilylation of enones using phenyldimethylsilane.[54] The following example shows the catalytic hydroboration of norbornene and its conversion to norbornanol **4** in a binary solvent system consisting of PFMC and toluene (Scheme 5). Interestingly, the reactions proceeded

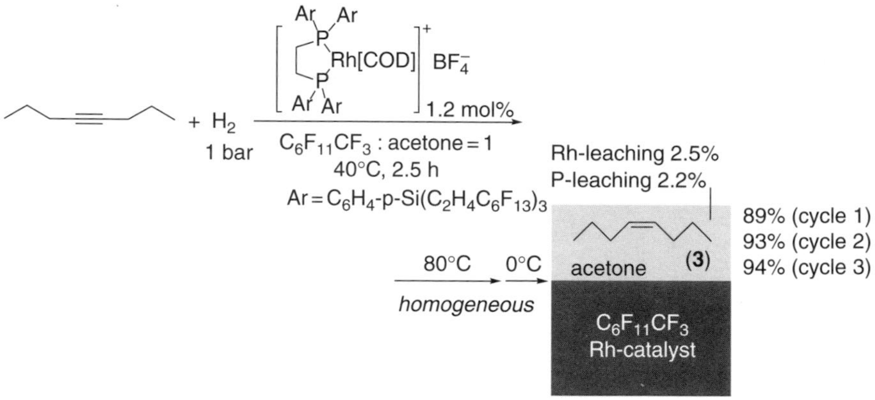

Scheme 4

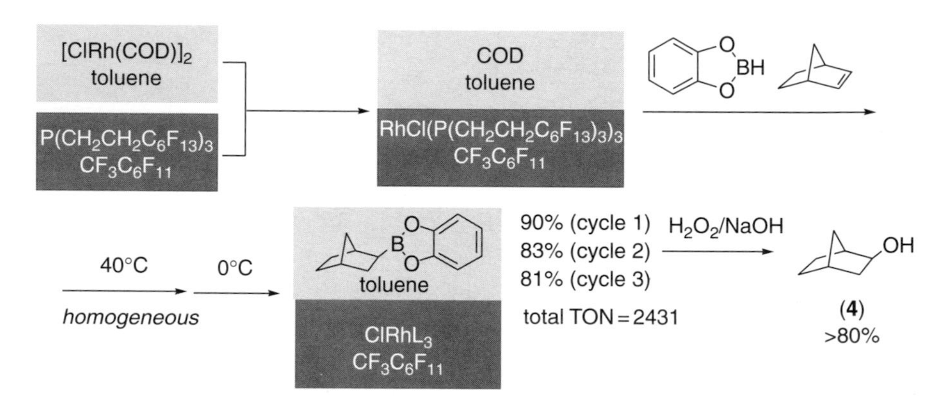

Scheme 5

more rapidly with PFMC alone, despite the fact that not all the reactants dissolved at the beginning of the reaction.

These fluorous Rh catalysts are also effective in the hydrosilylation of enones, which employs a similar binary solvent system (Scheme 6). Hexanes and ether can be used in place of toluene. Since these organic solvents are miscible with PFMC at room temperature, effective phase separation requires cooling to −30°C.

3.3.1.4 Catalytic oxidation reactions

A fluorous acetylacetonate ligand, $C_7F_{15}COCH_2COC_7F_{15}$, was introduced into metal-catalyzed aerobic oxidation reactions in fluorous media, which appears to be useful for two reasons: (1) catalyst recycling is easy; (2) the gas is highly soluble in the fluorous solvent employed.[55,56] For example, the fluorous Ni(acac)$_2$ complex was used as a catalyst for the auto-oxidation of aldehydes to carboxylic acids, where a mixed solvent system comprising toluene and perfluorodecalin was used.

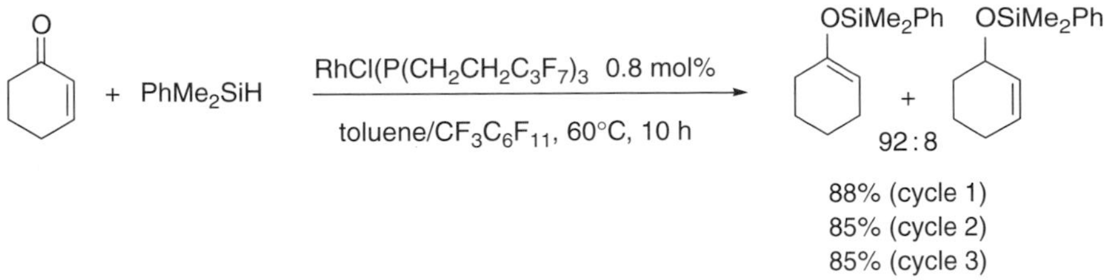

+ $PhMe_2SiH$ $\xrightarrow[\text{toluene}/CF_3C_6F_{11},\ 60°C,\ 10\ h]{RhCl(P(CH_2CH_2C_3F_7)_3\ 0.8\ mol\%}$ $OSiMe_2Ph$ + $OSiMe_2Ph$

92 : 8

88% (cycle 1)
85% (cycle 2)
85% (cycle 3)

Scheme 6

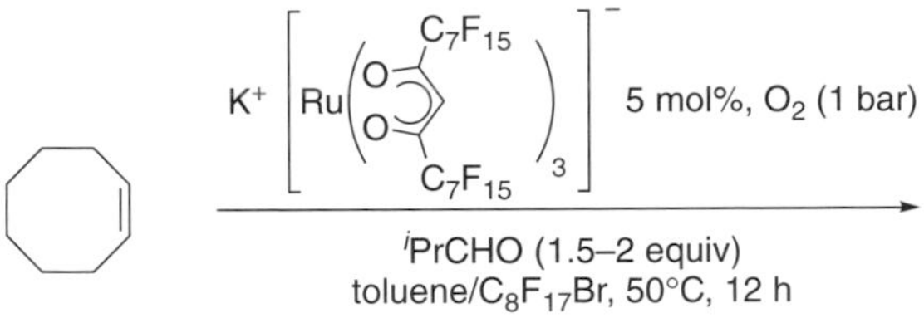

$K^+ [Ru(O,O\text{-}C_7F_{15}\text{acac}C_7F_{15})_3]^-$ 5 mol%, O_2 (1 bar)

^{i}PrCHO (1.5–2 equiv)
toluene/$C_8F_{17}Br$, 50°C, 12 h

85%

Scheme 7

The reaction mixture became homogeneous at 64°C. After the mixture was cooled to room temperature, phase separation occurred. Catalytic epoxidation of disubstituted alkenes was performed using a Ru-based fluorous acac complex in toluene/$C_8F_{17}Br$ (Scheme 7). The brown fluorous phase containing the Rh catalyst can be recycled several times in this epoxidation with no significant loss of activity.

A similar acac-type palladium complex catalyzes the Wacker oxidation of terminal alkenes, leading to methyl ketones in a binary solvent system comprising benzene and bromoperfluorooctane using *t*-BuOOH as an oxidant.[57] A nickel hexafluoroacetylacetonate catalyst was employed in a propylene dimerization, using a binary solvent system comprising toluene and perfluorodecalin.[56]

The Ru-catalyzed epoxidation of *trans*-stilbene in the presence of $NaIO_4$ was carried out using a bipyridyl ligand with a fluorous ponytail at the 4 and 4′ positions.[58] As illustrated by the first equation in Scheme 8, a triphasic system comprising water, dichloromethane and perfluorooctane was employed in the reaction. The reaction was complete in 15 min at 0°C and *trans*-stilbene oxide **5** was obtained from the dichloromethane layer in a 92% yield. The fluorous layer, containing the catalyst, could be recycled for four further runs without any addition of $RuCl_3$. The same perfluoroalkyl-substituted bipyridyl ligand was used successfully in the copper(I)-catalyzed TEMPO (2,2′,6,6′-tetramethylpiperidine *N*-oxyl)-oxidation of primary and secondary alcohols under aerobic conditions (Scheme 8, second equation).[59]

A fluorous version of the cobalt-porphyrin-catalyzed aerobic epoxidation of alkenes was examined in the presence of 2-methylpropanal (Scheme 9).[60] A fluorous porphyrin ring containing four 3,5-diperfluorooctylphenyl groups was prepared and complexed with $Co(OAc)_2$ to form the cobalt catalyst **6**. The reaction was carried out with vigorous stirring at room temperature in a biphasic system comprising

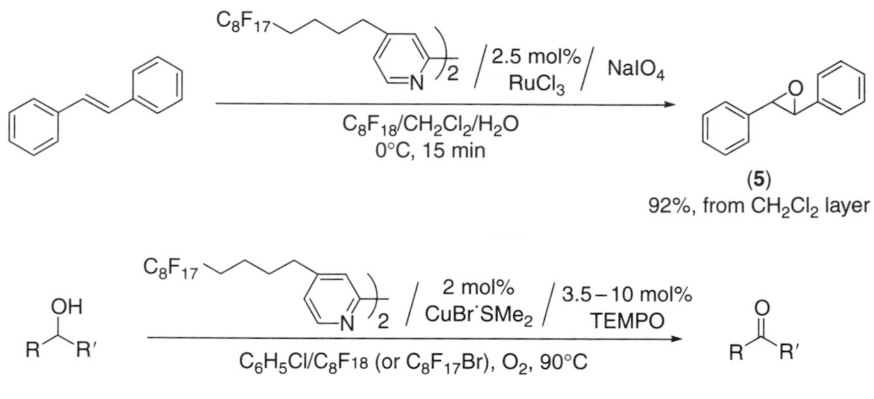

C_8F_{17} / 2.5 mol% $RuCl_3$ / $NaIO_4$
$C_8F_{18}/CH_2Cl_2/H_2O$
0°C, 15 min
(**5**)
92%, from CH_2Cl_2 layer

C_8F_{17} / 2 mol% $CuBr{\cdot}SMe_2$ / 3.5–10 mol% TEMPO
C_6H_5Cl/C_8F_{18} (or $C_8F_{17}Br$), O_2, 90°C

Scheme 8

$Co(OAc)_2{\cdot}4H_2O$
DMF, reflux
6
Ar = C_6H_3-3,5-$(CF_2(CF_2)_6CF_3)_2$

CH_3CN / **6** / C_6F_{14}
[Co]-catalyst
6 0.1 mol%, O_2
CH_3CN/C_6F_{14}, 25°C, 3 h
100% conversion
95% selectivity

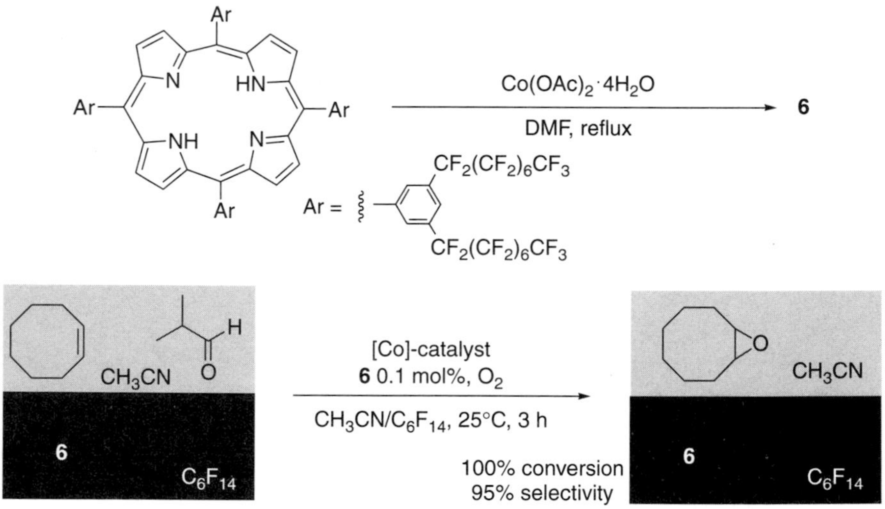

Scheme 9

acetonitrile and perfluorohexane. It is likely that the reaction takes place around the border surface of the two layers.

Selenium-catalyzed oxidation reactions have been reported by two different research groups, who introduced two fluorous ponytails into the aromatic ring of aryl butyl selenide (Scheme 10). Using 5 mol% of 2,4-bisperfluorooctylphenyl butyl selenide (**7**) as a catalyst, a variety of alkenes were oxidized by hydrogen peroxide to the corresponding epoxides.[61] The use of the 60% aqueous solution of hydrogen peroxide seems crucial, since a 30% solution resulted in the formation of an emulsion and lower yields. A ^{19}F-nuclear magnetic resonance (NMR) study

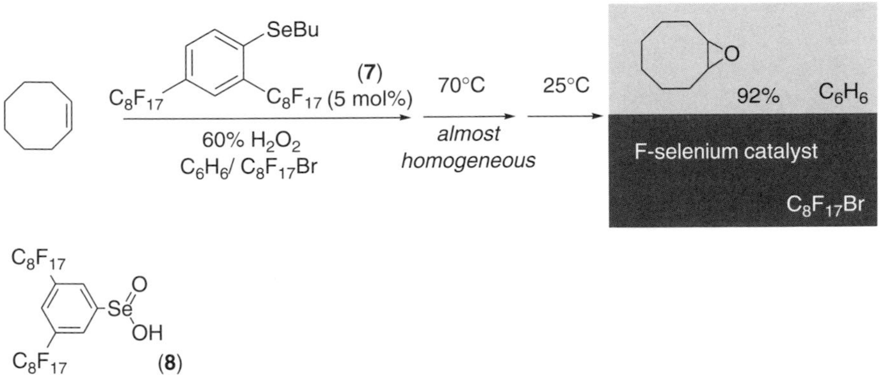

Scheme 10

revealed that leaching of the catalyst into the benzene layer is ~0.1%. The catalyst can be recycled more than ten times without any decrease in yields. The Baeyer–Villiger oxidation of cyclopentanone with hydrogen peroxide catalyzed by 2 mol% of 3,5-bis(perfluorooctyl)benzeneseleninic acid (**8**), generated *in situ* from the corresponding butyl selenide, was carried out using 60% hydrogen peroxide and hexafluoroisopropanol as a reaction medium.[62] The catalyst system gave an initial turnover frequency (TOF) of 75 h^{-1} and complete conversion to the δ-lactone in 2 h. Triphasic conditions (water/1,2-dichloroethane/perfluorodecalin) were also examined for the oxidation of *p*-nitrobenzaldehyde, allowing the easy separation and recycling of the catalyst phase. However, the activity of the catalyst decreased after several recyclings (90%, 71%, 86%, 78%, 50%).

Trifluoromethyl perfluorohexylethyl ketone is an efficient alternative catalyst for the epoxidation reactions of alkenes.[63] Such reactions were carried out using 1–5 mol% of this fluorous ketone as a catalyst and 1.5 equivalents of the oxone in a mixture of hexafluoroisopropanol and water. Thus, the new procedure circumvents the need for volatile fluorine compounds such as trifluoroacetone and hexafluoroacetone.

3.3.1.3 Coupling reactions

The Stille coupling reaction of aryl and benzyl halides with fluorous aryltin compounds permits the easy separation of reagent and product by means of a biphasic workup.[64] The reaction can be dramatically accelerated via the use of microwave irradiation.[65] The Stille coupling reactions were performed using a fluorous palladium complex **9** prepared from palladium dichloride and tri(*p*-perfluorononylphenyl)phosphine.[66] The reaction was conducted in a mixture of PFMC and DMF at 80°C for 3 h. After the mixture was cooled to room temperature, phase separation occurred. The lower fluorous layer, containing the catalyst, can be used repeatedly (Scheme 11). Related Pd catalysts with different fluorous ligands, such as tri(*m*-perfluorooctylphenyl)phosphine and

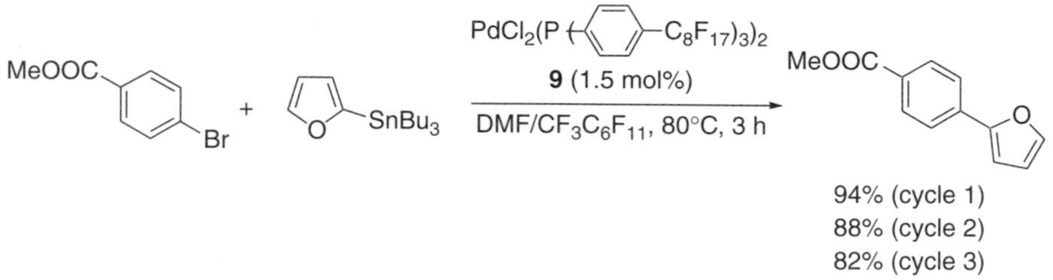

MeOOC (4-bromophenyl) + 2-furyl-$SnBu_3$ → [$PdCl_2(P(C_6H_4\text{-}C_8F_{17})_3)_2$ **9** (1.5 mol%), $DMF/CF_3C_6F_{11}$, 80°C, 3 h] → MeOOC-(4-(2-furyl)phenyl)

94% (cycle 1)
88% (cycle 2)
82% (cycle 3)

Scheme 11

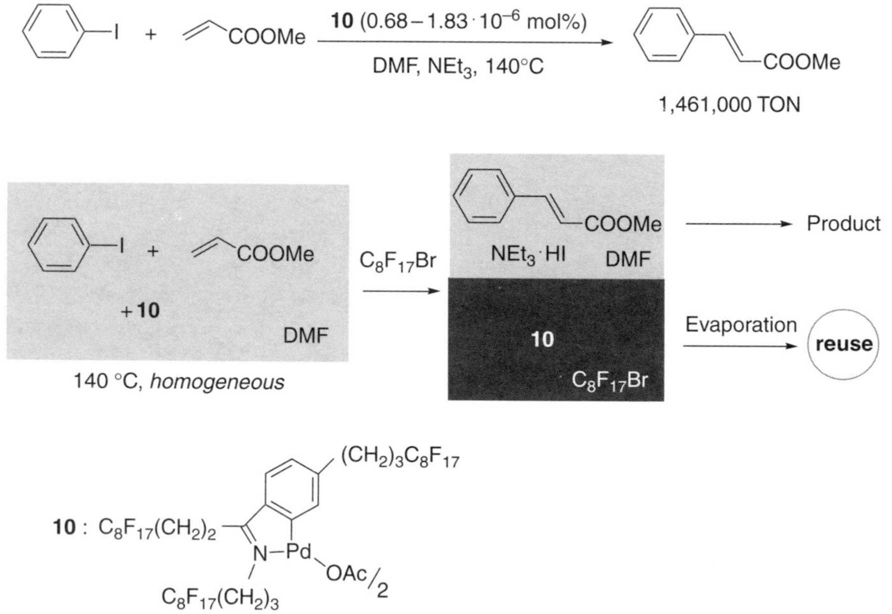

PhI + $\diagup$COOMe → [**10** (0.68–1.83·10^{-6} mol%), DMF, NEt_3, 140°C] → PhCH=CH–COOMe

1,461,000 TON

PhI + $\diagup$COOMe + **10**, DMF; 140 °C, *homogeneous* → $C_8F_{17}Br$ → upper phase: PhCH=CH–COOMe, $NEt_3\cdot HI$, DMF → Product; lower phase: **10**, $C_8F_{17}Br$ → Evaporation → reuse

10 : $C_8F_{17}(CH_2)_2$, $(CH_2)_3C_8F_{17}$, N–Pd, OAc/2, $C_8F_{17}(CH_2)_3$

Scheme 12

[*p*-(perfluorooctyl)ethylphenyl]phosphine also function well in Stille coupling reactions.

Using a fluorous palladacycle catalyst **10** originating from the corresponding fluorous Schiff base and palladium acetate, a fluorous Mizoroki–Heck reaction was achieved with an excellent turnover number (Scheme 12).[67] A homogeneous catalytic reaction system was obtained when DMF was used as the solvent. After the reaction, perfluorooctyl bromide was added to facilitate the separation of DMF (containing the products and amine salts) from the catalyst phase. The resulting lower fluorous layer was condensed under vacuum and the catalyst residue was used in a second run. In this reaction, the palladacycle catalyst appears to act as a source of palladium nanoparticles, which are thought to be the dominant active catalyst.

Fluorous triarylphosphine ligands, such as $P(Ph\text{-}p\text{-}OCH_2C_7F_{15})_3$ and $P(Ph\text{-}p\text{-}C_6F_{13})_3$, were examined in a Mizoroki–Heck reaction of methyl acrylate

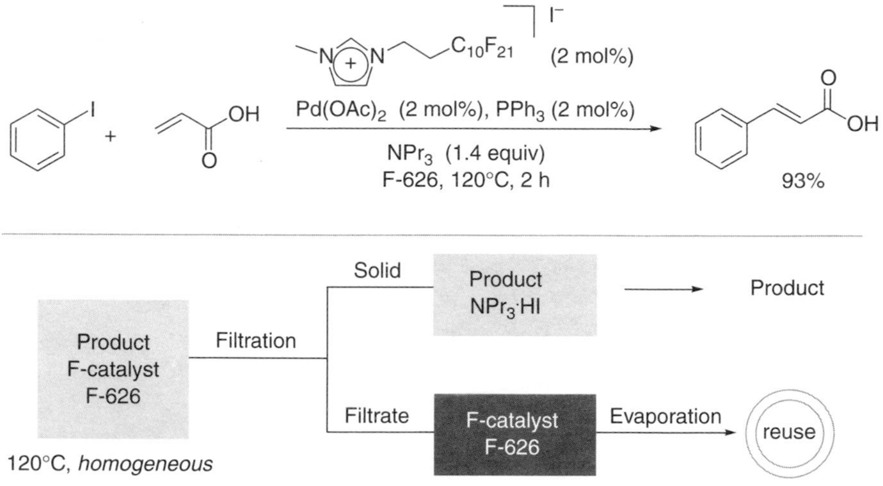

Scheme 13

with aryl iodides to give β-aryl acrylate.[68] These fluorous ligands were treated with $Pd_2(dba)_3$ or $Pd(OAc)_2$ to form the corresponding fluorous Pd complexes, which were found to be effective catalysts. Although the reaction can be conducted in acetonitrile as a solvent, D-100 (mainly perfluorooctane) was also added to achieve catalyst recycling. Thus, upon completion of the reaction, the acetonitrile layer containing the product and the amine salt was separated by decantation, and another acetonitrile solution of reactants was added to the fluorous catalyst layer.

An easy recycling method involving both catalyst and reaction medium was achieved in a Mizoroki–Heck arylation reaction of acrylic acid, using a fluorous carbene complex (prepared *in situ* from a fluorous ionic liquid and palladium acetate) as the catalyst and a fluorous ether solvent (F-626) as the reaction medium.[69] Because of the very low solubility of arylated carboxylic acids in F-626, the products precipitated during the course of the reaction. After separation of the products and amine salts by filtration, the filtrate, which contained the fluorous Pd catalyst, could be recycled for several runs (Scheme 13). The Mizoroki–Heck reaction was effectively promoted by a fluorous SCS pincer palladium, which is discussed in Section 3.4.5.

$(p\text{-}C_6F_{13}\text{-}C_6H_4)_3P$ has proven to be an excellent ligand for palladium-catalyzed cross-coupling reactions of arylzincbromide and vinylzincbromide with aryl iodides.[70] For example, the cross-coupling reaction of phenylzincbromide with 4-iodophenyl acetate in the presence of 1.5 mol% of tetrakis(triarylphosphine)palladium, prepared from $Pd(dba)_2$ and $(p\text{-}C_6F_{13}\text{-Ph})_3P$, gave biphenyl in a 93% yield (60°C, 0.5 h). In this reaction, a binary system of toluene and $C_8H_{17}Br$ was employed, which provided a homogeneous solution at 60°C. Upon cooling to 0°C, phase separation took place and the fluorous layer, which contained the catalyst, could be used in further experiments. The fluorous catalyst exhibited a much higher activity than $(PPh_3)_4Pd$, presumably as a result of the electron-withdrawing effect of the fluorous attachment, thus facilitating the reductive elimination.

Fluorous palladium-catalyzed substitution reactions of allylic acetates with nucleophiles were reported, in which $[(p\text{-}C_6F_{13}C_2H_4C_6F_{13})_3P]_4Pd$ was used as a catalyst and a 1 : 1 mixture of THF and perfluorocyclohexane as the reaction medium.[71] The fluorous version of Rh-catalyzed cyclopropanation with methyl diazoacetate has also been reported, for which $Rh_2(OCOC_7F_{15})_4$ and $Rh_2(OCOC_6H_4\text{-}p\text{-}C_6F_{13})_4$ were used as catalysts.[72] These catalysts could be extracted using PFMC and used several times without any significant loss of activity.

3.3.1.6 Fluorous acid and base catalysts

A fluorous trialkylphosphine was used as a base catalyst for Michael addition of alcohols (Scheme 14).[73] The fluorous phosphine exhibits a ~60- to 150-fold increase in solubility in octane between 20–80°C and 20–100°C, respectively. Utilizing the dramatic temperature dependence of the solubility of this compound, it could be recycled by cooling to lower temperatures. Thus, after completion of the reaction, the solution was cooled to −30°C and the precipitated catalyst isolated by decantation, permitting it to be recycled four times without any deterioration in yields.

The nickel complex **11**, with a fluorous Schiff base as ligand, was prepared from salicylaldehyde and 4-perfluorodecylaniline (Scheme 15).[74] This fluorous Ni complex was used as a catalyst for the conjugate addition of β-diketones to diethyl azodicarboxylate (DEAD). The mixture was stirred at 60°C in a mixture of

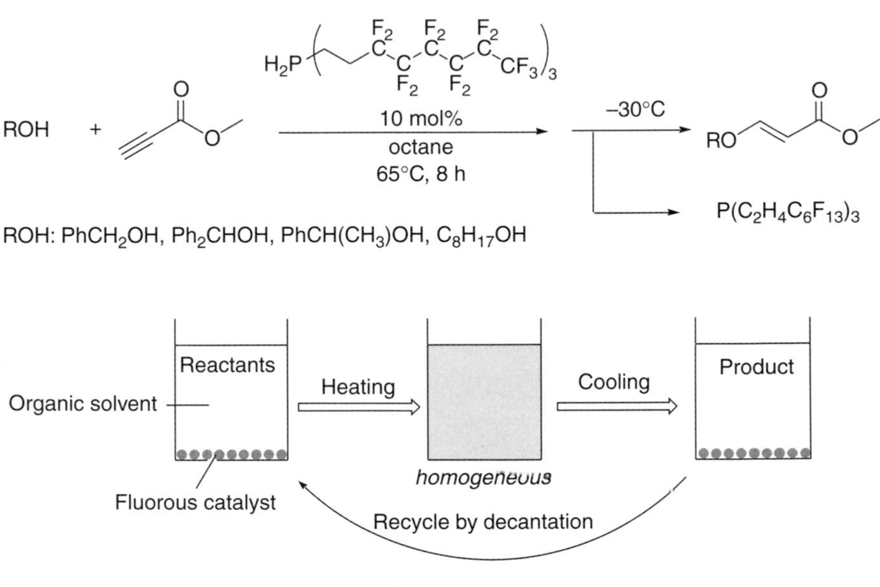

Scheme 14

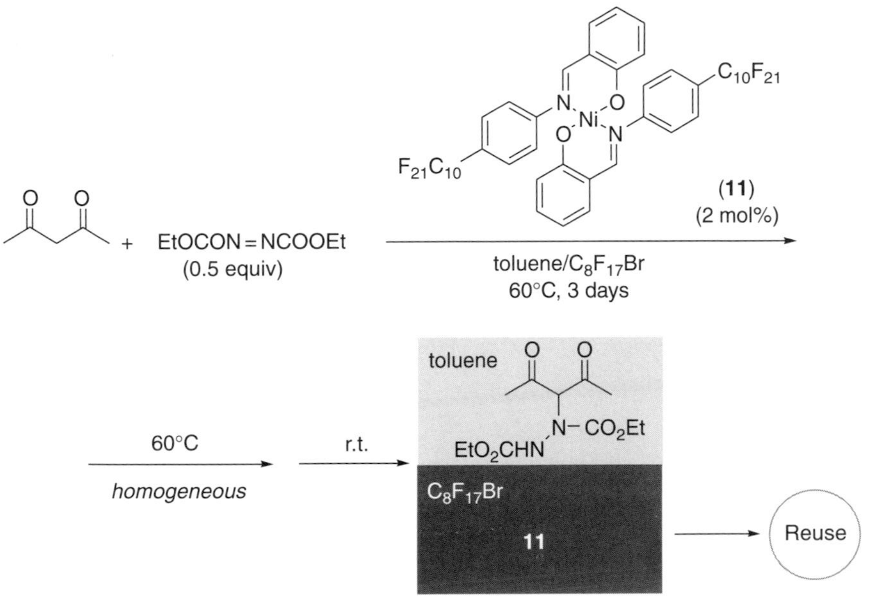

Scheme 15

toluene and 1-bromoperfluorooctane for 3 days to give a homogeneous solution. When the mixture was cooled, phase separation occurred. The organic layer contained the addition product, and the 1-bromoperfluorooctane layer, which contained the catalyst, was reused at least four times without any decrease in the yields of the adducts.

A series of fluorous arylboronic acids were designed to take advantage of the strong electron-withdrawing effect of perfluoroalkyl substituents, which would enhance acid activity along with immobility in the fluorous phase.[75] Among several fluorous arylboronic acids tested for a direct amide condensation between a carboxylic acid and an amine, 3,5-bis(perfluorodecyl)phenylboronic acid **12** was found to be an excellent catalyst. Although the catalyst was insoluble in *o*-xylene and toluene at room temperature, the amide condensation proceeded homogeneously under refluxing conditions. An example of a condensation between cyclohexanecarboxylic acid and benzylamine is given in Scheme 16. As in the case of Scheme 14 a biphasic workup is unnecessary for the separation of product and catalyst. After heating, the reaction mixture was left to stand at ambient temperature for 1 h to allow the catalyst to precipitate. The supernatant from the resulting mixture was removed by decantation, and the residual solid catalyst reused without the need to isolate it.

The two following examples incorporate the fluorous super Brønsted acid catalyst **13** for the acetalization of benzaldehyde with 1,3-propanediol and for the

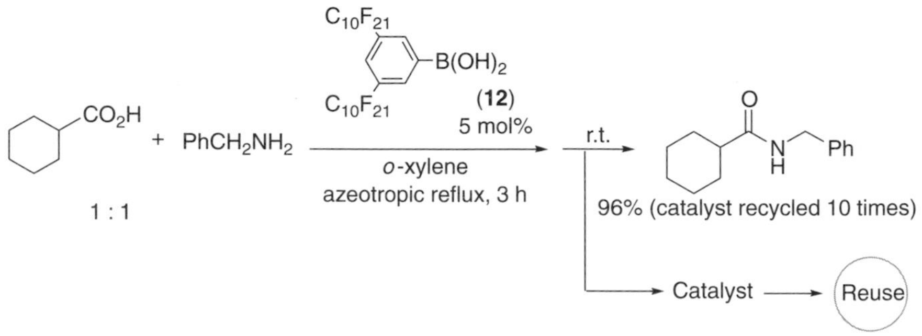

Scheme 16

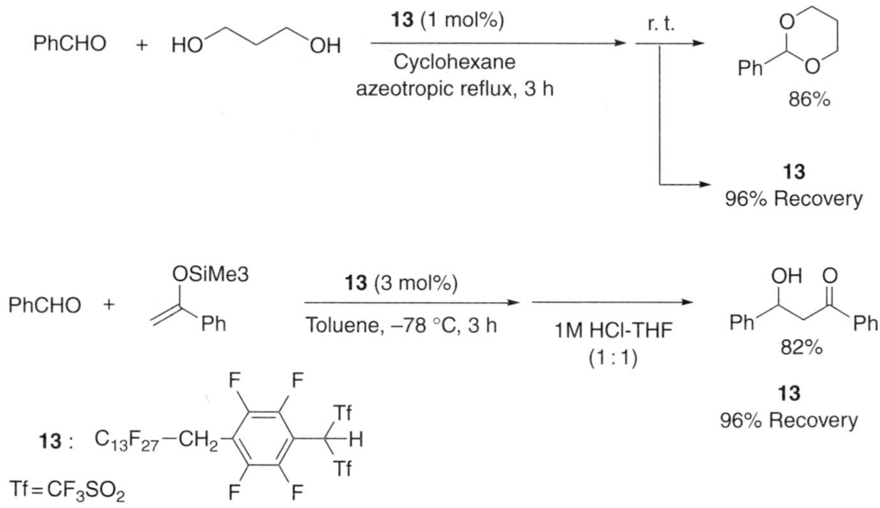

Scheme 17

Mukaiyama aldol reaction of an enol silyl ether with benzaldehyde (Scheme 17).[76] These reactions were conducted in a neat organic solvent, and the catalyst was precipitated after cooling the reaction mixture to room temperature, hence permitting its easy recovery. The same group also reported on the use of a polystyrene-bounded super Brønsted acid catalyst in a flow system.[77]

Lanthanide tris(perfluorooctanesulfonyl)methides, such as $Sc[C(SO_2C_8F_{17})_3]_3$ and $Yb[C(SO_2C_8F_{17})_3]_3$, can be immobilized in fluorous phases.[78] These fluorous Lewis acid catalysts are highly effective for the acylation of alcohols, Friedel–Crafts acylations, Mukaiyama aldol condensations and Diels–Alder reactions (Scheme 18). These reactions proceeded efficiently in an organic/fluorous

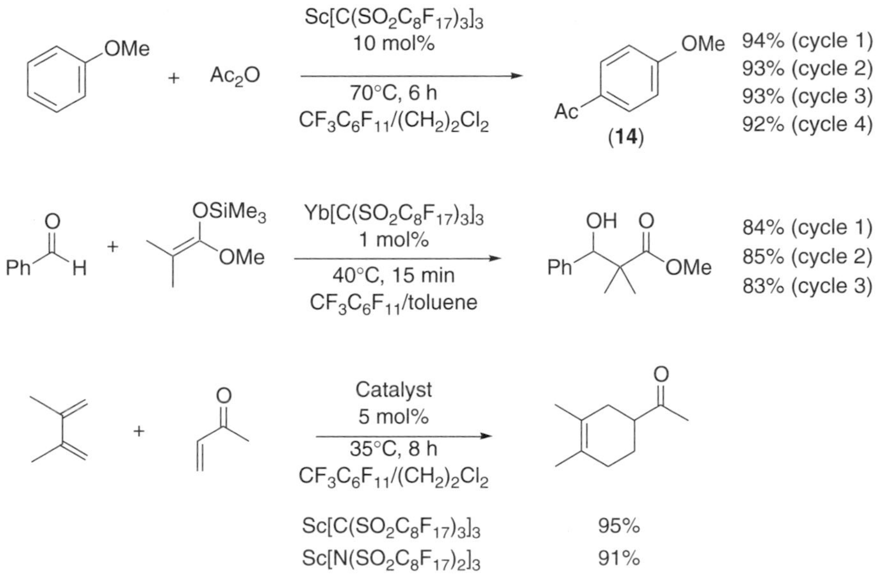

OMe + Ac_2O → $Sc[C(SO_2C_8F_{17})_3]_3$ 10 mol%, 70°C, 6 h, $CF_3C_6F_{11}/(CH_2)_2Cl_2$ → Ac, OMe (**14**) 94% (cycle 1) 93% (cycle 2) 93% (cycle 3) 92% (cycle 4)

Ph(C=O)H + $OSiMe_3$, OMe → $Yb[C(SO_2C_8F_{17})_3]_3$ 1 mol%, 40°C, 15 min, $CF_3C_6F_{11}$/toluene → Ph, OH O, OMe 84% (cycle 1) 85% (cycle 2) 83% (cycle 3)

+ O → Catalyst 5 mol%, 35°C, 8 h, $CF_3C_6F_{11}/(CH_2)_2Cl_2$ → O

$Sc[C(SO_2C_8F_{17})_3]_3$ 95%
$Sc[N(SO_2C_8F_{17})_2]_3$ 91%

Scheme 18

biphasic system as well as in neat organic solvents. For example, Friedel–Crafts acylation was conducted with the continuous recycling of the fluorous phase using a recycling apparatus; *p*-acetylanisole (**14**) was obtained in a high yield (87%) in the organic phase (1,2-dichloroethane), which automatically separated from the fluorous phase (perfluoromethylcyclohexane), which contains the catalysts. The ytterbium tris[bis(perfluorooctanesulfonyl)amide] complex ($Yb[N(SO_2C_8F_{17})_2]_3$) was shown to be immobilized as a solid when cooled to −20°C in 1,2-dichloroethane and could be recycled. The fluorous Lewis acid system was further modified by the inclusion of complexes of a cyclodextrin/cyclodextrin copolymer.[79] Using such catalysts, Mukaiyama aldol and Diels–Alder reactions could be performed in water. These fluorous lanthanide catalysts were well matched with supercritical CO_2,[80] which is discussed in Section 3.5.1.

A slightly different type of fluorous lanthanide complex, $Yb[C(SO_2C_6F_{13})_2(SO_2C_8F_{17})]_3$,[81] showed a modest catalytic activity in BTF, but was found to be an effective catalyst in perfluoromethyldecalin for the Friedel–Crafts acylation of arenes with an acid anhydride. For example, anisole reacted with acetyl anhydride in the presence of 10 mol% of the catalyst in perfluoromethyldecalin at 110°C for 2 h to give the product in an 82% yield. Furthermore, the catalyst was recovered in a 96% yield via extraction of the reaction mixture with hot perfluoromethyldecalin (~85°C). A 77% conversion was obtained in a second reaction with the recycled catalyst.

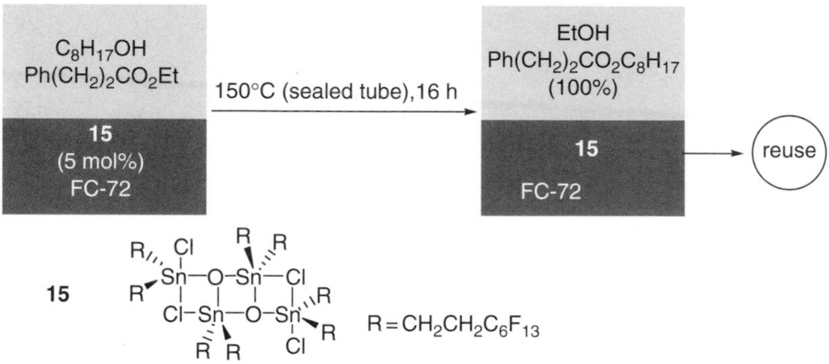

Scheme 19

In 1993, a perfluorocarbon was used as a reaction solvent for azeotropic transesterification reactions of carboxylic acid esters.[82] One of the ultimate goals of transesterification is the use of a 1 : 1 ratio of ester and alcohol.[83] Such a 1 : 1 transesterification was recently achieved using a fluorous version of the Otera method, which is based on tetraalkyldistannoxane.[84] For example, octanol and ethyl 3-phenylpropionate were added to an FC-72 solution containing fluorous stannoxane $[(C_6F_{13}CH_2CH_2)_2SnO]_n$ **15**. The mixture was then sealed and heated at 150°C for 16 h. When the mixture was cooled, phase separation occurred and octyl 3-phenylpropanoate was obtained by separation and evaporation of the upper layer (Scheme 19). The lower FC-72 layer, which contained the catalyst, was used more than 20 times without any loss in catalytic activity.[85]

3.3.2 Enantioselective catalysts for fluorous biphasic systems

3.3.2.1 Reduction

An asymmetric hydrogen transfer of ketones was reported using chiral perfluorinated ligands in a 2-propanol/*n*-perfluooctane biphasic system.[86] Several perfluorinated salen and diamine ligands were examined for the reaction catalyzed by the $[Ir(COD)Cl]_2$ complex; diamine **16** was found to be most effective (Scheme 20). The reaction was carried out at 70°C for 30 min and then the mixture was cooled to 0°C. The perfluorooctane solution was separated and used for the next reaction. The reactivity was almost the same as that of the first run, and the enantioselectivity was higher (79% ee). Two further recyclings of the fluorous layer yielded the product with enantioselectivities up to 59% ee, but a decrease in activity was observed.

3.3.2.2 Epoxidation

An enantioselective epoxidation in a fluorous biphasic system was reported in 1998.[87] The fluorous manganese complex **17** was tested for the enantioselective epoxidation of alkenes in the presence of molecular oxygen and sacrificial aldehydes. The characteristic ability of fluorocarbons to dissolve large quantities of

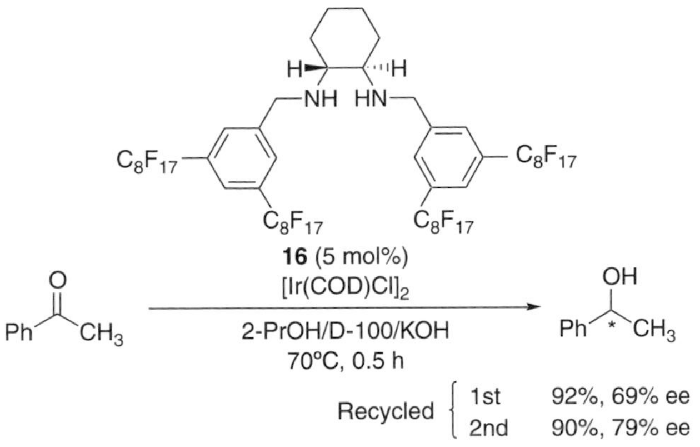

C_8F_{17} C_8F_{17} C_8F_{17} C_8F_{17}

H H NH HN

16 (5 mol%)
$[Ir(COD)Cl]_2$
2-PrOH/D-100/KOH
70°C, 0.5 h

Ph CH_3 (O) → Ph * CH_3 (OH)

Recycled { 1st 92%, 69% ee; 2nd 90%, 79% ee }

Scheme 20

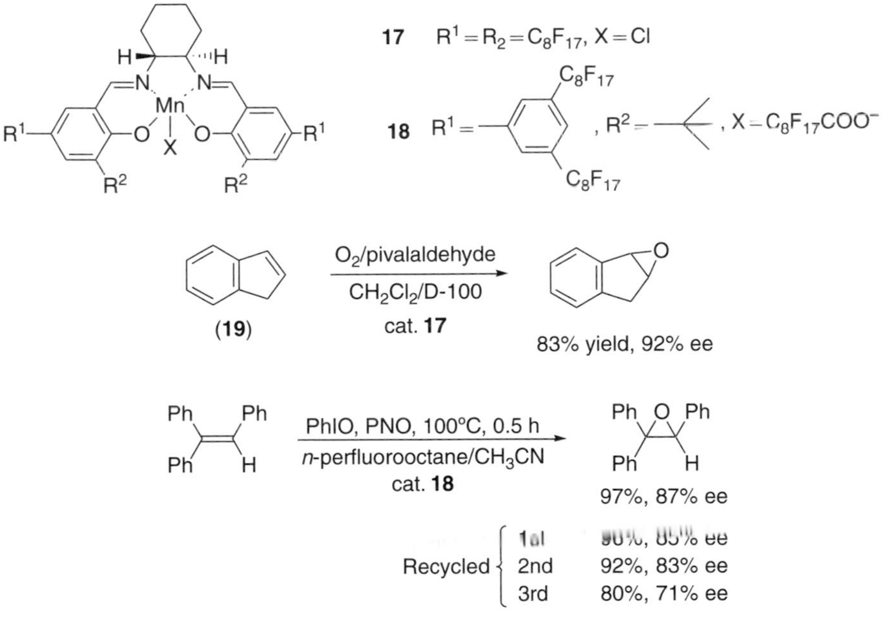

17 $R^1 = R_2 = C_8F_{17}$, X = Cl

18 R^1 = 3,5-(C_8F_{17})$_2C_6H_3$, R^2 = tert-butyl, X = $C_8F_{17}COO^-$

O_2/pivalaldehyde
CH_2Cl_2/D-100
cat. **17**

(**19**)

83% yield, 92% ee

PhIO, PNO, 100°C, 0.5 h
n-perfluorooctane/CH_3CN
cat. **18**

97%, 87% ee

Recycled { 1st 90%, 85% ee; 2nd 92%, 83% ee; 3rd 80%, 71% ee }

Scheme 21

molecular oxygen prompted the authors to exploit the epoxidation under atmospheric pressure of O_2 in D-100 (Ausimont, mainly *n*-perfluorooctane) and CH_2Cl_2. The epoxides were obtained in yields higher than 85%, even by using substantially smaller amounts of catalyst than in the original homogeneous conditions. The brown perfluorocarbon layer recovered by simple decantation was recycled in a second run without an appreciable decrease in activity. Indene **19** was epoxidized in 92% ee (Scheme 21).

However, <13% ee was attained with other substrates such as 1,2-dihydronaphthalene, styrenes and stilbenes. Second-generation fluorous chiral salen manganese complex **18** was then prepared (Scheme 21).[88] In the Mn complex, $C_7F_{15}COO^-$ was used as an axial base instead of Cl^-, improving its solubility in fluorocarbons. The partition coefficients of **18** were >1000 in both *n*-perfluorooctane/CH_3CN and *n*-perfluorooctane/toluene biphasic systems. PhIO, together with small amounts of pyridine *N*-oxide (PNO), was used as the oxidizing agent, and the epoxidation reaction was carried out in an *n*-perfluorooctane/CH_3CN biphasic system. Both chemical yield and enantioselectivity rose with temperature and reached a maximum at 100°C. The enantioselectivities were improved for 1,2-dihydronaphthalene (50% ee), triphenylethylene (87% ee) and other substrates.

3.3.2.3 Protonation

A catalytic enantioselective protonation of a samarium enolate was carried out with DHPEX (α, α-di[(s)-2-hydroxy-2-phenylethyl]-*o*-xylenedioxide) and a fluorous achiral proton source (Rfh_3-OH) in an FC-72/THF biphasic system.[89] A solution of Rfh_3-OH in FC-72 was added to the samarium enolate solution in THF within 1 min at −45°C, and then the reaction mixture was stirred at this temperature for 6 h to give the product in 89% ee. When the reaction was carried out in THF without the use of FC-72, Rfh_3-OH precipitated as fine crystals in the reaction mixture – the product being obtained in 90% ee (Scheme 22). Both reactions took place in biphasic systems, THF (liquid)/FC-72 (liquid) for the former, Rfh_3-OH (solid)/THF (liquid) for the latter. Rfh_3-OH was recovered quantitatively by liquid/liquid extraction with FC-72 in both cases. Trityl alcohol can be used instead of Rfh_3-OH for regenerating DHPEX: 67% and 93% ee were obtained after 1 and 26 h, respectively.

3.3.2.4 Et_2Zn or Et_3Al addition to aldehydes

Arylzinc thiolate complex **20** containing perfluoroalkyl chains was prepared as a model catalyst precursor for the enantioselective zinc-mediated 1,2-addition of Et_2Zn to benzaldehyde in fluorous biphasic systems.[90] The addition reaction to benzaldehyde was first carried out in hexane to give 94% ee compared with 72% ee for the original non-fluorous catalyst. When the reaction was carried out in an octane/perfluoromethylcyclohexane biphasic system, the catalyst immobilized in the fluorous phase could be used twice without loss of enantioselectivity, but the system was less efficient from the third run (Scheme 23).

A similar enantioselective addition of Et_2Zn to aromatic aldehydes was reported using Ti complexes of (*R*)-F_{13}BINOL (binaphthol) and (*R*)-F_{17}BINOL (Scheme 24).[91] Consecutive biphasic reactions were carried out at 0°C in toluene-hexane and FC-72 for 2 h using benzaldehyde. Chemical yields were >80%, and although the enantioselectivity gradually decreased with recycling, values >80% ee were obtained for the fifth run. Owing to its solubility in hexane, ~10% of F_{13}BINOL was recovered in the organic phase during every run. Employing

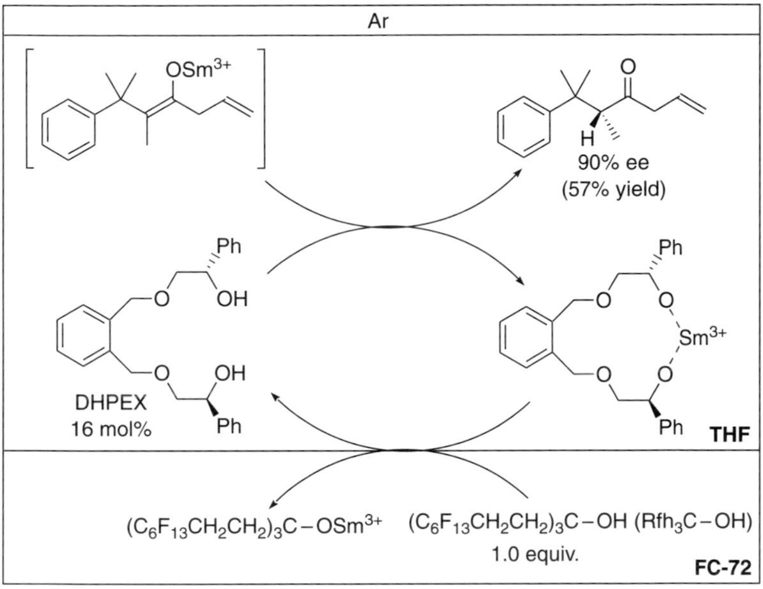

Scheme 22

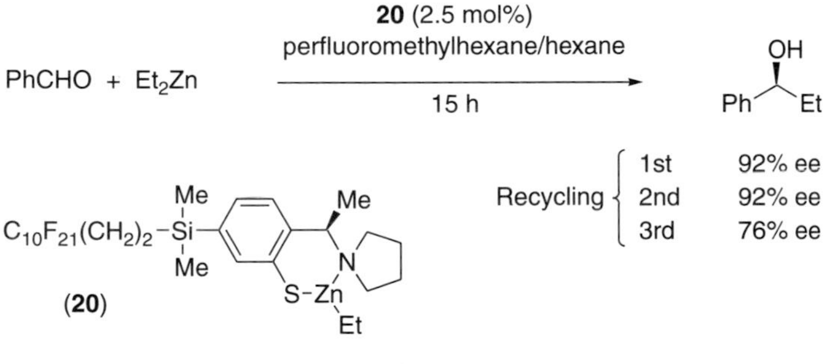

Scheme 23

F_{17}BINOL improved the immobilization (1% leaching), but the enantioselectivities were lowered slightly (78% ee).

Chiral 4, 4′, 6, 6′-tetraperfluorooctyl-BINOL **21** was applied to the titanium-catalyzed addition of Et_2Zn to aromatic aldehydes.[92] The reaction was carried out in a hexane/perfluoromethyldecalin biphasic system at 45°C for 1 h. At this temperature, the reaction mixture became homogeneous and, after the reaction, the two phases were separated by cooling the homogeneous solution to 0°C. The fluorous phase was used nine times to give constant chemical yields (70–80%) and enantioselectivities (54–58% ee). When a triethylaluminum solution in hexane was used instead of Et_2Zn solution, the reaction mixture became homogeneous at

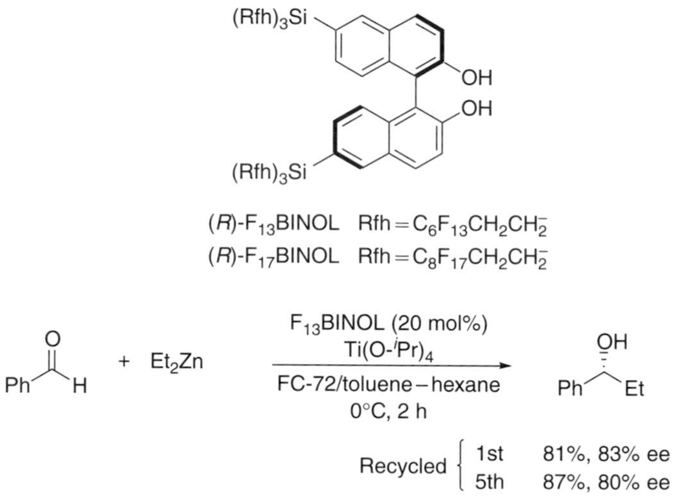

(Rfh)$_3$Si
OH
OH
(Rfh)$_3$Si

(*R*)-F_{13}BINOL Rfh = $C_6F_{13}CH_2CH_2^-$
(*R*)-F_{17}BINOL Rfh = $C_8F_{17}CH_2CH_2^-$

PhCHO + Et_2Zn → (F_{13}BINOL (20 mol%), Ti(O-^{i}Pr)$_4$, FC-72/toluene–hexane, 0°C, 2 h) → PhCH(OH)Et

Recycled 1st 81%, 83% ee
Recycled 5th 87%, 80% ee

Scheme 24

53°C and gave maximum chemical yield (76%) and enantioselectivity (77% ee). Consecutive reactions were examined and no significant difference was observed during six runs. Enantioselectivity reached a maximum of 82% ee, although the chemical yield and the enantiomeric excess were much lower from the seventh run (Scheme 25).

3.3.3 Heavy fluorous reagents

3.3.3.1 Fluorous tin hydrides

$(C_6F_{13}CH_2CH_2)_3SnH$ was synthesized on a 20–40 g scale in 80–83% yield via three steps from a perfluoroalkyl Grignard reagent (Scheme 26).[93] Despite its high molecular weight (1162), it is a clear, free-flowing liquid with a boiling point of 115°C at about 0.1 mm Hg. It lacks the unpleasant odor of alkyltin compounds – indeed, it has no detectable smell at all. It is freely soluble in fluorous solvents such as perfluorohexane (FC-72), but sparingly soluble or insoluble in most organic solvents. The analogous tin hydrides and their derivatives are the original reagents in fluorous synthetic chemistry and have been used successfully for catalytic reduction of alkylhalides, reductive cyclization, hydrostannation of multiple bonds, Diels–Alder reactions, nitrile oxide cycloaddition reactions and the following Stille coupling reaction, radical carbonylation reaction and reactions using fluorous tin azide.

3.3.3.2 The Stille coupling reaction

One of the coupling partners of a Stille coupling reaction is a trialkylorganotin compound. The alkyls on tin substituents are almost always methyl or butyl

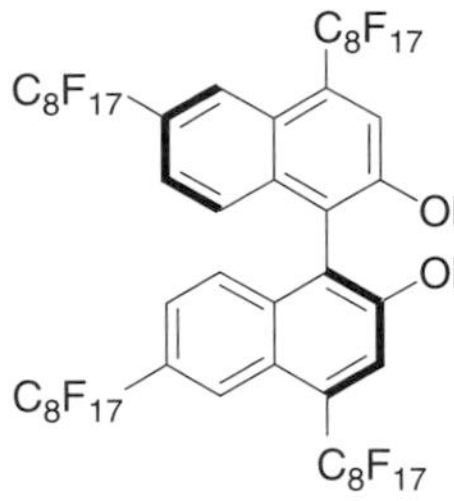

C_8F_{17} C_8F_{17} OH OH C_8F_{17} C_8F_{17} (**21**)

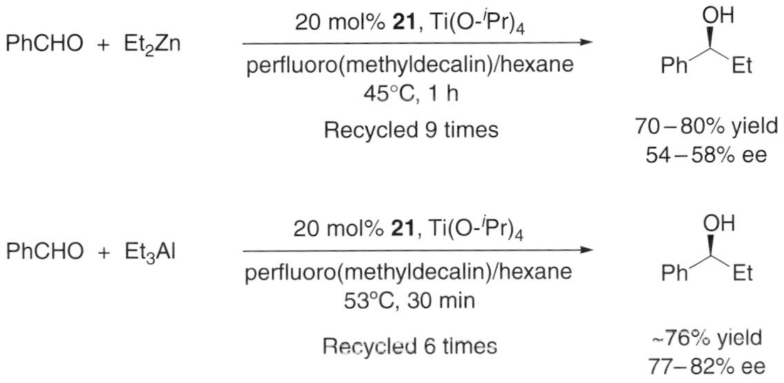

PhCHO + Et_2Zn → (20 mol% **21**, $Ti(O\text{-}^iPr)_4$; perfluoro(methyldecalin)/hexane, 45°C, 1 h; Recycled 9 times) → Ph–CH(OH)–Et, 70–80% yield, 54–58% ee

PhCHO + Et_3Al → (20 mol% **21**, $Ti(O\text{-}^iPr)_4$; perfluoro(methyldecalin)/hexane, 53°C, 30 min; Recycled 6 times) → Ph–CH(OH)–Et, ~76% yield, 77–82% ee

Scheme 25

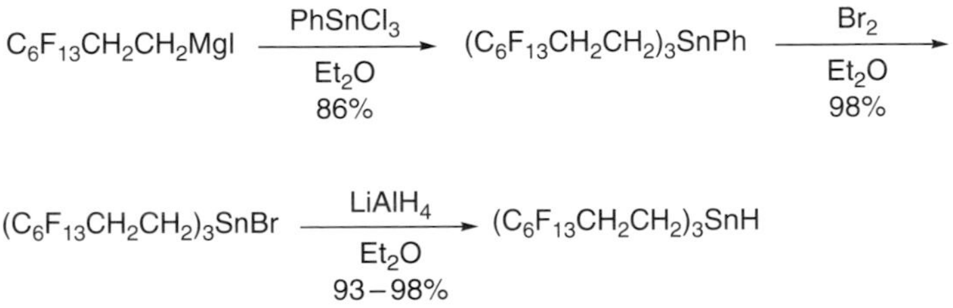

$C_6F_{13}CH_2CH_2MgI$ → ($PhSnCl_3$, Et_2O, 86%) → $(C_6F_{13}CH_2CH_2)_3SnPh$ → (Br_2, Et_2O, 98%) →

$(C_6F_{13}CH_2CH_2)_3SnBr$ → ($LiAlH_4$, Et_2O, 93–98%) → $(C_6F_{13}CH_2CH_2)_3SnH$

Scheme 26

groups: trimethyltin byproducts are easy to remove but toxic, whereas tributyltin byproducts are less toxic but difficult to remove. The reaction discussed here was selected to investigate the usefulness of using fluorous tin to solve these problems.[94] The reaction was carried out in DMF and THF in the presence of lithium chloride. After azeotropic evaporation with toluene at 75°C to remove some of the solvent, a three-phase extraction (water, dichloromethane and FC-72) was carried out. The dichloromethane phase was then washed three more times with water and FC-72 to remove DMF and fluorous compounds, giving the products in almost pure form. The crude fluorous tin chloride (99% yield) from the FC-72 phase was treated with phenylmagnesium bromide to provide the original tin reactant in a 96% yield after purification through a short column of neutral alumina. When the reactions were conducted using microwave irradiation, all were completed in <2 min.[95] The reactions using $PhSn(CH_2CH_2C_{10}F_{21})_3$ with single-mode microwave irradiation of 50 W finished within 6 min. The products were extracted easily using standard three-phase liquid extraction (Scheme 27).

3.3.3.3 Radical carbonylation reaction

The same strategy was applied to radical cabonylation with fluorous allyltin reagents. Propylene-spaced fluorous allyltin reagents **22** were tested as mediators for radical carbonylations: the four-components coupling reaction with RX, CO, alkenes and the fluorous allyltin produced the β-functionalized β-allylated ketones **23** (Scheme 28).[96] Biphasic workup (acetonitrile and FC-72) was successfully carried out to separate the products from the tin compounds. Comparison experiments

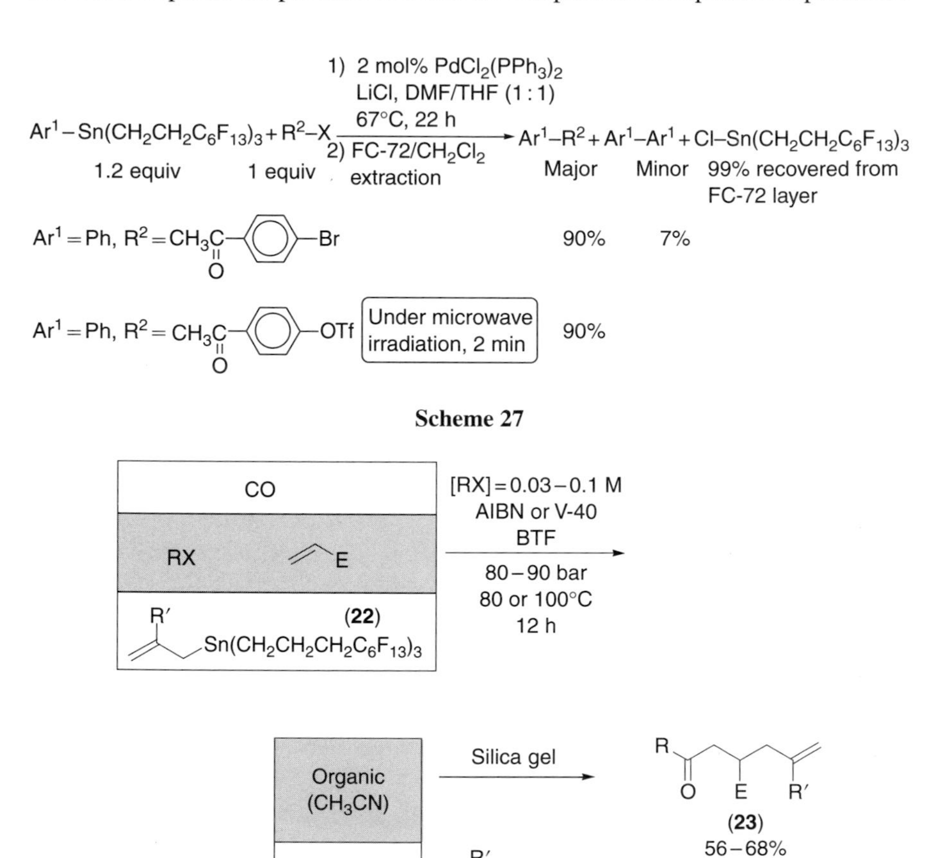

Scheme 27

Scheme 28

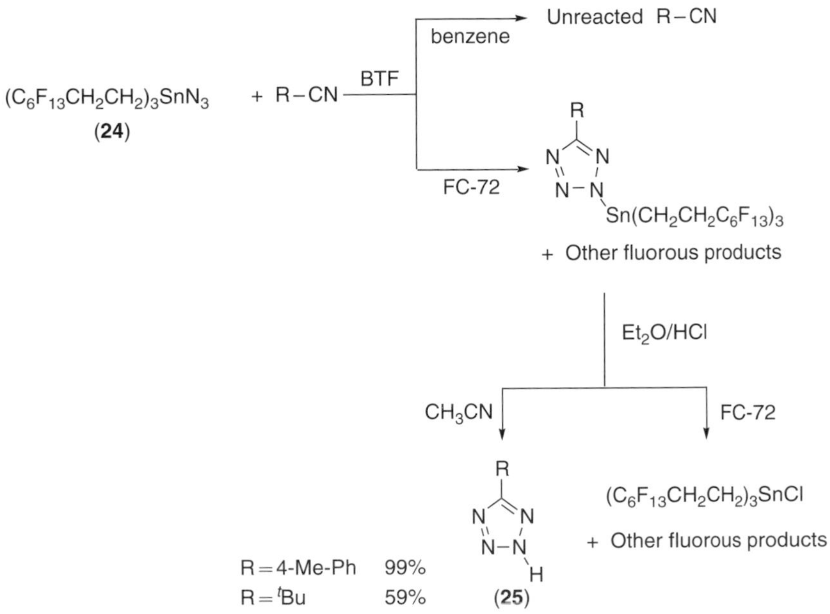

$(C_6F_{13}CH_2CH_2)_3SnN_3$ (**24**) + R–CN —BTF→

benzene → Unreacted R–CN

FC-72 → 5-R-tetrazolyl–$Sn(CH_2CH_2C_6F_{13})_3$ + Other fluorous products

Et_2O/HCl

CH_3CN → 5-R-1H-tetrazole (**25**)

FC-72 → $(C_6F_{13}CH_2CH_2)_3SnCl$ + Other fluorous products

R = 4-Me-Ph 99%

R = ^{t}Bu 59%

Scheme 29

suggested that the fluorous allyltin was less reactive than allylbutyltin for this radical chain reaction. The fluorous tin iodide was recovered quantitatively and treated with allylmagnesium bromide in ether to provide the allyltin reagent. The strategy was also applied to a catalytic hydroxymethylation of organic halides with $(C_6F_{13}CH_2CH_2)_3SnH$, CO and $NaBH_3CN$.[97]

3.3.3.4 Fluorous tin azide

Fluorous tin azide **24** lacks the strong unpleasant odor of trialkyltin compounds.[98] It showed little solubility in organic solvents such as benzene, toluene and acetonitrile, but good solubility in FC-72, ethers and BTF. It seemed to be stable with respect to heating at the temperatures typically used for reactions. The reaction with R–CN was carried out in BTF at reflux for 12 h. The solvent was then evaporated and the residue was treated with benzene/FC-72. The residue from the FC-72 layer was briefly exposed to ethereal HCl prior to partitioning between acetonitrile and FC-72. 5-Substituted tetrazole **25** was isolated from the organic phase and the corresponding tin chloride $(C_6F_{13}CH_2CH_2)_3SnCl$ was recovered from the fluorous phase, and then reconverted to tin azide by treatment with NaN_3 (Scheme 29).

3.3.3.5 Fluorous sulfide and sulfoxide

$C_6F_{13}CH_2CH_2S(=O)CH_3$ and $C_4F_9CH_2CH_2S(=O)CH_3$ were prepared and used for Swern oxidation.[99] Both are white and odorless crystalline products, but whereas

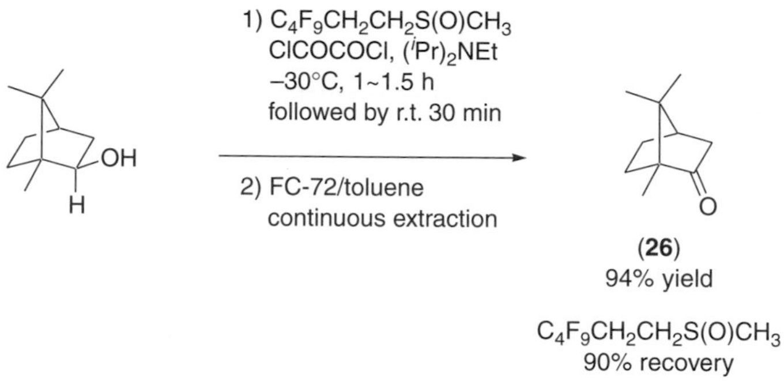

1) $C_4F_9CH_2CH_2S(O)CH_3$
ClCOCOCl, $(^iPr)_2NEt$
−30°C, 1~1.5 h
followed by r.t. 30 min

2) FC-72/toluene
continuous extraction

(26)
94% yield

$C_4F_9CH_2CH_2S(O)CH_3$
90% recovery

Scheme 30

the C_6F_{13} derivative is insoluble in dichloromethane below −30°C, the C_4F_9 analogue is soluble down to −45°C. The fluorous C_4F_9-DMSO was activated with oxalyl chloride in dichloromethane at −30°C. After completion of the reaction, the mixture was treated with water and dichloromethane and finally evaporated. The residue was partitioned between toluene and FC-72 in a continuous extractor for 4 h. The ketone **26** was recovered from the toluene phase, and treatment of the FC-72 phase with H_2O_2 returned the sulfoxide ready for reuse (Scheme 30).

The Corey–Kim reaction employs the sulfide and not the sulfoxide, and hence it was possible to work with $C_6F_{13}CH_2CH_2SCH_3$.[100] $C_8F_{17}CH_2CH_2SCH_3$ was also synthesized and employed as a ligand of borane.[101] Pyrophoric borane is commonly used for THF or dimethylsulfide complexes; usually the latter is preferred because of its greater stability and longevity. The solid mixture of fluorous sulfide and borane is completely odorless and shows no tendency to ignite in air. It is indefinitely stable under a nitrogen atmosphere in the refrigerator. The solid can be weighed on a simple laboratory balance. Since this solid is soluble in dichloromethane, the hydroboration reaction was carried out under nitrogen in a biphasic mixture of FC-72 and dichloromethane. After the reaction the organic layer was subjected to oxidative workup with alkaline H_2O_2. The fluorous sulfide and the complex were recovered mainly from the FC-72 phase. Fluorous diaryl diselenides were also prepared and used for elimination reactions.[102]

3.3.3.6 Other fluorous reagents

Fluorous DEAD was prepared and used in a Mitsunobu reaction.[103] Fluorous triaryl phosphine was used for a Wittig reaction.[104] Fluorous boronates were prepared and used in functional transformations and a Suzuki–Miyaura coupling reaction.[105]

Fluorous carbodiimide was used for dipeptide and ester synthesis.[106] The reaction was carried out in CH_2Cl_2 and FC-72, and the fluorous urea byproduct was extracted in the FC-72 using perfluoroheptanoic acid. The acid formed a stable complex with the fluorous urea and the solubility of the complex in FC-72 was dramatically increased compared with the fluorous urea.

3.3.4 Heavy fluorous protecting groups

When reagents or products contain >60% by weight of fluorine atoms, they generally partition into fluorous solvents upon separation via fluorous and organic liquid/liquid extraction. When the fluorous content is much lower than 60% (light fluorous molecules), more effective separation must be used, such as solid-phase extraction with reverse-phase silica gel. The fluorous solid-phase extraction and the fluorous high performance liquid chromatography (HPLC) method are reviewed in Section 3.4. In this section, heavy fluorous protecting groups that can be isolated using fluorous liquid/liquid extractions are described.

3.3.4.1 Trifluoroalkylsilyl protecting group

$Si(CH_2CH_2C_6F_{13})_3$ was initially employed as a protecting group of allyl alcohols for a reaction with nitrile oxides under standard Mukaiyama or Huisgen conditions.[107] Later, this fluorous synthetic method was applied to Ugi and Biginelli multicomponent condensations. Benzoic acid bearing $Si(CH_2CH_2C_{10}F_{21})_3$ at the *para* position **27** was prepared and used as a starting material for the reactions.[108] The Ugi four-components reaction was carried out with large excesses of reactants other than the fluorous benzoic acid (Scheme 31). After the reaction, the reaction mixture was diluted with benzene and then extracted three times with FC-72. The product **28** that was obtained from the FC-72 layer (together with the unreacted acid) was desilylated with tetrabutylammonium fluoride (TBAF) in THF at room temperature within 30 min. After evaporation of the solvent, the crude product was purified by two-phase extraction using benzene and FC-72. The yields of final products from the benzene layer were higher than 71% with one exception (32%), and the average purity was ~90%. A three-component Biginelli reaction was performed in a similar way (Scheme 32). These fluorous procedures, which do not require any chromatographic procedure, allow a rapid preparation of Ugi and Biginelli products in comparable yields to the standard protocols. In contrast to most solid-phase reactions, the fluorous reactions can be run in a homogeneous phase and can be followed by thin layer chromatography (TLC) or other standard tools. The products are single entities that can be characterized by standard spectroscopic methods.

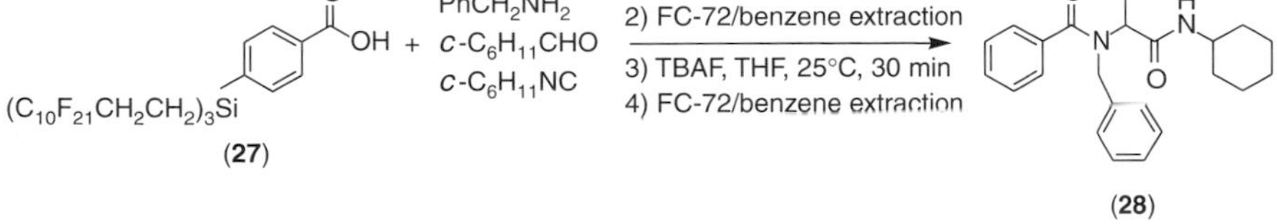

$(C_{10}F_{21}CH_2CH_2)_3Si$–C₆H₄–COOH **(27)** + $PhCH_2NH_2$ + c-$C_6H_{11}CHO$ + c-$C_6H_{11}NC$ → 1) CF_3CH_2OH, 90°C, 48 h; 2) FC-72/benzene extraction; 3) TBAF, THF, 25°C, 30 min; 4) FC-72/benzene extraction → **(28)** 84%, purity > 95%

Scheme 31

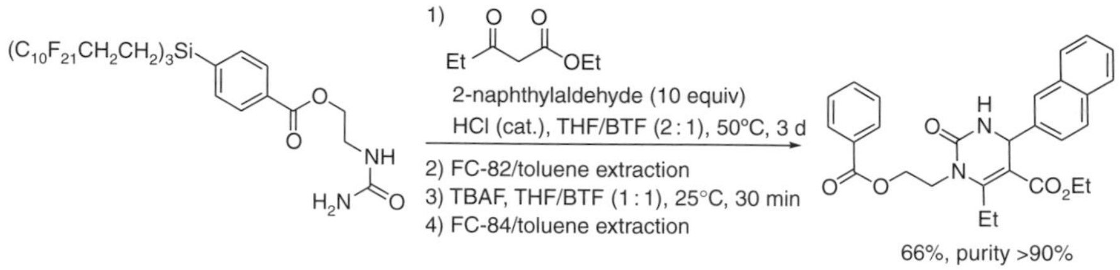

Scheme 32

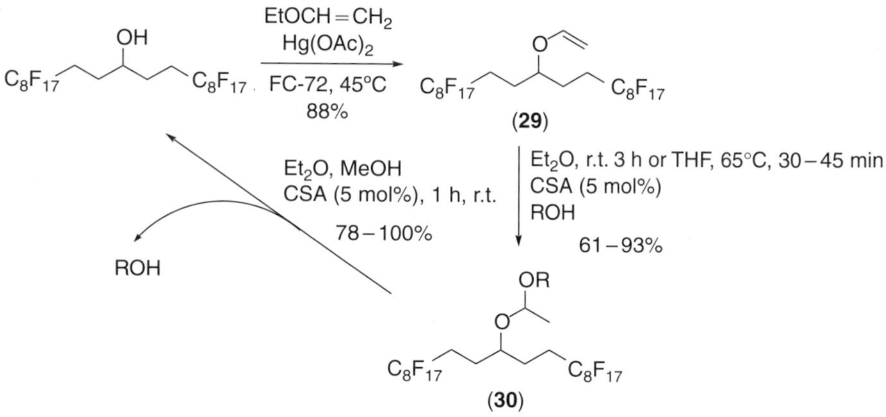

Scheme 33

Benzyl bromide bearing $Si(CH_2CH_2C_6F_{13})_3$ at its *para* position was also prepared and used for a fluorous approach to synthesizing disaccharides.[109]

3.3.4.2 Fluorous alcohol protective group

A highly fluorous alkoxy ethyl ether was synthesized and used as a protective group of alcohols.[110] Treatment of an ether solution of 1 equivalent of a primary alcohol and 3 equivalents of the fluorous vinyl ether **29** with 5 mol% of camphorsulfonic acid (CSA) for 3 h at room temperature provided the desired protected alcohols **30** in 84–93% yields (Scheme 33). The products and the excess fluorous vinyl ether were extracted with FC-72 and then separated by column chromatography. Secondary and even tertiary alcohols were similarly protected in good yields in THF at 65°C for 30–45 min. The protected fluorous acetals were treated in a 1 : 1 solution of ether and methanol with 5 mol% of CSA for 1 h for deprotection. After completion of the reaction, the products were isolated in pure form by simple three-phase extraction (reaction mixture/saturated aqueous $NaHCO_3$/FC-72).

Fluorous tertiary alcohols were prepared and applied as protecting groups of carboxylic acids.[111] The fluorous alcohols were reacted with phenylacetic acid and several linear aliphatic acids using DCI and DMAP at 40°C in BTF for 15–21 h to give the corresponding esters in 60–92% yields. Their partition coefficients (KD) between FC-72 or perfluoromethylcyclohexane and ten organic solvents were examined; it was observed that the KD values had to be higher than 4 to extract >99% of the esters into the fluorous phase using three extractions, and that perfluoromethylcyclohexane has stronger extraction power than FC-72.

3.3.4.3 Fluorous carboxylic acid protecting group

A novel fluorous protective group was prepared and employed for the synthesis of an oligosaccharide.[112] A bisfluorous chain-type propanoyl group (Bfp) was prepared from β-alanine ethyl ester (three steps) and introduced into the three hydroxyl functions of the mannose derivative to give the corresponding tri-Bfp derivative **31** in 87%. After removing the trityl group, the fluorous glycosyl acceptor was used for the synthesis of a disaccharide. The reaction of the acceptor with 5 equivalents of the glucose derivative in the presence of TMS-OTf in ether followed by FC-72/toluene extraction afforded the disaccharide **32** in 75% yield from the FC-72 layer (Scheme 34). The TBDPS (*tert*-butyldiphenylsilyl) group was removed by treatment with HF/pyridine in THF to give the pure fluorous disaccharide acceptor **33**, which was extracted with FC-72 from water and toluene in a three-phase extraction. The disaccharide was coupled with the glucose derivative under similar Schmidt's conditions to the first step. Similar reactions and extraction workups led to a precursor of tetrasaccharide **34**. Although the precursor tetrasaccharide with

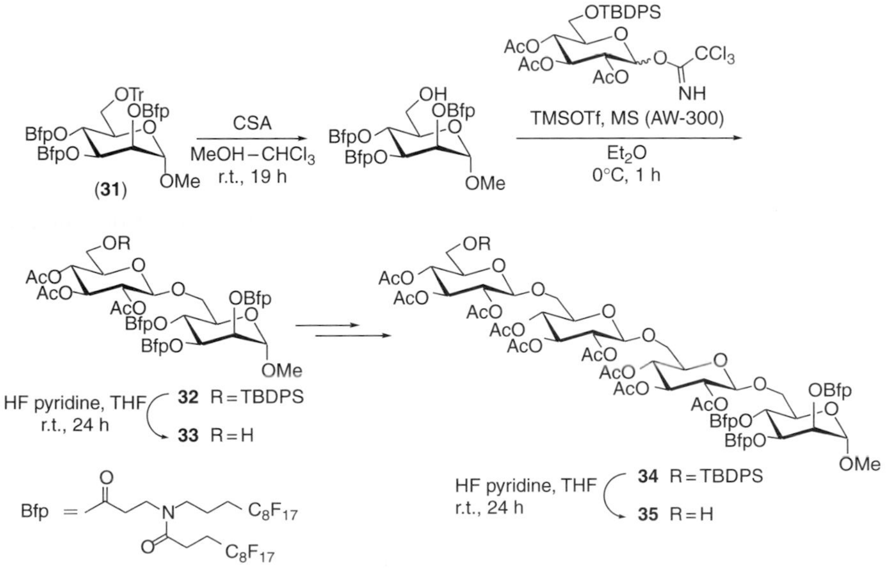

Scheme 34

TBDPS was not extracted with FC-72, the deprotected tetrasaccharide **35** was easily extracted with FC-72. Each synthetic intermediate was easily purified by simple FC-72 and organic solvent extraction and monitored as a single compound by NMR, mass spectroscopy and TLC, in contrast to the solid-phase synthesis.

This strategy was successfully applied to a practical synthesis of a trisaccharide part of globotriaosyl ceramide.[113] A highly fluorinated carboxylic acid (Hfb) was also prepared and used for rapid and large-scale synthesis of a thyrotropin-releasing hormone (TRH).[114]

3.4 Light fluorous compounds and fluorous silica gel

The first wave of research into fluorous compounds focused exclusively on using fluorous solvents as separation media. The first publication on the use of fluorous silica gel in 1997 marked a dividing point,[115] as the field began to diverge into what are now often called its *heavy* and *light*[116,117] fluorous branches.

3.4.1 Heavy and light fluorous molecules and separation strategy

In a simple view, both heavy and light fluorous molecules can be divided into an organic domain that controls the reaction chemistry and a fluorous domain that controls the separation chemistry. This view coincides with the principles of "strategy level separations,"[118] which dictate that reactions should be purified only by simple workup-level procedures whenever possible. In the "ideal separation," the target products of a reaction partitions into a phase that is different from all of the other reaction components, thereby allowing rapid and in many cases environment friendly isolation. The fluorous ponytails (permanent domains) or tags (temporary domains) on both heavy and light fluorous molecules allow them to partition into a fluorous phase under suitable workup conditions.

Heavy fluorous techniques use fluorous reaction components that have a large number of fluorines. Heavy fluorous molecules can have as few as ~39 fluorines, but it is not uncommon for them to have >50 or even >100. Such a high fluorine content gives heavy fluorous molecules unusual properties, and they can be separated from reaction mixtures by simple separation techniques such as extraction with a fluorinated solvent or even just filtration (see Section 3.3). However, the unusual properties that facilitate separations can complicate reactions since heavy fluorous molecules sometimes do not behave well under standard reaction conditions, and a search for suitable solvents and reaction conditions is a prerequisite.

Light fluorous molecules typically have 21 fluorines or fewer, and partition coefficients are often too low for efficient extraction into fluorinated solvents.[116] However, such molecules can be separated from organic molecules by a fluorous solid-phase extraction and from each other as well as organic molecules by fluorous chromatography.[119] These simple separation techniques combined with the lower molecular weights and higher organic solubilities of light fluorous molecules make light fluorous techniques convenient and attractive for drug discovery in combinatorial and parallel settings as well as in larger-scale applications.

3.4.2 *Solid-phase extractions with fluorous silica gel*

Silica gel with a fluorocarbon bonded phase, hereafter called fluorous silica gel, is an unusual material that separates molecules based primarily on fluorine content.[119a] The separation of light fluorous molecules containing 13–21 fluorines from small organic molecules containing few or no fluorines can often be accomplished by a simple solid-phase extraction, as Figure 3.3 illustrates. A reaction mixture containing organic and fluorous reaction components is charged to a short column of fluorous silica gel and eluted first with a fluorophobic solvent such as methanol or acetonitrile (often with 5–20% water). Organic molecules do not stick to the "Teflon coated" columns under these conditions; they instead elute with or near the solvent front. The fluorous fraction that is retained during the first pass can be eluted in a second pass with a fluorophilic solvent such as THF.

Fluorous solid-phase extractions are readily conducted in parallel (or serial), with or without automation. All reaction mixtures are processed identically and behave in substantially the same way, even though all the products are different. Careful fractionation and analysis are not required; there is one organic fraction and one fluorous fraction. Because the separation is so easy, loading levels can be high. Loose fluorous silica gel and cartridges suitable for solid-phase extraction have recently been commercialized under the trademark Fluoro*Flash*™ by Fluorous Technologies, Inc.[120]

The unique separation characteristics of fluorous silica gel are illustrated in Figure 3.3. The cartoon shows the separation of an organic compound from a

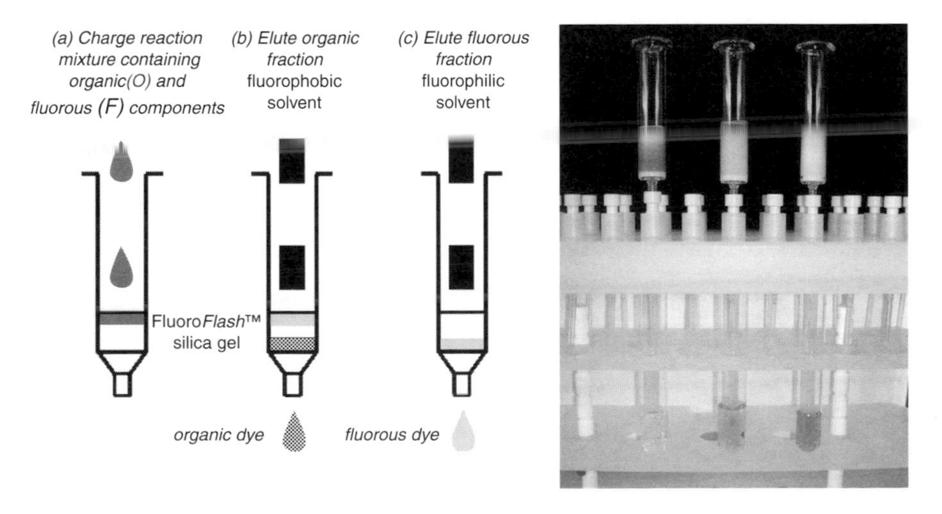

Figure 3.3

fluorous compound by solid-phase extraction over Fluoro*Flash*™ silica. The photograph shows the stages of a solid-phase extraction experiment conducted with an organic dye (Solvent Blue™) and a fluorous dye (F-orange I™, C_7F_{15}tag). These aminoanthraquinone dyes have about the same polarity and cannot be separated with normal-phase or reverse-phase silica gel. However, their separation with Fluoro*Flash*™ silica ($SiCH_2CH_2C_8F_{17}$ bonded phase) is straightforward. In the left Fluoro*Flash*™ column, the organic dye separates immediately after the elution with a fluorophobic solvent (80% methanol/water) begins. The fluorous dye remains immobilized on the column, even with extensive flushing with aqueous methanol (center column). However, it elutes immediately after washing with ~1 mL of THF (right column).

It is also possible to perform flash chromatographic and HPLC separations with fluorous silica gel. The HPLC separations are the cornerstone of the new technique of fluorous mixture synthesis,[121,122] a discovery-oriented technique that is not covered here. The following sections provide representative examples of the use of light fluorous reaction components, including fluorous reagents, catalysts, scavengers and protecting groups.

3.4.3 Light fluorous reagents

The use of fluorous reagents to transform traditional small organic precursors into products is straightforward, and it is a good starting point for newcomers to the field. A substrate or library of substrates is treated with a fluorous analogue of a standard organic reagent under reaction conditions similar or identical to those used for the non-fluorous reagent. After the reaction is complete, the mixtures are bifurcated into organic and fluorous fractions using fluorous solid-phase extraction. This simple technique removes the excess reagent and reagent-derived byproducts from the desired organic product(s).

The fluorous Mitsunobu reaction is illustrative of this approach, and a typical example is shown in Scheme 35.[123] Coupling of hydroxy ester **37** and 3,5-dinitrobenzoic acid **36** is carried out using a standard procedure in THF, but fluorous phosphine **39** and a fluorous azodicarboxylate **38** are used in place of the standard triphenylphosphine and DEAD reagents. Two-stage filtration of the reaction mixture through fluorous silica gives the pure substitution product **40** from the (first) organic fraction, and the (second) fluorous fraction is a mixture of the oxidized triarylphosphine **41** and the reduced hydrazide **42**. These two products are readily separated by using standard flash chromatography, and the resulting pure fluorous products **41** and **42** can be reconverted to the original reagents **38** and **39** in high yields by reduction or oxidation, respectively.

This procedure is very convenient for parallel synthesis and avoids the tedious chromatographies that are usually needed to separate spent Mitsunobu reagents from desired products. If excess reagents are used, then they can also be removed by the fluorous separation. But the reagents are soluble in THF, and large excesses are not

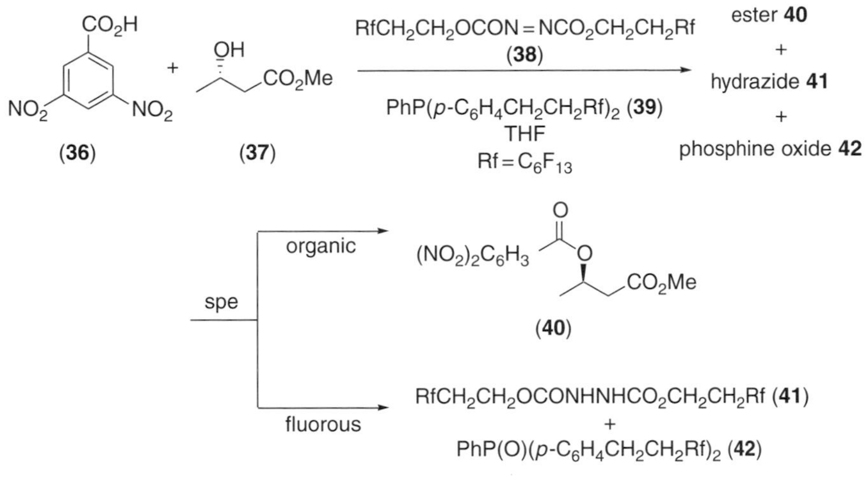

CO_2H / NO_2 / NO_2 (36) + OH / CO_2Me (37)

$RfCH_2CH_2OCON{=}NCO_2CH_2CH_2Rf$ (38)

$PhP(p\text{-}C_6H_4CH_2CH_2Rf)_2$ (39)

THF

$Rf = C_6F_{13}$

ester 40 + hydrazide 41 + phosphine oxide 42

spe

organic: $(NO_2)_2C_6H_3$ O O CO_2Me (40)

fluorous: $RfCH_2CH_2OCONHNHCO_2CH_2CH_2Rf$ (41) + $PhP(O)(p\text{-}C_6H_4CH_2CH_2Rf)_2$ (42)

Scheme 35

needed to drive the reaction to completion. This is in contrast to polymer-bound reagents, where the use of large excesses is commonly recommended.

A growing number of useful fluorous reagents are now known, including phosphines,[124] tin reagents,[125] sulfides and selenides,[126] and diacetoxyiodosobenzene oxidants (Nagashima, T., Fluorous Technologies Inc., and Lindsley, C., Merck, unpublished results),[127] to name a few. These can all be used to conduct standard organic reactions, and then the target product can be partitioned into the first fraction of the solid-phase extraction, and the excess reagents and byproducts can be partitioned into the second fraction. This provides not only for isolation of pure products, but also for recovery or reagents for recycling. The reuse of spent reagents is still uncommon in organic synthesis, yet fluorous methods help to make it straightforward.

3.4.4 Light fluorous catalysts

As chemical synthesis moves from discovery to production, scales increase and the use of catalytic rather than stoichiometric quantities of reagents is increasingly advantageous from both the economic and environmental standpoints. The vast majority of fluorous catalysts prepared to date are best classified as heavy fluorous catalysts, and they are removed from the reaction mixture by liquid/liquid separation techniques. On the one hand, fluorous silica gel provides another option for these catalysts, which is to use a solid/liquid separation instead. On the other hand, fluorous silica gel enables the use of light fluorous catalysts, such as the palladium catalyst shown in Scheme 36. Mizoroki–Heck reactions are promoted by standard conductive heating (oil bath) or microwave heating.[128] After cooling and solid-phase extraction,

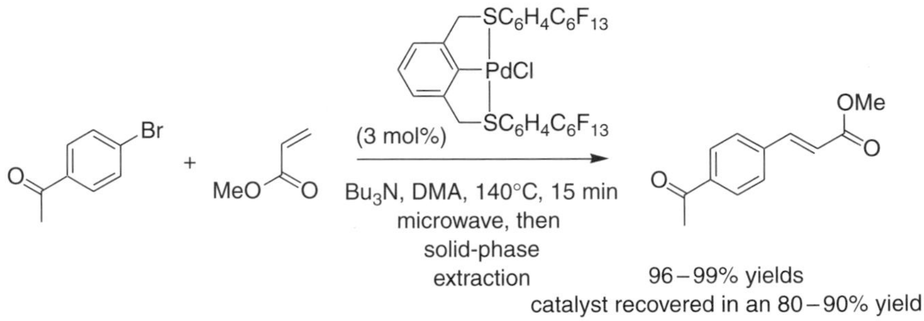

$SC_6H_4C_6F_{13}$
PdCl
$SC_6H_4C_6F_{13}$
(3 mol%)
Br
+
MeO O
OMe
O
O
O
Bu_3N, DMA, 140°C, 15 min
microwave, then
solid-phase
extraction
96–99% yields
catalyst recovered in an 80–90% yield

Other Catalysts

$^FDPP = (C_4F_9CH_2CH_2C_6H_4)_2PCH_2CH_2CH_2P(C_6F_4CH_2CH_2C_4F_9)_2$

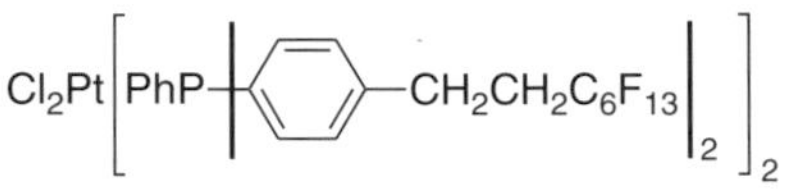

$Cl_2Pt[PhP[C_6H_4CH_2CH_2C_6F_{13}]_2]_2$

Scheme 36

the target product is isolated from the first fraction and the catalyst is recovered from the second fraction (Unpublished results of K. Fisher and G. Moura (University of Pittsburgh), and T. Nagashima (Fluorous Technologies, Inc.)). The catalyst is then purified by recrystallization (~80–90% recovery) and reused. Several other recently introduced light fluorous catalysts are shown in the lower part of Scheme 36.[116b,129]

3.4.5 Light fluorous scavengers

The use of fluorous reagents ensures only that the final organic product will be fluorous free; this does not mean that it will be pure. For example, if an excess of an organic reactant is used, then it will contaminate the product in the organic fraction, as will any unreacted substrate or (non-fluorous) side-product. These types of undesired impurities can readily be removed by fluorous scavenging,[130] that is, post-reaction treatment with a fluorous scavenger designed to consume likely contaminants of the desired product rapidly.

Scheme 37 illustrates one example from a 16-member library where fluorous thiol scavenger **46** quickly purified the crude products of amine alkylation reactions.[131,132] Amine **43** was reacted with a modest excess (2 equiv.) of alkylating **44** to ensure rapid and complete conversion to **45**. The excess halide **44** was then scavenged by addition of thiol **46**. Quick aqueous workup to remove the salts followed by two-stage fluorous solid-phase extraction to remove the quenched product **47** and the unreacted thiol **46** provided the pure alkylation product **45** in an excellent yield and purity. Rapid quenching is an attractive feature of this procedure, and control experiments showed that only about one-tenth the reaction time was needed for **46** compared with a related polymer-bound thiol.

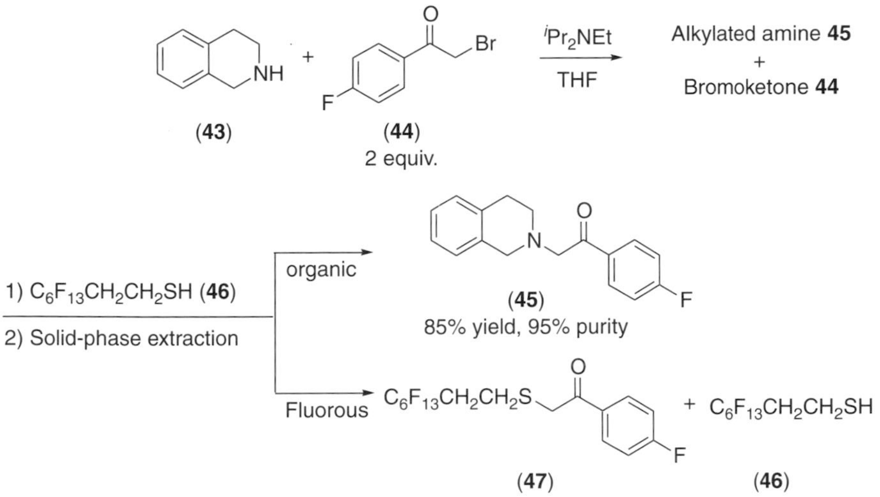

(43) + (44) 2 equiv. $\xrightarrow[\text{THF}]{^{i}Pr_2NEt}$ Alkylated amine **45** + Bromoketone **44**

1) $C_6F_{13}CH_2CH_2SH$ (**46**)
2) Solid-phase extraction

organic → (**45**) 85% yield, 95% purity

Fluorous → $C_6F_{13}CH_2CH_2S$– (**47**) + $C_6F_{13}CH_2CH_2SH$ (**46**)

Scheme 37

In a discovery setting, the technique of scavenging is not usually considered to be environmentally friendly – in an optimized reaction, neither the excess reagent nor the scavenger would be needed. But the need in discovery settings is to isolate products for testing, not to optimize reaction conditions. Here, the fluorous scavenging technique may indeed help since the waste generated using a slight excess of reagent and fluorous scavenger could well be less than the solvent waste generated if careful chromatography were needed to purify the products. And scavenging may not be limited to discovery chemistry. In process chemistry, it could be used to remove trace impurities such as metals that cannot be tolerated and whose removal might otherwise consume large amounts of resources. An assortment of other fluorous scavengers, nucleophilic and electrophilic scavengers are commercially available.[130–133]

3.4.6 *Light fluorous protecting groups*

Although it is now used more broadly to describe all fluorous techniques, the term *fluorous synthesis* originally referred to methods where a fluorous substrate is used to produce a fluorous-tagged product (by analogy to *solid-phase synthesis*). This technique was introduced in a heavy fluorous setting in early 1997,[134] but light fluorous synthesis quickly emerged as the preferred mode of operation. Here, a single fluorous protecting group can be used to render a library of substrates fluorous, and then standard organic and inorganic reagents and reactants are used to promote reactions. At the end of the synthesis, the fluorous group is removed to yield the final organic products. The use of fluorous protecting groups is probably better suited to discovery chemistry than process chemistry. In the discovery setting, fluorous groups

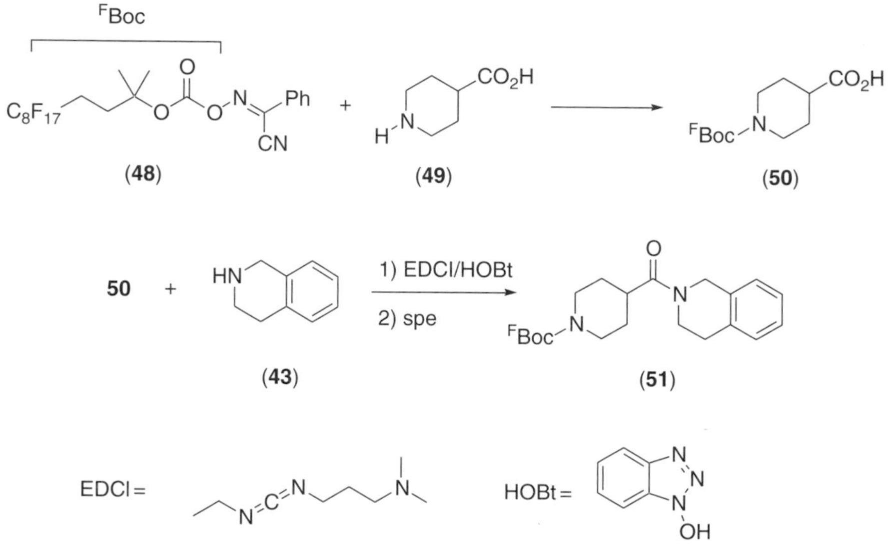

Scheme 38

could be considerably more environmentally friendly than solid-phase techniques since no large excesses of reagents and reactants are used. Also, the solvent volumes used for solid-phase extraction separations are typically less than those needed for washings in solid-phase synthesis or chromatographies in traditional synthesis.

Fluorous tagging of substrates is ideally suited to expediting parallel synthesis, as illustrated by the simple amide coupling chemistry shown in Scheme 38. Only a single example is shown, but 16 and 96 compound libraries were made using this approach.[135] ^{F}Boc-protected amino acid **50** is readily prepared from the fluorous Boc-On reagent **48** and amino acid **49**. The free acid of **50** can be coupled with amine **43** under standard conditions to make ^{F}Boc amide **51**. This is readily separated from the excess reagents, reactants and derived byproducts by two-stage fluorous solid-phase extraction. When desired, the ^{F}Boc group of **51** can be removed to give the free amine ready for further chemistry (in this case, additional *N*-alkylation reactions were conducted to diversify the library products). Likewise, the residual tag can be recovered and reused.

An assortment of light fluorous protecting groups are commercially available (^{F}CBz[136], ^{F}diisopropyl silane[137], ^{F}PMB[138], ^{F}Marshall ester[139]) or can be readily prepared (acetals,[140] alkoxysilanes[141]). Also, the attachment of fluorous tags to target molecules is not limited to protecting groups, and displaceable tags have also been used in parallel synthesis settings.[142]

The introduction and subsequent commercialization of fluorous silica gel have enabled the light branch of fluorous chemistry. The applicability of this chemistry covers all types of reaction components (substrates, reagents, reactants, catalysts, scavengers) and cuts across the synthesis spectrum from small-scale discovery chemistry to large-scale process chemistry. The simple solid-phase extraction or flash chromatographic separations are generally much less solvent intensive than traditional chromatographies, and the fluorous reaction components can routinely be recovered and reused. Likewise, washing and reuse of the fluorous silica gel is common practice. The immobilization of the fluorous separation medium on the silica can be advantageous relative to the use of volatile fluorous solvents, whose release should be controlled owing to their environmental persistence. Synthesis and separation techniques based on fluorous silica gel will become increasingly useful as more and more light fluorous reaction components become available.

3.5 Fluorous reactions in supercritical carbon dioxide ($scCO_2$) and fluorous triphasic reactions

3.5.1 Fluorous reactions in $scCO_2$

The favorable features of carbon dioxide as a green reaction medium are being increasingly appreciated by synthetic chemists[143] and are highlighted in Chapter 4. The key issues in using carbon dioxide (either liquid or supercritical) as a solvent are what kinds of reactions can be run in carbon dioxide and how to run them. Solvents with carbonyl groups are rarely used for organic reactions because of their reactivity. Furthermore, carbon dioxide is relatively non-polar and does not have excellent dissolving power for broad classes of organic substrates, reagents, reactants and catalysts. Fluorous chemistry, on the other hand, offers a number of potentially general solutions to the solubility problem because highly fluorinated compounds frequently have excellent solubility in carbon dioxide. In a simple view, carbon dioxide is a "hybrid" solvent,[144] like benzotrifluoride or trichlorotrifluoroethane, that has the ability to dissolve under suitable conditions of temperature and pressure both organic compounds and fluorous compounds. Accordingly, it is an attractive reaction solvent in fluorous chemistry.

Two groups concurrently introduced fluorous reaction components for use in reactions in supercritical carbon dioxide in 1997.[145,146] Leitner and coworkers were looking for general ways to solubilize catalysts in supercritical CO_2.[145] They added the standard rhodium precursor Rh(hfacac)(η^4-C_8F_{12}) (hfacac = $CF_3COCHCOCF_3$) to fluorous phosphine **52a** and showed that the resulting catalyst was soluble in supercritical CO_2 and smoothly promoted the typical hydroformylation reaction shown in Scheme 39. Control experiments with the standard triphenylphosphine **52b** provided much poorer results, presumably because the catalyst and perhaps the derived intermediates along the catalytic cycle have little or no solubility in the reaction medium.

In contrast, Beckman and Curran were looking for solvents that would dissolve both organic and fluorous reaction components and suggested supercritical CO_2 as a general solution to fluorous reaction and separation problems.[146] The setting was

scCO$_2$; CO/H$_2$; Rh/L(1/6) → CHO + CHO

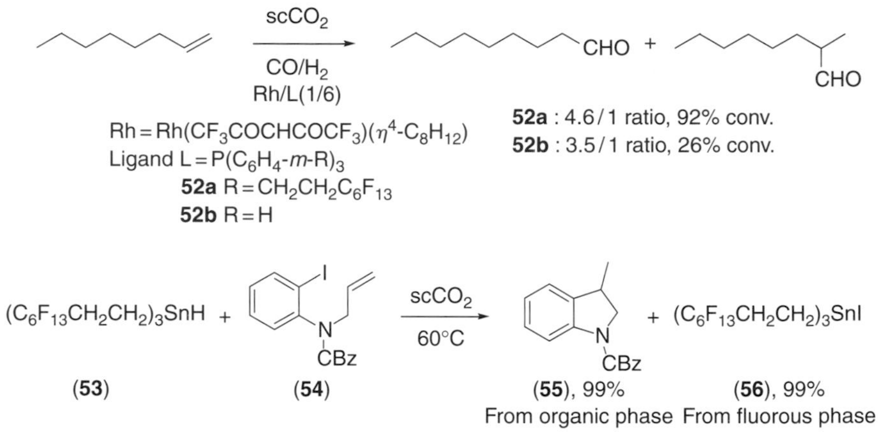

Rh = Rh(CF$_3$COCHCOCF$_3$)(η^4-C$_8$H$_{12}$)
Ligand L = P(C$_6$H$_4$-*m*-R)$_3$
52a R = CH$_2$CH$_2$C$_6$F$_{13}$
52b R = H

52a : 4.6 / 1 ratio, 92% conv.
52b : 3.5 / 1 ratio, 26% conv.

(C$_6$F$_{13}$CH$_2$CH$_2$)$_3$SnH (**53**) + (**54**) —scCO$_2$, 60°C→ (**55**), 99% From organic phase + (C$_6$F$_{13}$CH$_2$CH$_2$)$_3$SnI (**56**), 99% From fluorous phase

Scheme 39

radical chemistry mediated by the fluorous tin hydride **53**. For example, reductive cyclization of **54** in supercritical CO_2 was followed by partitioning of the crude reaction product between dichloromethane and perfluorohexane. Cyclized product **55** was recovered from the dichloromethane phase in a 99% yield and tin iodide **56** was recovered from the fluorous phase also in a 99% yield (Scheme 39).

Each of these papers proposed extensions that have since been realized. The concept of adding fluorous ponytails to ligands or other reaction components to promote reactions in supercritical CO_2 has been extended to a number of other systems[142] and appears to have broad generality. It is thus one of several important emerging strategies for conducting reactions in CO_2. The projected ability to use the CO_2 phase to effect both reaction and separation has also been realized in several important studies. For example, fluorous lanthanide catalysts with good solubility in supercritical CO_2 and poor solubility in liquid CO_2 have been used to conduct both reaction and separation stages of alcohol acylation and Friedel–Crafts acylation (Figure 3.4).[148] Water has been used in biphasic reactions with fluorous rhodium catalysts.[149] This is highly effective for water-soluble products, which can be removed from the reactor without depressurization.

Fluorous compounds are also potentially useful as additives to promote organic reactions in carbon dioxide. For example, a fluorous alcohol R_fCH_2OH assists asymmetric hydrogenations with non-fluorous ruthenium BINAP catalysts,[150] and a fluorous aryl alkyl ether ($C_8F_{17}C_6H_4$-*p*-$OC_{12}H_{25}$) does so in scandium-triflate-catalyzed aldol and Friedel–Crafts reactions.[151] These additives are presumed to act as solubilizers or emulsifiers to promote contact among the various reaction components. Since they are fluorous, they can be readily recovered from the otherwise organic reaction mixtures for reuse.

In a recent new direction, researchers have capitalized on the ability of some organic solvents to absorb large amounts of CO_2 when pressurized with the gas.[152]

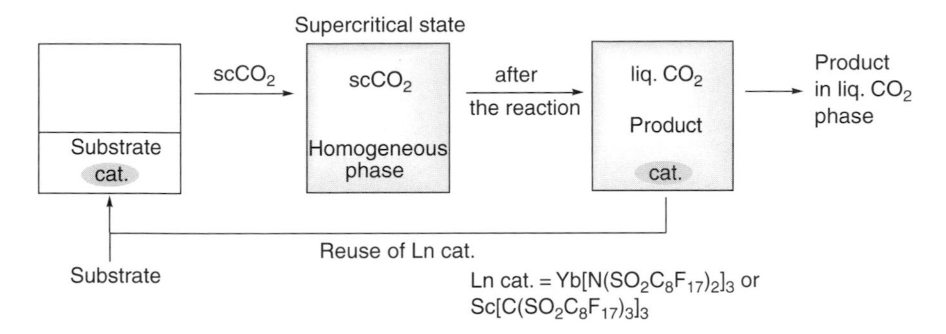

Figure 3.4

Fluorous organometallic complexes can be induced to dissolve in liquid hydrocarbons by CO_2 pressurization and then precipitated in a controlled fashion by depressurization. Slow depressurization to induce crystallization is a promising method for obtaining crystals of suitable quality for X-ray diffraction. Extension to reactions where catalysts or reagents are dissolved by pressurization and then precipitated by depressurization can also be envisioned.

3.5.2 Fluorous triphasic reactions

Fluorous biphasic and triphasic reactions are at once similar and different. Like fluorous biphasic reactions (see Section 3.3), fluorous triphasic reactions use a fluorous reaction solvent. However, whereas biphasic reactions use heavy fluorous molecules, triphasic reactions use light fluorous molecules or sometimes no fluorous molecules at all. The reaction and separation occur simultaneously in triphasic reactions. Indeed, the reaction drives the separation in most triphasic processes, whereas a separation follows a reaction in biphasic methods.

Triphasic reactions were introduced in 2001 in a detagging setting.[153] Molecules bearing fluorous tags were detagged with simultaneous purification of non tagged impurities (Figure 3.5). For example, fluorous tagged 2-napthylethanol derivative **57** was doped with 20–100 mol% 1-napthylethanol **58** to simulate an untagged impurity. These mixtures were then added to the source side of a U-tube with organic solvents (typically MeOH) floating on both sides of the tube above FC-72. The desilylation reagent H_2SiF_6 was added to the receiving side. After 1–3 days, the desilylated product **59** was isolated from the receiving side of the U-tube in excellent yields and purities, and the untagged impurity **58** was recovered from the source side.

Presumably, the tagged alcohol (**57**) partitions into and then passes through the fluorous phase to the receiving side. Detagging by H_2SiF_6 then strands alcohol **59** on that side. Impurity **58** has a very low partition coefficient into FC-72 and remains on the source side throughout the reaction. In essence, the chemical energy released by the detagging reaction drives the separation. The paper envisioned a number of ways to use fluorous triphasic reactions, and one of these, enzyme kinetic

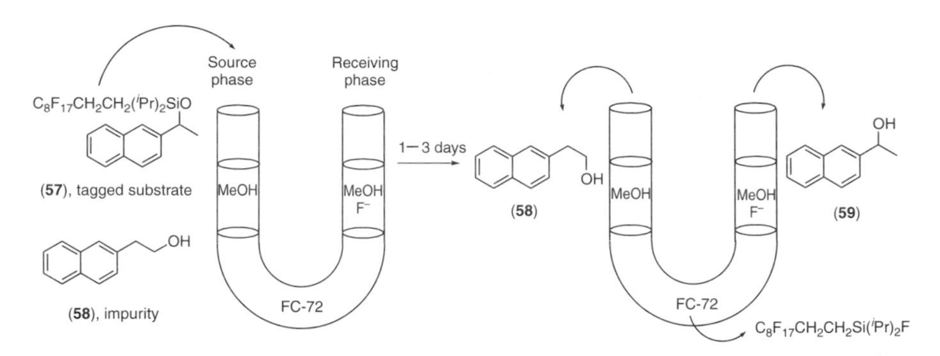

Figure 3.5

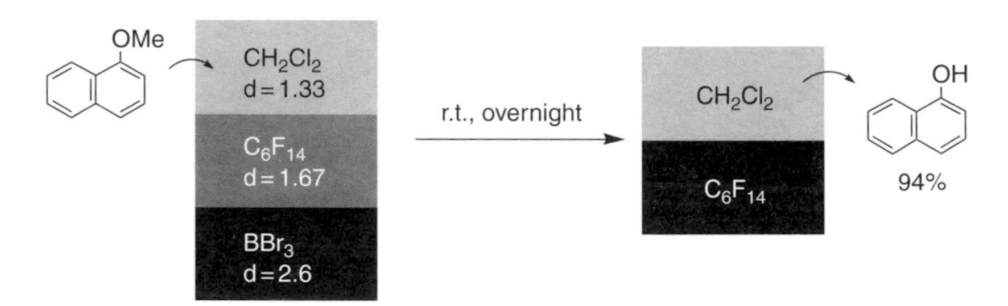

Figure 3.6

resolution followed by a fluorous triphasic separation, has been demonstrated in proof-of-principle experiments.[154]

In a recent new direction, "phase vanishing" reactions have been introduced that use a fluorous reaction solvent as a barrier or membrane but otherwise employ no fluorous reaction components.[155] For example, aryl methyl ethers can be demethylated by BBr_3 in a reaction that starts out triphasic (Figure 3.6). The substrate and reaction solvent float on top of FC-72 and the BBr_3 sinks to the bottom. Although it is not miscible with FC-72, BBr_3 has some solubility and it gradually diffuses up to encounter and demethylate the aryl methyl ether. As the reaction nears completion, the BBr_3 phase vanishes, hence the name. No cooling is needed, nor are syringe pumps or other mechanical equipment used to regulate the addition in this exothermic reaction; the FC-72 serves as the chemical regulator. Similar reactions can be carried out with Br_2 and other reagents that are denser than FC-72.[156] In addition, less dense reagents such as $SOCl_2$ or $SOBr_2$ can be used in a U-tube setup and also constitute a simple reactor for multiple parallel reactions fed by a single reagent source.[157]

Although a few triphasic reactions have been conducted with water, mostly in the molecular recognition/transport field, synthetic chemists are only just beginning to recognize the potential and advantages of this class of reaction and separation process. This could well be because water is not a suitable solvent for many types of organic reactions. The use of fluorous solvents in triphasic reactions and separations opens up new options.

3.6 Experimental part

3.6.1 Asymmetric epoxidation of alkenes under fluorous biphasic conditions

(Ref. 88) In a 10 mL Schlenk vessel placed in a thermoregulated bath at 100°C, 1 mL of a 0.2*M* solution of alkene in CH_3CN containing *o*-dichlorobenzene (0.1*M*, internal standard for gas chromatography (GC)) and 0.2 mL of a 0.25*M* solution of pyridine *N*-oxide in CH_3CN were added under nitrogen to 1 mL of a solution of catalyst in *n*-perfluorooctane. PhIO (67 mg, 0.3 mmol) was quickly added under a nitrogen stream. The two-phase mixture was magnetically stirred at 1300 ± 50 rpm and cooled to room temperature at the end of the reaction. The brown fluorous layer was separated, washed with CH_3CN (2×0.5 mL) and reused in further runs. The combined organic layers were washed with saturated aqueous Na_2SO_3 (1 mL) and brine (1 mL) and dried ($MgSO_4$). Epoxide yield and enantiomeric excess were determined by gas chromatographic analysis of the organic solution (Cyclodex-B chiral column).

3.6.2 Asymmetric addition of Et_2Zn to benzaldehyde catalyzed by F_{13}BINOL-Ti in an FC-72/organic biphasic system

(Ref. 7013 91b) Ti(O-*i*Pr)$_4$ (341 mg, 1.20 mmol) was added to a solution of (*R*)-F_{13}BINOL (485 mg, 0.20 mmol) in FC-72 (5 mL) under argon at room temperature. After stirring for 30 min, 1*M* Et_2Zn hexane solution (3.0 mL, 3.0 mmol) was added to the reaction mixture and the mixture was stirred for another 10 min. The reaction mixture was cooled to 0°C and a solution of benzaldehyde (110 mg, 1.03 mmol) in toluene (3 mL) was added. After vigorous stirring for 2 h at that temperature, the organic phase was separated with a cannula and quenched with 1 N hydrochloric acid (6 mL). The mixture was extracted with ether (15 mL $\times$ 4). The combined organic layer was washed with brine (10 mL $\times$ 3), dried over anhydrous $MgSO_4$ and concentrated *in vacuo*. The residue was dissolved in ether (2 mL). ((1*H*,1*H*,2*H*,2*H*)-perfluorooctyl)dimethylsilyl bound silica gel (FRP silica gel; 1 g) was added to the solution, which was then evaporated to dryness. The powder obtained was loaded on a column of FRP silica gel (5 g) and then eluted successively with CH_3CN (30 mL) and FC-72 (40 mL). The CH_3CN fraction was evaporated *in vacuo* and purified by preparative TLC (hexane/EtOAc = 4 : 1) to give 1-phenyl-1-propanol (116 mg, 83% yield) in 82% ee as a colorless oil. (*R*)-F_{13}BINOL (50 mg, 10%) was recovered from the FC-72 fraction. To the fluorous phase Ti(O-*i*Pr)$_4$, Et_2Zn solution and benzaldehyde solution were added, and the reaction was carried out in the same way.

3.6.3 *Ugi four-components condensation*

(Ref. 108) 4-Tris((2-(perfluorodecyl)ethyl)silyl)benzoic acid (26.2 mg, 0.015 mmol), the amine (0.25 mmol), the aldehyde (0.25 mmol) and the isonitrile (0.25 mmol) were added to a sealed tube with CF_3CH_2OH (0.3 mL). (For some examples, the preformed imine was used.) The suspension was heated under argon to 90°C for 48 h. After removal of the solvent, the residue was diluted with FC-72 (15 mL) and washed with benzene (15 mL). The benzene layer was additionally extracted twice with FC-72 (15 mL). The combined fluorous phases were evaporated to yield the perfluorosilylated amino acid amide. For desilylation, the amino acid amide was dissolved at 25°C in THF (2 mL), TBAF (1*M* in THF, 0.022 mL, 0.022 mmol) was added and the resulting solution was stirred at 25°C for 30 min. After removal of the solvent, the residue was diluted with benzene (30 mL) and washed twice with FC-72 (15 mL). Et_2O (30 mL) was added to the organic layer, which was washed with 0.1 N HCl, saturated aqueous Na_2CO_3 and brine (15 mL each). The organic phase was dried ($MgSO_4$) and evaporated to yield the benzoylated amino acid amide. The purity was checked using GC analysis.

3.6.4 *Synthesis of a tetrasaccharide using a fluorous protective group*

(Ref. 112) 4-(Dimethylamino)pyridine (139 mg, 1.14 mmol) and dicyclohexylcarbodiimide (353 mg, 1.71 mmol) were added to a solution of 6-*O*-trityl-α-methylmannoside (124 mg, 0.285 mmol) and Bfp (1.05 g, 1.03 mmol) in dry CH_2Cl_2 (25 mL). After stirring for 2 h at room temperature, toluene (20 mL) was added to the reaction mixture. The reaction mixture was extracted three times with FC-72 (20 mL $\times$ 3). The combined FC-72 layers were evaporated. The almost pure 2,3,4-tri-*O*-Bfp-6-*O*-trityl-α-methylmannoside (865 mg, 87%) was used in the next step without further purification. The trityl group of this compound was removed by treatment with camphorsulfonic acid in methanol-ether at room temperature for 19 h to produce 2,3,4-tri-*O*-Bfp-α-methylmannoside in an 88% yield. Molecular sieves AW-300 (700 mg) were added to a solution of the compound (142 mg, 442 μmol) and 2,3,4-*O*-acetyl-6-*O*-TBDPS-trichloroacetylimidoyl glucoside (152 mg, 221 μmol) in dry ether (5 mL), under argon. After stirring for 1 h at room temperature, TMS-OTf (8 μL, 442 μmol) was added to the reaction mixture at 0°C. After stirring for 20 min at 0°C, the reaction mixture was filtered. The filtrate was added to saturated $NaHCO_3$, and extracted three times with EtOAc. The EtOAc layers were washed with brine, dried over anhydrous Na_2SO_4 and evaporated. The residue was partitioned between methanol and FC-72. The FC-72 layer was evaporated to give the disaccharide (124 mg, 75%). The almost pure compound was used in the next step without further purification. To a solution of this compound (95 mg, 25.4 mmol) in dry THF (2 mL) was added 70% HF-pyridine (180 μL) at room temperature. After stirring for 8 h at room temperature, the reaction mixture was partitioned between toluene, saturated $NaHCO_3$ and FC-72. The FC-72 layer was washed with brine, dried over anhydrous Na_2SO_4 and evaporated to give the deprotected product at 6-*O*-TBDPS (85 mg, 95%). The almost pure compound was used in the next step without further purification. Similar reactions were repeated to provide the final tetrasaccharide.

3.6.5 Large-scale preparation of tris(3,3,4,4,5,5,6,6,7,7,8,8,8-tridecafluorooctyl)tin hydride

(Ref. 93a) A 1 L, three-necked flask was dried in an oven. Bromo tris(3,3,4,4,5,5,6,6,7,7,8,8,8-tridecafluoro)tin (13.8 g, 11.1 mmol) was dissolved in dry ether (275 mL) and transferred to the dried three-necked flask equipped with a thermometer, stirring bar and an outlet to argon. The solution was cooled to 0°C. A 1 *M* solution of lithium aluminum hydride in ether (11.1 mL, 11.1 mmol) was added dropwise over 45 min to the solution, the addition rate being adjusted to maintain a temperature between 0°C and 1°C. The reaction mixture was stirred for 6 h at 0°C. Water (75 mL) was slowly added (initially dropwise) while stirring to the ice-cold mixture. After addition of sodium potassium tartrate (20%) (250 mL), the mixture was transferred to a 1 L separatory funnel. The ethereal layer was separated, and the aqueous layer was extracted three times with ether (3 × 100 mL). The combined extracts were dried with magnesium sulfate and vacuum-filtered into a 1 L round-bottomed flask. The solvent was evaporated under reduced pressure. The crude product was distilled under reduced pressure (0.02 mm Hg) at 133–140°C to provide 11.3 g (9.69 mmol, 87%) of the pure product as an oil.

3.6.6 The Stille coupling reaction with fluorous tin reactant under microwave irradiation

(Ref. 95b) In a screw-capped Pyrex tube were placed tris[2-(perfluorohexyl) ethyl]organotin (0.24 mmol), organohalide or organic triflate (0.20 mmol), bis(triphenylphosphine)palladium(II) chloride (2.8 g, 0.004 mmol), lithium chloride (25.4 mg, 0.60 mmol) and DMF (1.0 mL). The tube was capped, and the suspension was purged with nitrogen. The content of the tube was mixed with a whirlmixer and thereafter rapidly heated using microwave power (MicroWell 10 single-mode microwave cavity producing continuous irradiation (2450 MHz) from Labwell AB, Uppsala, Sweden). The microwave power was adjusted to give full conversion in <2 min (60–70 W). After the reaction mixture was cooled to room temperature, toluene (2 mL) was added, and most of the DMF was azeotropically evaporated under reduced pressure at 50–70°C. The residue was then partitioned in a three-phase extraction between water (10 mL, top), dichloromethane (20 mL, middle) and FC-84 (perfluoroheptane) (10 mL, bottom). The three layers were separated, and the dichloromethane layer was washed three additional times with water (3 × 10 mL) and FC-84 (3 × 10 mL) and evaporated under reduced pressure. The resulting crude product was purified by chromatography (silica column) to afford pure coupling product (>95% pure by gas chromatographic mass spectrometry). No attempt was made to recover the fluorous tin chloride from the FC-84 phase.

3.6.7 Fluorous Swern oxidation

(Ref. 99) To a well-stirred solution of anhydrous CH_2Cl_2 (5 mL) under argon at −30°C was added oxalyl chloride (0.14 mL, 1.6 mmol). $C_4F_9CH_2CH_2S(=O)CH_3$ (1.0 g, 3.2 mmol) was then added dropwise and the reaction mixture was stirred for an additional 20 min. Isoborneol (0.153 g, 1.0 mmol) dissolved in CH_2Cl_2 (5 mL) was then added to this solution, followed after an additional 0.5–1 h by $(i\text{-Pr})_2NEt$ (0.88 mL, 5.0 mmol). The reaction mixture was then allowed to warm to room temperature and stirred for a period of 30 min before it was quenched with H_2O, washed with ammonium chloride (5 mL), extracted with CH_2Cl_2 (10 mL) and carefully concentrated under aspirator vacuum in a cold water bath. The reaction mixture was then dissolved in toluene (6 mL) and extracted continuously with FC-72 (15 mL) in a cooled continuous extractor for 4 h. After decantation, concentration of the toluene layer and chromatography on silica gel yielded camphor (0.142 g, 94%). The FC-72 phase, containing a mixture of the fluorous sulfoxide and sulfide was then stirred with methanol (3 mL) and H_2O_2 (0.23 mL of 30%, 2 mmol) for 1 h, after which it was diluted with H_2O (5 mL) and extracted with CH_2Cl_2 (10 mL) in a three-phase system. Concentration of the CH_2Cl_2 layer allowed the recovery of the fluorous sulfoxide (0.9 g, 90%).

3.6.8 Preparation of 4-methylacetophenone in the presence of a fluorous acetylacetonate palladium complex

(Ref. 57) A Schlenk flask was charged with 4-methoxystyrene (1.0 mmol), *t*-BuOOH (1.5 mmol, 3*M* solution in benzene) and benzene (0.5 mL). The fluorous acetylacetonate palladium catalyst ($(C_7F_{15}COCH_2COC_7F_{15})_2Pd$, 0.05 mmol, 5 mol%) dissolved in $C_8F_{17}Br$ (1.0 mL) was added. The heterogeneous reaction mixture was then heated to 56°C, leading to a homogeneous solution. The reaction was complete after 3 h, as indicated by GC analysis. The reaction mixture was cooled to 25°C, leading to the formation of two phases. The orange fluorous phase was separated and washed with benzene three times, and was then ready for use in further runs. The organic phase was diluted with ether (10 mL) and washed with an aqueous solution of $Na_2S_2O_5$ to destroy the excess of *t*-BuOOH, then with brine. After drying ($MgSO_4$) and concentration of the solution, the crude oil was purified by flash column chromatography (pentane : ether = 9 : 1), yielding pure 4-methoxyacetophenone (76%) as a colorless solid.

3.6.9 The Mizoroki–Heck reaction using a fluorous palladacycle catalyst

(Ref. 67) A Schlenk tube was sequentially charged with DMF (6 mL), iodobenzene (5.02 mmol), methyl acrylate (6.28 mmol), NEt_3 (1.4 mL) and a solution of the fluorous palladacycle catalyst **10** in $CF_3C_6H_5$ (0.229 m*M*; 0.015 mL, 3.44×10^{-6} mmol). The tube was connected to a condenser and placed in a 140°C oil bath. The solution was vigorously stirred (14 h), removed from the bath to cool and diluted with ether. A standard basic aqueous workup and flash chromatography on silica gel (hexane : EtOAc = 9 : 1) gave *trans*-methyl cinnamate in 98% yield.

3.6.10 Direct amide condensation using a fluorous phenylboronic acid

(Ref. 75) A dried 10 mL-flask fitted with a 5 mL pressure-equalized addition funnel containing 4 Å molecular sieves surmounted by a reflux condenser was charged with cyclohexanecarboxylic acid (1.0 mmol), benzylamine (1.0 mmol) and the 3,5-bisperfluorodecylphenyl boronic acid catalyst (0.05 mmol) in *o*-xylene (5 mL). The mixture was heated at azeotropic reflux with removal of water to provide a homogeneous solution. After 3 h, the resulting mixture was cooled to room temperature to precipitate the catalyst. The liquid phase was decanted and the residual solid catalyst was reused without isolation. The liquid phase was concentrated under reduced pressure, and the residue was purified by column chromatography on silica gel (hexane : EtOAc = 3 : 1) to give the corresponding amide in 96% yield.

3.6.11 Esterification of alcohol with acetic anhydride using a fluorous scandium catalyst

(Ref. 78) Cyclohexanol (2 mmol) and acetic anhydride (2 mmol) were added to a mixture of $CF_3C_6H_5$ (5 mL) and toluene (5 mL). To the resultant mixture was added $Sc[C(SO_2C_8F17)_3]_3$ (0.02 mmol, 1 mol%), and the solution was stirred at 30°C for 20 min. The mixture obtained was allowed to stand still at room temperature, so that the reaction mixture separated into an upper phase of toluene and a lower phase of $CF_3C_6H_5$. Cyclohexyl acetate (98% yield) was obtained from the upper phase after evaporation under reduced pressure and silica gel chromatography. To the lower phase containing the scandium catalyst were again added toluene (5 mL), cyclohexanol (2 mmol) and acetic anhydride (2 mmol), followed by stirring at 30°C for 20 min. After reaction, the same workup procedure gave the product in quantitative yield.

References

[1] For historical views and definitions of fluorous chemistry, see: Symposium-in-Print on fluorous chemistry, Gladysz, J.; Curran, D. P., Eds.; *Tetrahedron* **2002**, *58*, 3823.

[2] Lead reviews: (a) Horváth, I. T. *Acc. Chem. Res.* **1998**, *31*, 641. (b) Curran, D. P. *Angew. Chem., Int. Ed. Engl.* **1998**, *37*, 1175. (c) Barthel-Rosa, L. P.; Gladysz, J. A. *Coord. Chem. Rev.* **1999**, 190–192, 587. (d) Cornils, B. *Angew. Chem., Int. Ed. Engl.* **1997**, *36*, 2057. (e) Kitazume, T. *J. Fluorine Chem.* **2000**, *105*, 265. (f) Furin G. G. *Russ. Chem. Rev.* **2000**, *69*, 491. (g) Curran, D. P. In *Stimulating Concepts in Chemistry*; Vögtle, F.; Stoddard, J. F.; Shibasaki, M., Eds.; Wiley-VCH: New York, 2000. (h) de Wolf, E.; van Koten, G.; Deelman, B.-J. *Chem. Soc. Rev.* **1999**, *28*, 37.

[3] (a) Ogawa, A.; Curran, D. P. *J. Org. Chem.* **1997**, *62*, 450–51. (b) Maul, J. J.; Ostrowski, P. J.; Ublacker, G. A.; Linclau, B.; Curran, D. P. In *Topics in Current Chemistry, Modern Solvents in Organic Synthesis*, Vol. 206; Knochel P.; Springer-Verlag: Berlin, 1999, pp. 80–104.

[4] Matsubara, H.; Yasuda, S.; Sugiyama, H.; Ryu, I.; Fujii, Y.; Kita K. *Tetrahedron* **2002**, *58*, 4071.
[5] Curran, D. P.; Hadida, S. *J. Am. Chem. Soc.* **1996**, *118*, 2531.
[6] Curran, D. P.; Luo, Z.; Degenkolb, P. *Bioorg. Med. Chem. Lett.* **1998**, *8*, 2403.
[7] Elsner, P.; Wigger-Alberti, W.; Pantini, G. *Dermatology* **1998**, *197*, 141.
[8] (a) Fujii, Y.; Furugaki, H.; Kajihara, Y.; Kita, K.; Morimoto, H.; Uno, M. US Patent 6011071, 2000. (b) Fujii, Y.; Furugaki, H.; Tamura, E.; Yano, S.; Kita, K. *Bull. Chem. Soc. Jpn.* **2005**, *78*, 456.
[9] (a) Clark, L. C.; Gollan, F. *Science* **1966**, 152, 1755. (b) Riess, J. G.; Le Blanc, M. *Pure Appl. Chem.* **1982**, *54*, 2383.
[10] Handa, T.; Mukerjee, P. *J. Phys. Chem.* **1981**, *85*, 3816.
[11] Horváth, I. T.; Rábai, J. *Science* **1994**, *266*, 72.
[12] Luo, Z.; Zhang, Q.; Oderaotoshi, Y.; Curran, D. P. *Science* **2001**, *291*, 1766.
[13] Yoshida, J.-I.; Itami, K. *Chem. Rev.* **2002**, *102*, 3693.
[14] Gladysz, J. A.; Curran, D. P. *Tetrahedron* **2002**, *58*, 3823.
[15] Jiao, H.; Le Stang, S.; Soós, T.; Meier, R.; Kowski, K.; Rademacher, P.; Jafarpour, L.; Hamard, J.-B.; Nolan, S. P.; Gladysz, J. A. *J. Am. Chem. Soc.* **2002**, *124*, 1516 and earlier work cited therein.
[16] de Wolf, E.; Mens, A. J. M.; Gijzeman, O. L. J.; van Lenthe, J. H.; Jenneskens, L. W.; Deelman, B.-J.; van Koten, G. *Inorg. Chem.* **2003**, *42*, 2115.
[17] Reichardt, C. *Solvents and Solvent Effects in Organic Chemistry*; Wiley-VCH: Weinheim, 2003.
[18] Zhu, D.-W. *Synthesis* **1993**, 953.
[19] (a) Sianesi, D.; Marchionni, G.; De Pasquale, R. J. In *Organofluorine Chemistry Principles and Commercial Applications*, Banks, R. E.; Smart, B. E.; Tatlow, J. C., Eds.; Plenum Press: New York, 1994; Chapter 20. (b) Ohsaka, Y. In *Organofluorine Chemistry Principles and Commercial Applications*; Banks, R. E.; Smart, B. E.; Tatlow, J. C., Eds.; Plenum Press: New York, 1994, Chapter 21.
[20] Monk, T. G. *J. Crit. Illness* **2000**, *15*, No. 9 [Supplement; Symposium on novel hematologic therapies], S10–S17.
[21] (a) Merrigan, T. L.; Bates, E. D.; Dorman, S. C.; Davis, J. H., Jr. *Chem. Commun.* **2000**, 2051. (b) van den Broeke, J.; Winter, F.; Deelman, B.-J.; van Koten, G. *Org. Lett.* **2002**, *4*, 3851.
[22] Maul, J. J.; Ostrowski, P. J.; Ublacker, G. A.; Linclau, B.; Curran, D. P. *Top. Curr. Chem.* **1999**, *206*, 79.
[23] Matsubara, H.; Yasuda, S.; Sugiyama, H.; Ryu, I.; Fujii, Y.; Kita, K. *Tetrahedron* **2002**, *58*, 4071.
[24] (a) Filler, R. In *Fluorine-Containing Molecules*; Liebman, J. F.; Greenberg, A.; Dolbier, W. R., Jr., Eds.; VCH: Weinheim, 1988; Chapter 2. (b) Alkorta, I.; Rozas, I.; Elguero, J. *J. Org. Chem.* **1997**, *62*, 4687 and references therein.
[25] Collings, J. C.; Roscoe, K. P.; Robins, E. G.; Batsanov, A. S.; Stimson, L. M.; Howard, J. A. K.; Clark, S. J.; Marder, T. B. *New J. Chem.* **2002**, *26*, 1740 and references therein.
[26] Freed, B. K.; Biesecker, J.; Middleton, W. J. *J. Fluorine Chem.* **1990**, *48*, 63.
[27] Smart, B. E. In *Organofluorine Chemistry Principles and Commercial Applications*; Banks, R. E.; Smart, B. E.; Tatlow, J. C., Eds.; Plenum Press: New York, 1994, Chapter 3.
[28] Rocaboy, C.; Gladysz, J. A. *Tetrahedron* **2002**, *58*, 4007.
[29] Alvey, L. J.; Rutherford, D.; Juliette, J. J. J.; Gladysz, J. A. *J. Org. Chem.* **1998**, *63*, 6302.
[30] Wende, M.; Gladysz, J. A. *J. Am. Chem. Soc.* **2003**, *125*, 5861.
[31] Ishihara, K.; Kondo, S.; Yamamoto, H. *Synlett* **2001**, 1371.
[32] (a) Benesi, H. A.; Hildebrand, J. H. *J. Am. Chem. Soc.* **1948**, *70*, 3978. (b) Scott, R. L. *J. Am. Chem. Soc.* **1948**, *70*, 4090. (c) Hildebrand, J. H.; Cochran, D. R. F. *J. Am. Chem. Soc.* **1949**, *71*, 22. (d) Hildebrand, J. H.; Fisher, B. B.; Benesi, H. A. *J. Am. Chem. Soc.* **1950**, *72*, 4348. (e) *Solubilities of Inorganic and Organic Compounds*, Vol. 1, Part 2; Stephen, H.; Stephen, T., Eds.; Pergamon Press: New York, 1963, pp. 1027, 1028, 1086, 1394, 1412, 1472, 1473.
[33] Serratrice, G.; Delpuech, J.-J.; Diguet, R. *Nouv. J. Chem.* **1982**, *6*, 489.
[34] See Riess, J. G.; LeBlanc, M. *Pure Appl. Chem.* **1982**, *54*, 2388.

[35] Guillevic, M.-A.; Rocaboy, C.; Arif, A. M.; Horváth, I. T.; Gladysz, J. A. *Organometallics* **1998**, *17*, 707.

[36] (a) Klement, I.; Lütjens, H.; Knochel, P. *Angew. Chem. Int. Ed. Engl.* **1997**, *36*, 1454; *Angew. Chem.* **1997**, *109*, 1605. (b) Laboratory observations from Gladysz' group.

[37] Atkins, P. W. In *Physical Chemistry*, 3rd Edn.; W. H. Freeman and Company: New York, 1986, pp. 197–198.

[38] Juliette, J. J. J.; Rutherford, D.; Horváth, I. T.; Gladysz, J. A. *J. Am. Chem. Soc.* **1999**, *121*, 2696.

[39] Kiss, L. E.; Kövesdi, I.; Rábai, J. *J. Fluorine Chem.* **2001**, *108*, 95.

[40] Gladysz, J. A.; Curran, D. P.; Horváth, I. T., Eds.; Handbook of Fluorous Chemistry; Wiley-VCH: Weinheim, 2004.

[41] Curran, D. P.; Ferritto, R.; Hua, Y. *Tetrahedron Lett.* **1998**, *39*, 4937.

[42] (a) Richter, B.; de Wolf, E.; van Koten, G.; Deelman, B.-J. *J. Org. Chem.* **2000**, *65*, 3885. (b) de Wolf, E.; Richter, B.; Deelman, B.-J.; van Koten, G. *J. Org. Chem.* **2000**, *65*, 5424. (c) Richter, B.; Spek, A. L.; van Koten, G.; Deelman, B.-J. *J. Am. Chem. Soc.* **2000**, *122*, 3945. (d) de Wolf, E.; Speets, E. A.; Deelman, B.-J.; van Koten, G. *Organometallics* **2001**, *20*, 3686.

[43] Huque, F. T. T.; Jones, K.; Saunders, R. A.; Platts, J. A. *J. Fluorine Chem.* **2002**, *115*, 119.

[44] (a) Poliakoff, M.; Fitzpatrick, J. M.; Farren, T. R.; Anastas, P. T. *Science* **2002**, *297*, 807. (b) DeSimone, J. M. *Science* **2002**, *297*, 799.

[45] Clayton, J. W., Jr. *Fluorine Chem. Rev.* **1967**, *1*, 197.

[46] Hogue, C. *Chem. Eng. News* **2003**, *April 21 issue*, 9.

[47] Wuebbles, D. J.; Calm, J. M. *Science* **1997**, *278*, 1090.

[48] (a) Ravishankara, A. R.; Solomon, S.; Turnipseed, A. A.; Warren, R. F. *Science* **1993**, *259*, 194. (b) See also Ravishankara, A. R.; Turnipseed, A. A.; Jensen, N. R.; Barone, S.; Mills, M.; Howard, C. J.; Solomon, S. *Science* **1994**, *263*, 71.

[49] (a) Giesy, J. P.; Kannan, K. *Environ. Sci. Technol.* **2001**, *35*, 1339. (b) Giesy, J. P.; Kannan, K.; Jones, P. D. *Sci. World* **2001**, *1*, 627.

[50] Horváth, I. T.; Kiss, G.; Cook, R. A.; Bond, J. E.; Stevens, P. A.; Rabái, J.; Mazeleski, E. J. *J. Am. Chem. Soc.* **1998**, *120*, 3133.

[51] (a) Foster, D. F.; Adams, D. J.; Gudmunsen, D.; Stuart, A. M.; Hope, E. G.; Hamilton, D.J. *Chem. Commun.* **2002**, 722. (b) Foster, D. F.; Adams, D. J.; Gudmunsen, D.; Adams, D. J.; Stuart, A. M.; Hope, E. G.; Cole-Hamilton, D. J. *Tetrahedron* **2002**, *58*, 3901.

[52] Richter, B.; Spek, A. L.; van Koten, G.; Deelman, B.-J. *J. Am. Chem. Soc.* **2000**, *122*, 3945.

[53] de Wolf, E.; Spek, A. L.; Kuipers, B. W. M.; Philipse, A. P.; Meeldijk, J. D.; Bomans, P. H. H.; Frederik, P. M.; Deelman, B.-J.; van Koten, G. *Tetrahedron* **2002**, *58*, 3922.

[54] (a) Juliette, J. J. J.; Horváth, I. T.; Gladysz, J. A. *Angew. Chem., Int. Ed. Engl.* **1997**, *36*, 1610. (b) Juliette, J. J. J.; Rutherford, D.; Horváth, I. T.; Gladysz, J. A. *J. Am. Chem. Soc.* **1999**, *121*, 2696. (c) Dinh, L. V.; Gladysz, J. A. *Tetrahedron Lett.* **1999**, *40*, 8995.

[55] Klement, I.; Lutjens, H.; Knochel, P. *Angew. Chem., Int. Ed. Engl.* **1997**, *36*, 1454.

[56] Benvenuti, F.; Carlini, C.; Marchionna, M.; Galletti, [illegible] *J. Mol. Catal. (A)*, [illegible]

[57] Betzemeier, B.; Lhermitte, F.; Knochel, P. *Tetrahedron Lett.* **1998**, *39*, 6667.

[58] Quici, S.; Pozzi, G. *Tetrahedron Lett.* **1999**, *40*, 3647.

[59] (a) Ragagnin, G.; Betzemeier, B.; Quici, S.; Knochel, P. *Tetrahedron* **2002**, *58*, 3985. (b) Betzemeier, B.; Cavazzini, M.; Quici, S.; Knochel, P. *Tetrahedron Lett.* **2000**, *41*, 4343.

[60] Pozzi, G.; Montanari, F.; Quici, S. *Chem. Commun.* **1997**, 69.

[61] Betzemeier, B.; Lhermitte, F.; Knochel, P. *Synlett* **1999**, 489.

[62] Ten Brink, G.-J.; Vis, J.-M.; Arends, I. W. C. E.; Sheldon, R. A. *Tetrahedron* **2002**, *58*, 3977.

[63] (a) Legros, J.; Crousse, B.; Bourdon, J.; Bonnet-Delpon, D.; Bégué, J.-P. *Tetrahedron Lett.* **2001**, *42*, 4463. (b) Legros, J.; Crousse, B.; Bonnet-Delpon, D.; Bégué, J.-P. *Tetrahedron* **2002**, *58*, 3993.

[64] (a) Curran, D. P.; Hoshino, M. *J. Org. Chem.* **1996**, *61*, 6480. (b) Hoshino, M.; Degenkolbe, P.; Curran, D. P. *J. Org. Chem.* **1997**, *62*, 8341.

[65] Larhed, M.; Hoshino, M.; Hadida, S.; Curran, D. P.; Hallsberg, A. *J. Org. Chem.* **1997**, *62*, 5583.

[66] Schneider, S.; Bannwarth, W. *Angew. Chem., Int. Ed. Engl.* **2000**, *39*, 4142.

[67] Rocaboy, C.; Gladysz, J. A. *Org. Lett.* **2002**, *4*, 1993.
[68] Moineau, J.; Pozzi, G.; Quici, S.; Sinou, D. *Tetrahedron Lett.* **1999**, *40*, 7683.
[69] Fukuyama, T.; Arai, M.; Matsubara, H.; Ryu, I. *J. Org. Chem.* **2004**, *69*, 8105.
[70] Betzemeier, B.; Knochel, P. *Angew. Chem., Int. Ed. Engl.* **1997**, *36*, 2623.
[71] Kling, R.; Sinou, D.; Pozzi, G.; Chopin, A.; Quignard, F.; Busch, S.; Kainz, S.; Koch, D.; Leitner, W. *Tetrahedron Lett.* **1998**, *39*, 9439.
[72] Endres, A.; Maas, G. *Tetrahedron Lett.* **1999**, *40*, 6365.
[73] Wende, M.; Meier, R.; Gladysz, J. A. *J. Am. Chem. Soc.* **2001**, *123*, 11490.
[74] Meseguer, M.; Moreno-Manas, M.; Vallribera, A. *Tetrahedron Lett.* **2000**, *41*, 4093.
[75] Ishihara, K.; Kondo, S.; Yamamoto, H. *Synlett* **2001**, 1371.
[76] Ishihara, K.; Hasegawa, A.; Yamamoto, H. *Synlett* **2002**, 1299.
[77] Ishihara, K.; Hasegawa, A.; Yamamoto, H. *Synlett* **2002**, 1296.
[78] (a) Mikami, K.; Mikami, Y.; Matsumoto, Y.; Nishikido, J.; Yamamoto, F.; Nalkajima, H. *Tetrahedron Lett.* **2001**, *42*, 289. (b) Mikami, K.; Mikami, Y.; Matsuzawa, H.; Matsumoto, Y.; Nishikido, J.; Yamamoto, F.; Nalkajima, H. *Tetrahedron* **2002**, *58*, 4015.
[79] Nishikido, J.; Nanbo, M.; Yoshida, A.; Nakajima, H.; Matsumoto, Y.; Mikami, K. *Synlett* **2002**, 1613.
[80] Nishikido, J.; Kamishima, M.; Matsuzawa, H.; Mikami, K. *Tetrahedron* **2002**, *58*, 8345.
[81] (a) Barrett, A. G.; Braddock, D. C.; Catterick, D.; Henschke, J. P.; McKinnell, R. M. *Synlett* **2000**, 847. (b) Barrett, A. G.; Bouloc, N.; Braddock, D. C.; Catterick, D.; Chadwick, D.; White, A. J. P.; Williams, D. J. *Tetrahedron* **2002**, *58*, 3835.
[82] Zhu, D.-W. *Synthesis* **1993**, 953.
[83] (a) Otera, J.*Chem. Rev.* **1993**, *93*, 1449. (b) Otera, J. *Angew. Chem., Int. Ed. Engl.* **2001**, *40*, 2044.
[84] Xing, J.; Toyoshima, S.; Orita, A.; Otera, J. *Angew. Chem., Int. Ed. Engl.* **2001**, *40*, 3670.
[85] Otera, J. *Acc. Chem. Res.* **2002**, *37*, 288.
[86] (a) Mailland, D.; Pozzi, G.; Quici, S.; Sinou, D. *Tetrahedron* **2002**, *58*, 3971. (b) Maillard, D.; Nguefak, C.; Pozzi, G.; Quici, S.; Valade, B.; Sinou, D. *Tetrahedron Asymmetry* **2000**, *11*, 2881.
[87] Pozzi, G.; Cinato, F.; Montanari, F.; Quici, S. *Chem. Commun.* **1998**, 877.
[88] Cavazzini, M.; Manfredi, A.; Montanari, F.; Quici, S.; Pozzi, G. *Chem. Commun.* **2000**, 2171.
[89] Takeuchi, S.; Nakamura, Y.; Ohgo, Y.; Curran, D. P. *Tetrahedron Lett.* **1998**, *39*, 8691.
[90] Kleijn, H.; Rijnberg, E.; Jastrzebski, J. T. B. H.; van Koten, G. *Org. Lett.* **1999**, *1*, 853.
[91] (a) Nakamura, Y.; Takeuchi, S.; Ohgo, Y.; Curran, D. P. *Tetrahedron Lett.* **2000**, *41*, 57. (b) Nakamura, Y.; Takeuchi, S.; Okumura, K.; Ohgo, Y.; Curran, D. P. *Tetrahedron* **2002**, *58*, 3963.
[92] (a) Tian, Y.; Chan, K. S. *Tetrahedron Lett.* **2000**, *41*, 8813. (b) Tian, Y.; Yang, Q. C.; Mak, T. C. W.; Chan, K. S. *Tetrahedron* **2002**, *58*, 3951.
[93] (a) Crombie, A.; Kim, S.-Y.; Hadida, S.; Curran, D. P. *Org. Synth.* **2002**, *72*, 1. (b) Curran, D. P.; Hadida, S.; Kim, S.-Y.; Luo, Z. *J. Am. Chem. Soc.* **1999**, *121*, 6607. (c) Hadida, S.; Super, M.; Beckman, E. J.; Curran, D. P. *J. Am. Chem. Soc.* **1997**, *119*, 7406. (d) Curran, D. P.; Hadida, S. *J. Am. Chem. Soc.* **1996**, *118*, 2531.
[94] (a) Hoshino, M.; Degenkolb, P.; Curran, D. P. *J. Org. Chem.* **1997**, *62*, 8341. (b) Curran, D. P.; Hoshino, M. *J. Org. Chem.* **1996**, *61*, 6480.
[95] (a) Olofsson, K.; Kim, Y.-S.; Larhed, M.; Curran, D. P., Hallberg, A. *J. Org. Chem.* **1999**, *64*, 4539. (b) Larhed, M.; Hoshino, M.; Hadida, S.; Curran, D. P., Hallberg, A. *J. Org. Chem.* **1997**, *62*, 5583.
[96] Ryu, I.; Niguma, T.; Minakata, S.; Komatsu, M., Luo, Z.; Curran, D. P. *Tetrahedron Lett.* **1999**, *40*, 2367.
[97] Ryu, I.; Niguma, T.; Minakata, S.; Komatsu, M.; Hadida, S.; Curran, D. P. *Tetrahedron Lett.* **1997**, *38*, 7883.
[98] Curran, D. P.; Hadida, S.; Kim, S.-Y. *Tetrahedron* **1999**, *55*, 8997.
[99] Crich, D.; Neelamkavil, S. *J. Am. Chem. Soc.* **2001**, *123*, 7449.
[100] Crich, D.; Neelamkavil, S. *Tetrahedron* **2002**, *58*, 3865.

[101] Crich, D.; Neelamkavil, S. *Org. Lett.* **2002**, *4*, 4175.
[102] (a) Crich, D.; Neelamkavil, S.; Sartillo-Piscil, F. *Org. Lett.* **2000**, *2*, 4029. (b) Crich, D.; Barba, G. R. *Org. Lett.* **2000**, *2*, 989. (c) Crich, D.; Hao, X.; Lucas, M. *Tetrahedron* **1999**, *55*, 14261.
[103] Dobbs, A. P.; McGregor-Johnson, C. *Tetrahedron Lett.* **2002**, *43*, 2807.
[104] Galante, A.; Lhoste, P.; Sinou, D. *Tetrahedron Lett.* **2001**, *42*, 5425.
[105] Chen, D.; Qing, F.-L.; Huang, Y. *Org. Lett.* **2002**, *6*, 1003.
[106] Palomo, C.; Aizpurua, J. M.; Loinaz, I.; Fernandez-Berridi, M. J.; Irusta, L. *Org. Lett.* **2001**, *3*, 2361.
[107] (a) Studer, A.; Hadida, S.; Ferritto, R.; Kim, Y.-S.; Jeger, P.; Wipf, P.; Curran, D. P. *Science* **1997**, *275*, 823. (b) Studer, A.; Curran, D. P. *Tetrahedron* **1997**, *53*, 6681.
[108] Studer, A.; Jeger, P.; Wipf, P.; Curran, D. P. *J. Org. Chem.* **1997**, *62*, 2917.
[109] Curran, D. P.; Ferritto, R.; Hua, Y. *Tetrahedron Lett.* **1998**, *39*, 4937.
[110] (a) Rover, S.; Wipf, P. *Tetrahedron Lett.* **1999**, *40*, 5667. (b) Wipf, P.; Reeves, J. *Tetrahedron Lett.* **1999**, *40*, 5139.
[111] Pardo, J.; Cobas, A.; Guitian, E.; Castedo, L. *Org. Lett.* **2001**, *3*, 3711.
[112] Miura, T.; Hirose, Y.; Ohmae, M.; Inazu, T. *Org. Lett.* **2001**, *3*, 3947.
[113] Miura, T.; Inazu, T. *Tetrahedron Lett.* **2003**, *44*, 1819.
[114] Mizuno, M.; Goto, K.; Miura, T.; Hosaka, D.; Inazu, T. *Chem. Commun.* **2003**, 972.
[115] Curran, D. P.; Hadida, S.; He, M. *J. Org. Chem.* **1997**, *62*, 6714.
[116] (a) Curran, D. P.; Luo, Z. Y. *J. Am. Chem. Soc.* **1999**, *121*, 9069. (b) Zhang, Q.; Luo, Z.; Curran, D. P. *J. Org. Chem.* **2000**, *65*, 8866.
[117] For a comprehensive overview of fluorous silica gel and light fluorous chemistry, see Curran, D. P. In *The Handbook of Fluorous Chemistry*; Gladysz, J.; Horvath, I.; Curran, D. P., Eds.; Wiley-VCH: New York, 2004, pp. 101, 128.
[118] (a) Curran, D. P. *Chemtracts – Org. Chem.* **1996**, *9*, 75. (b) Curran, D. P. *Angew. Chem., Int. Ed. Eng.* **1998**, *37*, 1175. (c) Curran, D. P. In *Stimulating Concepts in Chemistry*; Vögtle, F., Stoddard, J. F., Shibasaki, M., Eds.; Wiley-VCH: New York, 2000. (d) Yoshida, J.; Itami, K. *Chem. Rev.* **2002**, *102*, 3693. (e) Tzschucke, C. C.; Markert, C.; Bannwarth, W.; Roller, S.; Hebel, A.; Haag, R. *Angew. Chem., Int. Ed.* **2002**, *41*, 3964.
[119] (a) Review: Curran, D. P. *Synlett* **2001**, 1488. (b) Curran, D. P.; Hadida, S.; Studer, A.; He, M.; Kim, S.-Y.; Luo, Z.; Larhed, M.; Hallberg, M.; Linclau, B. In *Combinatorial Chemistry: A Practical Approach*, Vol. 2; Fenniri, H., Ed.; Oxford University Press: Oxford, 2001.
[120] Fluorous Technologies, Inc. is on the web at www.fluorous.com. DPC holds an equity interest in this company.
[121] Reviews: (a) Zhang, W. *Chem. Rev.* **2004**, *104*, 2531. (b) Ref. [40], chapters 7 & 8. (c) Curran, D. P. *Synlett* **2001**, 1488. Lead references: (d) Zhang, Q.; Lu, H.; Richard, C.; Curran, D. P. *J. Am. Chem. Soc.* **2004**, *126*, 36. (e) Zhang, W.; Luo, Z.; Chen, C. H.-T.; Curran, D. P. *J. Am. Chem. Soc.* **2002**, *124*, 10443. (f) Luo, Z.; Zhang, Q.; Oderaotoshi, Y.; Curran, D. P. *Science* **2001**, *291*, 1766.
[122] β-CD columns for separation of fluorous and non-fluorous mixture synthesis, and "racemic synthesis": (a) Matsuzawa, H.; Mikami, K. *Synlett* **2002**, [illegible]. (b) Matsuzawa, H.; Mikami, K. *Tetrahedron Lett.* **2003**, *44*, 6227. (c) Nakamura, Y.; Takeuchi, S.; Okumura, K.; Ohgo, Y; Matsuzawa, H.; Mikami, K. *Tetrahedron Lett.* **2003**, *44*, 6221. (d) Curran, D. P.; Dandapani, S.; Werner, S.; Matsugi, M. *Synlett* **2004**, 1545. (e) Mikami, K.; Matsuzawa, H.; Takeuchi, S.; Nakamura, Y.; Curran, D. P. *Synlett* **2004**, 2713.
[123] (a) Dandapani, S.; Curran, D. P. *Tetrahedron* **2002**, *58*, 3855. (b) Dobbs, A. P.; McGregor-Johnson, C. *Tetrahedron Lett.* **2002**, *43*, 2807. (c) Dandapani, S.; Curran, D. P. *Tetrahedron* **2004**, *69*, 8751.
[124] Many fluorous phosphines are known. For recent leading references, see Ref. [2b] and Clarke, M. L. *J. Organomet. Chem.* **2003**, *665*, 65.
[125] Selected references: (a) Hoshino, M.; Degenkolb, P.; Curran, D. P. *J. Org. Chem.* **1997**, *62*, 8341. (b) Ryu, I.; Niguma, T.; Minakata, S.; Komatsu, M.; Luo, Z. ; Curran, D. P. *Tetrahedron Lett.* **1999**, *40*, 2367. (c) Curran, D. P.; Hadida, S.; Kim, S.-Y. *Tetrahedron* **1999**, *55*, 8997. (d) Curran, D. P.; Hadida, S.; Kim, S.-Y.; Luo, Z. *J. Am. Chem. Soc.* **1999**, *121*, 6607. (e) Bucher, B.; Curran, D. P. *Tetrahedron Lett.* **2000**, *41*, 9617.

[126] (a) Crich, D.; Hao, X. L.; Lucas, M. *Tetrahedron* **1999**, *55*, 14261. (b) Crich, D.; Neelamkavil, S. *J. Am. Chem. Soc.* **2001**, *123*, 7449. (c) Crich, D.; Neelamkavil, S. *Org. Lett.* **2002**, *4*, 4175.

[127] Rocaboy, C.; Gladysz, J. A. *Chem. Eur. J.* **2003**, *9*, 88.

[128] (a) Larhed, M.; Moberg, C.; Hallberg, A. *Acc. Chem. Res.* **2002**, *35*, 717. (b) Lindeberg, G.; Larhed, M.; Hallberg, A. US Patent 6136157, **2000**.

[129] Vallin, K. S.; Zhang, Q.; Larhed, M.; Curran, D. P.; Hallberg, A. *J. Org. Chem.* **2003**, *68*, 6639.

[130] Linclau, B.; Singh, A. K.; Curran, D. P. *J. Org. Chem.* **1999**, *64*, 2835.

[131] Zhang, W.; Curran, D. P.; Chen, C. H. T. *Tetrahedron* **2002**, *58*, 3871.

[132] (a) Lindsley, C. W.; Zhao, Z.; Leister, W. H. *Tetrahedron Lett.* **2002**, *43*, 4225. (b) Lindsley, C. W.; Zhao, Z. J.; Leister, W. H.; Strauss, K. A. *Tetrahedron Lett.* **2002**, *43*, 6319.

[133] Zhang, W.; Chen, C. H. T.; Nagashima, T. *Tetrahedron Lett.* **2003**, *44*, 2065.

[134] Studer, A.; Hadida, S.; Ferritto, R.; Kim, S. Y.; Jeger, P.; Wipf, P.; Curran, D. P. *Science* **1997**, *275*, 823.

[135] Luo, Z.; Williams, J.; Read, R. W.; Curran, D. P. *J. Org. Chem.* **2001**, *66*, 4261.

[136] (a) Filippov, D. V.; van Zoelen, D. J.; Oldfield, S. P.; van der Marel, G. A.; Overkleeft, H. S.; Drijfhout, J. W.; van Boom, J. H. *Tetrahedron Lett.* **2002**, *43*, 7809. (b) Curran, D. P.; Amatore, M.; Guthrie, D.; Campbell, M.; Go, E.; Luo, Z. *J. Org. Chem.* **2003**, *68*, 4643.

[137] Zhang, Q.; Rivkin, A.; Curran, D. P. *J. Am. Chem. Soc.* **2002**, *124*, 5774.

[138] Curran, D. P.; Furukawa, T. *Org. Lett.* **2002**, *4*, 2233.

[139] Chen, C. H. T.; Zhang, W. *Org. Lett.* **2003**, *5*, 1015.

[140] (a) Wipf, P.; Reeves, J. T. *Tetrahedron Lett.* **1999**, *40*, 4649. (b) Wipf, P.; Reeves, J. T. *Tetrahedron Lett.* **1999**, *40*, 5139.

[141] Rover, S.; Wipf, P. *Tetrahedron Lett.* **1999**, *40*, 5667.

[142] (a) Wipf, P.; Methot, J. L. *Org. Lett.* **1999**, *1*, 1253. (b) Zhang, W. *Org. Lett.* **2003**, *5*, 1011.

[143] Leitner, W. *Acc. Chem. Res.* **2002**, *35*, 746.

[144] (a) Ogawa, A.; Curran, D. P. *J. Org. Chem.* **1997**, *62*, 450. (b) Maul, J. J.; Ostrowski, P. J.; Ublacker, G. A.; Linclau, B.; Curran, D. P. In *Topics in Current Chemistry, "Modern Solvents in Organic Synthesis,"* Vol. 206; Knochel, P., Ed.; Springer-Verlag: Berlin, **1999**, pp. 80–104.

[145] Kainz, S.; Koch, D.; Baumann, W.; Leitner, W. *Angew. Chem., Int. Ed. Eng.* **1997**, *36*, 1628. The fluorous terminology was not used therein, but its applicability is obvious in retrospect.

[146] Hadida, S.; Super, M. S.; Beckman, E. J.; Curran, D. P. *J. Am. Chem. Soc.* **1997**, *119*, 7406.

[147] Examples: (a) Shezad, N.; Oakes, R. S.; Clifford, A. A.; Rayner, C. M. *Tetrahedron Lett.* **1999**, *40*, 2221. (b) Osuna, A. M. B.; Chen, W. P.; Hope, E. G.; Kemmitt, R. D. W.; Paige, D. R.; Stuart, A. M.; Xiao, J. L.; Xu, L. J. *J. Chem. Soc., Dalton Trans.* **2000**, 4052. (c) Kani, I.; Omary, M. A.; Rawashdeh-Omary, M. A.; Lopez-Castillo, Z. K.; Flores, R.; Akgerman, A.; Fackler, J. P. *Tetrahedron* **2002**, *58*, 3923. (d) Osswald, T.; Schneider, S.; Wang, S.; Bannwarth, W. *Tetrahedron Lett.* **2001**, *42*, 2965.

[148] Nishikido, J.; Kamishima, M.; Matsuzawa, H.; Mikami, K. *Tetrahedron* **2002**, *58*, 8345.

[149] McCarthy, M.; Stemmer, H.; Leitner, W. *Green Chem.* **2002**, *4*, 501.

[150] Xiao, J.; Nefkens, S. C. A.; Jessop, P. G.; Ikariya, T.; Noyori, R. *Tetrahedron Lett.* **1996**, *37*, 2813.

[151] Komoto, I.; Kobayashi, S. *Org. Lett.* **2002**, *4*, 1115.

[152] Jessop, P. G.; Olmstead, M. M.; Ablan, C. D.; Grabenauer, M.; Sheppard, D.; Eckert, C. A.; Liotta, C. L. *Inorg. Chem.* **2002**, *41*, 3463.

[153] Nakamura, H.; Linclau, B.; Curran, D. P. *J. Am. Chem. Soc.* **2001**, *123*, 10119.

[154] Luo, Z.; Swaleh, S. M.; Theil, F.; Curran, D. P. *Org. Lett.* **2002**, *4*, 2585.

[155] Ryu, I.; Matsubara, H.; Yasuda, S.; Nakamura, H.; Curran, D. P. *J. Am. Chem. Soc.* **2002**, *124*, 12946.

[156] Matsubara, H.; Yasuda, S.; Ryu, I. *Synlett* **2003**, 247.

[157] Nakamura, H.; Usui, T.; Kuroda, H.; Ryu, I.; Matsubara, H.; Yasuda, S.; Curran, D. P. *Org. Lett.* **2003**, *5*, 1167.

4 Supercritical carbon dioxide

Christopher M. Rayner and R. Scott Oakes (Sections 4.1 and 4.2); Toshiyasu Sakakura and Hiroyuki Yasuda (Sections 4.3 and 4.4)

4.1 Historical background

Ever-increasing demands for sustainability and new environmental regulations have led to intense research in the area of alternative solvents. Supercritical fluids (SCFs), and supercritical carbon dioxide ($scCO_2$) in particular, are emerging as among the most promising sustainable technologies to have been developed in recent years. They are not new and can be traced back to original reports dated as long ago as 1822, when Baron Cagniard de LaTour[1] first observed the disappearance of two distinct gas and liquid phases into one single phase by increasing the temperature, thus demonstrating the existence of the critical point of alcohol using equipment originally designed by Denys Papin in 1680.[2] A fascinating historical introduction to the development of SCFs is provided in a recent text on the area.[3]

One of the best known and most influential developments in the use of $scCO_2$ was based on the work of Kurt Zosel in the early 1960s, where $scCO_2$ played a key role in the extraction of caffeine from coffee, which has obvious benefits to the consumer over alternative, solvent-based processes.[4] This is now carried out industrially on a huge scale, producing many thousands of tons of decaffeinated coffee per year. Liquid and supercritical CO_2 are routinely used for the extraction of natural products, particularly hops and essential oils[5], and this is now considered to be a relatively mature area of CO_2 technology.

Of particular relevance to this chapter is the use of CO_2 in polymer synthesis, in the manufacture of polymethylmethacrylate and polystyrene (Xerox)[6,7] and for the production of fluoropolymers (DuPont).[8] One of the main drivers for the latter was the phasing out of the chlorofluorocarbons (CFCs) used in the original process. The main advantage to this application is not necessarily the avoidance of the use of CFCs (although this is important), but the superior polymer processing properties made possible by the relative volatility of CO_2 and its ease of removal.

Other important large-scale uses of CO_2 are emerging, such as dry cleaning[9] utilizing liquid CO_2 as an alternative to perchloroethylene, for which commercial franchises are now in operation in the United States and Europe. Related technology is also being applied to other cleaning applications, including degreasing, spray painting[9,10:152], coating of various surfaces[11-13], dyeing[14-18] and foaming.[19,20]

Some industrial applications have been reviewed [13]. It is now clear from these examples that high-pressure CO_2 technology is viable on an industrial scale. This also suggests that, in the future, similar developments for chemical synthesis – the basis of this chapter – can be expected.

Several reviews on the use of supercritical fluids for reaction chemistry have been published, describing topics such as homogeneous catalysis[21–23], heterogeneous catalysis[24] and SCFs as solvent replacements in chemical synthesis[25] and in synthetic organic chemistry.[26] Other reviews are referred to in the relevant part of the text. There have also been two particularly useful books published in the field: one describes many aspects of chemical synthesis using SCFs[63], and the second concentrates more on the physical aspects.[27]

4.2 Physical properties

4.2.1 Fundamentals of supercritical fluids

A supercritical fluid is defined as a substance above its critical temperature (T_c) and critical pressure (P_c). This definition should include the clause "but below the pressure required for condensation into a solid," which is, however, commonly omitted because the pressure required to condense an SCF into a solid is generally impracticably high.[27] In addition, the important benefits and distinctions of SCFs are most apparent around the critical point, which is the region of the phase diagram where most studies are carried out. The critical point represents the highest temperature and pressure at which the substance can exist as a vapor and liquid in equilibrium. The phenomenon can be easily explained with reference to the phase diagram for pure carbon dioxide (Figure 4.1). This shows the areas where carbon dioxide exists

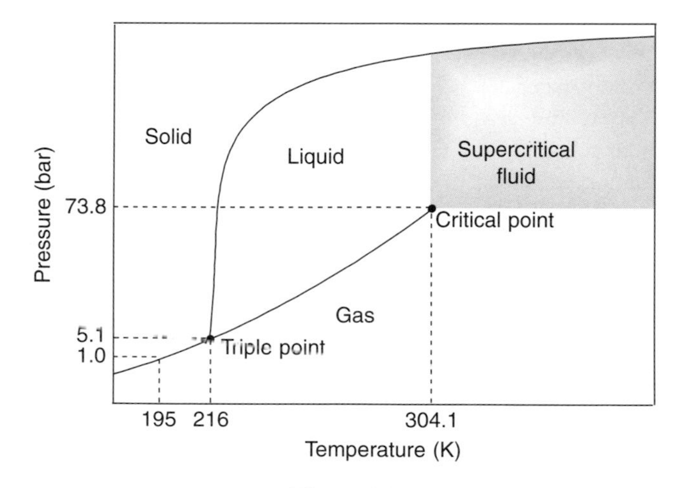

Figure 4.1

as a gas, liquid, solid or SCF. The curves represent the temperatures and pressures where two phases coexist in equilibrium (at the triple point, the three phases coexist). The gas–liquid coexistence curve is known as the boiling curve. Moving upward along this curve, both temperature and pressure increase: the liquid becomes less dense owing to thermal expansion, and the gas becomes denser as the pressure rises. Eventually, the densities of the two phases converge and become identical, the distinction between gas and liquid disappears and the boiling curve comes to an end at the critical point. The critical point for carbon dioxide occurs at a pressure of 73.8 bar and a temperature of 31.1°C. Other important supercritical fluids for reaction chemistry[3,27] include fluoroform (CHF_3, $T_c = 25.9$°C, $P_c = 48.2$ bar) and water ($T_c = 374.0$°C, $P_c = 220.6$ bar). Although the high critical parameters for water significantly limit its application as a reaction medium, it has important applications nonetheless.

The phase behavior of a SCF around the critical point can be demonstrated visually in an autoclave with a window, in which the meniscus between liquid and gas can be seen to disappear as the critical point is reached (Figure 4.2).[28] Figure 4.2a shows a two-phase liquid–gas system with a clearly defined meniscus. When the temperature and pressure of the system increase, the difference between the densities of the two phases decreases and the meniscus is less well defined (Figure 4.2b). Finally, in Figure 4.2c, no meniscus is present as the system is now a single homogeneous SCF.

It should be emphasized that the phase diagram shown in Figure 4.1 is for pure CO_2 alone. The addition of reagents or co-solvents can change its solvent properties significantly, and the phase diagram for the reaction mixture, and associated critical parameters (T_c, P_c), may differ considerably from that of pure CO_2. Indeed, in many literature examples, the description of CO_2 as being supercritical might not be the case, especially for solutions with significant concentrations of additional reagents; the term "supercritical" should therefore be read with caution. The term "dense phase CO_2" is often used, particularly when there is some uncertainty regarding the actual phase of a mixture. As $scCO_2$ is a relatively weak solvent, it is also often the case that two-phase mixtures of reagents and $scCO_2$ are obtained, which can greatly affect the outcome and rationalization of a reaction. It is therefore crucial

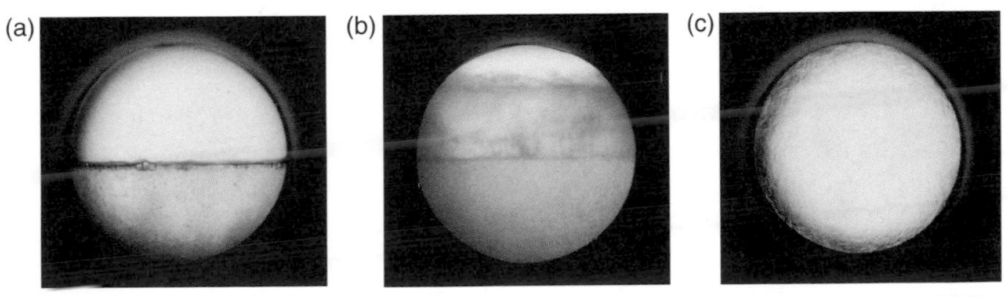

Two-phase liquid–gas system — Meniscus less well defined — Homogeneous SCF

Figure 4.2[28]

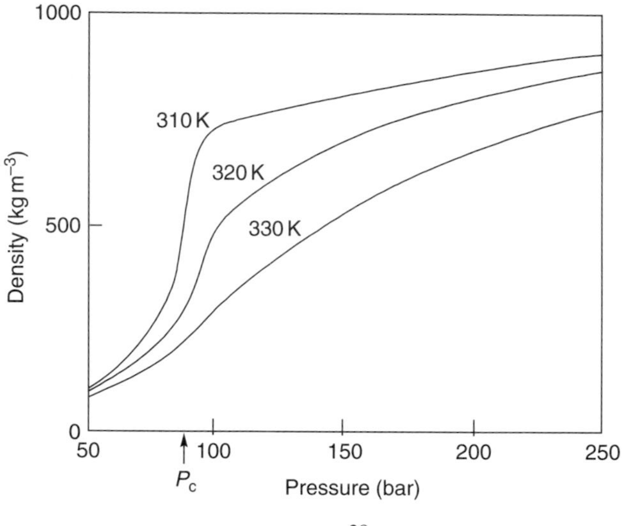

Figure 4.3[28]

to attempt to verify the homogeneity of the reaction mixture, as there is always a significant probability that the reaction may occur in a neat layer under an atmosphere of $scCO_2$. In some cases, this may be beneficial, but in others the use of high-pressure apparatus for a reaction taking place in neat conditions is, of course, not justified.

One of the most important properties of an SCF is that its physical properties can vary dramatically as a result of relatively small changes in temperature and pressure. This effect is most pronounced around the critical point[27], where the density of carbon dioxide is approximately 0.46 g/mL. If the pressure is doubled, the density of the fluid increases dramatically, reaching a density comparable to that of liquid carbon dioxide (Figure 4.3).

In the case of carbon dioxide, the solvent power is closely related to the density. In some cases this can be exploited, as modification of the density of the reaction medium can exert some control over reaction pathways. A number of reactions have been reported in which product selectivity is controlled by variation in SCF density, in some cases with results far superior to anything possible using conventional solvents (see below).

Table 4.1 is a comparison of typical values for the physical properties of a pure substance in different phases, including an SCF around the critical point.[3] It can be seen that the density of an SCF is approximately two orders of magnitude higher than that of the gas, but still less than half that of a conventional liquid phase. The viscosity and diffusivity – which are also temperature and pressure dependent – are in general at least an order of magnitude lower and higher, respectively, than for the liquid phase. If a chemical reaction is particularly fast, the diffusion can be the limiting factor; an increase in the diffusivity can then lead to reaction rate enhancement. This

Table 4.1[5]

Property	Gas	SCF	Liquid
Density (g/mL)	10^{-3}	0.4	1
Viscosity (Pa · s)	10^{-5}	10^{-4}	10^{-3}
Diffusivity (cm^2/s)	0.1	10^{-3}	$10^{-5}/10^{-6}$

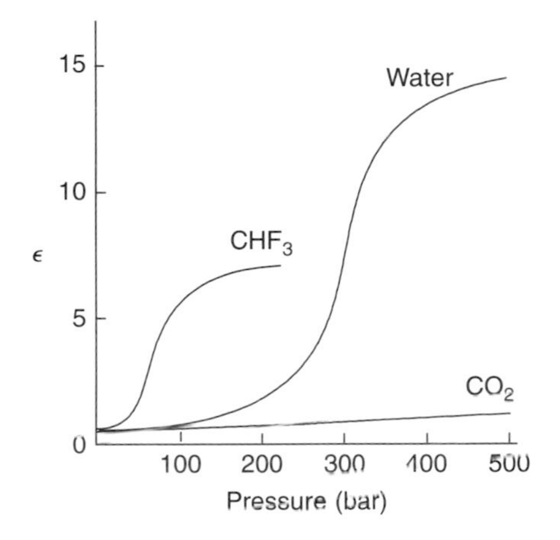

Figure 4.4[27]

principle is particularly applicable to unimolecular fission reactions, where increased diffusion allows an increase in radical separation rather than recombination, and highly efficient bimolecular processes such as free radical or enzymatic reactions.[29]

The majority of SCFs also show a sharp increase in the dielectric constant (ε) with increasing pressure in the compressible region (around the critical point). This behavior reflects, to some extent, the change in density. The magnitude of the increase depends on the nature of the SCF: whereas the dielectric constant varies little with pressure for non-polar substances such as $scCO_2$, dramatic increases are observed for more polar SCFs such as water or fluoroform (Figure 4.4).[27]

Supercritical fluids also show interesting thermal conductivity properties.[27,30] There is a general increase with density, as expected. There is also an enhancement of thermal conductivity around the critical region, and this is quite marked even 12 K above the critical temperature.

SCFs, like gases, have no surface tension: they diffuse rapidly to occupy the entire volume of a system. This also means that if other gases are introduced, they will also diffuse to fill the entire volume and mix perfectly. Unlike the solubility of gases in liquid solvents, which is relatively low and decreases as temperature increases, gases are totally miscible with SCFs and can be said to have perfect solubility. The concentration of hydrogen in a supercritical mixture of hydrogen

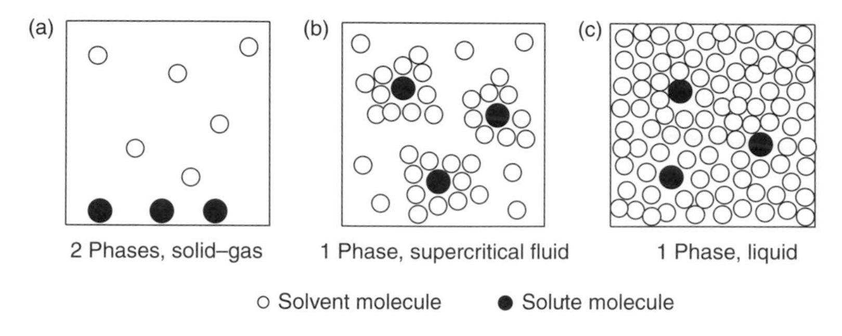

Figure 4.5

(85 bar) and carbon dioxide (120 bar) at 50°C is 3.2*M*, whereas the concentration of hydrogen in tetrahydrofuran under the same pressure is merely 0.4*M*.[31] There is, therefore, potential to much improve chemical processes where gaseous reagents have traditionally been used in solution phase, particularly if catalysts and substrates also have appreciable solubility in the reaction mixture.

The solubility of liquid and solid solutes in SCFs is surprisingly high, especially when SCFs are compressed to liquid-like densities. As solubility is related to density, the solubility of this medium has the added benefit of being tunable, and hence the solubility of an organic solute can be controlled directly: the reaction can be controlled by precipitation of a product, or the product can be purified by selective precipitation. Furthermore, the ability to dissolve gaseous, liquid and solid materials, and hence to homogenize reaction mixtures in the SCF phase, is particularly powerful.[32] This phenomenon is potentially extremely important to the bulk chemical industry for such reactions as hydrogenation, hydroformylation and oxidation, where traditionally the solubility of the gaseous reagent in the liquid solvent has been rate limiting.

An additional feature of SCFs, which enhances solubilities at moderate densities, is solute–solvent clustering.[33] Figure 4.5a shows a two-phase system in which a solid solute is under the pressure of a gas. Interactions between the molecules are very weak, and no significant solvation occurs. In Figure 4.5c, the system is now in the liquid phase, offering high solvation, and we therefore see a single phase. Figure 4.5b represents the intermediate case of an SCF. The bulk density is moderate, higher than the gas phase but much lower than that of a liquid. Solvent clustering around the solute molecules does, however, mean that the local density is relatively high and therefore a moderate solvation is offered; we observe a one-phase system for low-concentration solutions.[27,34]

4.2.2 *Enhancement of solubilities in* $scCO_2$

Since carbon dioxide is a non-polar molecule, non-polar solutes are much more soluble in it than polar molecules. However, because of its large molecular quadrupole, $scCO_2$ has a higher affinity for polar solutes than a true non-polar solvent such as pentane.[27] Solubility can still be a problem, but it can be enhanced in a number of ways.

- Increasing the bulk density of the SCF
- Adding a co-solvent
- Modifying the solute

Increasing the bulk density of the SCF (usually by an increase in pressure) is simple but not always desirable. The solvent power can be increased by the addition of a co-solvent to modify the characteristics of the SCF to be more like the substrate. Modifiers (e.g. MeOH, toluene) can therefore be added to increase or decrease polarity, or to increase affinity for aromatic species. However, the more modifier that is added, the further scCO_2 moves away from being the ideal green solvent. Perhaps less obvious is the fact that reagents themselves may also in effect act as co-solvents, enhancing the solubility of other, more polar solutes in the reaction medium. Hence, although measuring the solubility of reagents in pure CO_2 is helpful, it can act only as an indicator – it is the solubility of a reagent in the reaction mixture as a whole that is the most crucial factor. It should also be appreciated that this may change significantly as the reaction proceeds and the reaction mixture composition changes, which may lead to changes in phase behavior (e.g. phase separation).

A common approach has been to modify the solute (or choose a more suitable solute) to make it more CO_2-philic. Molecules with hydrocarbon chains show greater solubility than aromatic or polar substrates. Of special interest, however, is the greater solubility of organic fluorocarbons and siloxanes and related compounds in scCO_2 compared with the corresponding hydrocarbons. The nature of this increased solubility is not yet fully understood; however, computational and nuclear magnetic resonance (NMR) studies have suggested that the explanation might be an increase in solute–solvent Van der Waals interactions as the fluorocarbon character is increased, combined with a lower solute–solute attraction.[35,36] The effect of the increased solubility of fluorocarbon species has been used to good effect in the design of CO_2-philic surfactants, chelating agents and ligands in order to enhance the solubility of polymers, metals and catalysts, respectively.[3,37,38] Recent reports on cheaper, more sustainable approaches to solubility enhancement are likely to have a significant impact in the future.[39–41]

4.3 Applications as reaction media

4.3.1 Background

Organic syntheses in supercritical CO_2 emerged in the mid-1980s after supercritical CO_2 was successfully applied to food extraction (e.g. decaffeinated coffee) and material processing (e.g. the rapid expansion of supercritical solutions (RESS) method). Pioneering work concerning enzymatic hydrolysis and transesterification first appeared in 1985.[42] A few years later, Diels–Alder reactions[43] and the photo-isomerization of alkenes[44] in scCO_2 were reported. In the early 1990s, radical polymerization of fluorous alkenes was investigated by DeSimone[45], and halogenation[46] and hydroformylation[47] were studied by other groups. In 1994, Noyori and Ikariya reported that CO_2 hydrogenation was dramatically accelerated under supercritical conditions[48], which resulted in the explosive propagation of this technology. Since the late 1990s, $scCO_2$ has been applied to almost every organic reaction. This chapter highlights recent progress in homogeneous catalysis. Several other excellent, relevant reviews are also available.[3,22,23,26,49]

4.3.2 Hydroformylation

The total miscibility of $scCO_2$ with many reaction gases, including hydrogen and carbon monoxide, can avoid rate limitations by diffusion processes, which are often encountered in gas/liquid phase reactions. Thus, the enhancement of reaction rates may be expected using $scCO_2$ as solvent for reactions with gaseous reagents, such as hydroformylation and hydrogenation (Section 4.3.3). The first hydroformylation in $scCO_2$ was reported in 1991.[47] The cobalt carbonyl-catalyzed hydroformylation of propylene proceeded cleanly at 80°C in $scCO_2$ ($d = 0.5$ g/mL) to yield n-butyraldehyde **1** in 88% (Scheme 1), which compares favorably with the value obtained in benzene at higher CO and H_2 pressures. The equilibrium and the rate for forming the catalytic intermediate, $HCo(CO)_4$, in $scCO_2$ were comparable to those in apolar solvents such as methylcyclohexane. The activation energy for the cobalt-catalyzed hydroformylation of propylene in $scCO_2$ (23.3±1.4 kcal/mol) is also comparable to the value in conventional organic solvents (e.g. hexane, toluene).[50] For the stoichiometric hydroformylation of 3,3-dimethyl-1,2-diphenylcyclopropene **2** with $MnH(CO)_5$, it was suggested that the cage effects in $scCO_2$, neat olefin and hexane are identical and that the aldehyde is formed by a non-radical pathway in $scCO_2$ (Scheme 2).[51]

In 1997, it was demonstrated that [Rh(hfacac)(cod)] (hfacac = hexafluoroacetylacetonate $CF_3COCHCOCF_3$) is an effective unmodified rhodium catalyst for the

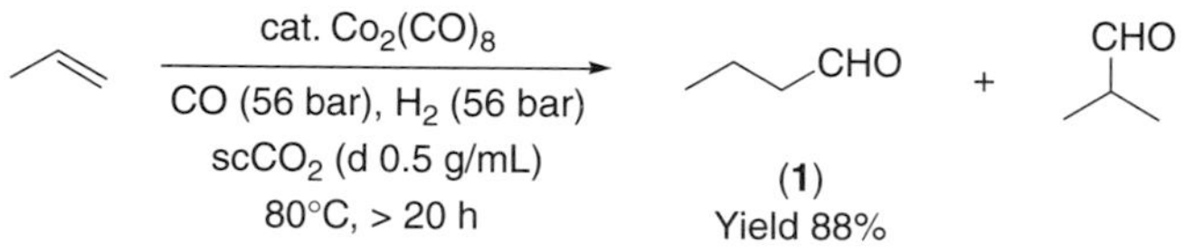

cat. $Co_2(CO)_8$
CO (56 bar), H_2 (56 bar)
$scCO_2$ (d 0.5 g/mL)
80°C, > 20 h
CHO
(1)
Yield 88%
+
CHO

Scheme 1

Ph Ph
(2)
+ $MnH(CO)_5$
$scCO_2$ (200 bar), 60°C
Ph Ph
H CHO

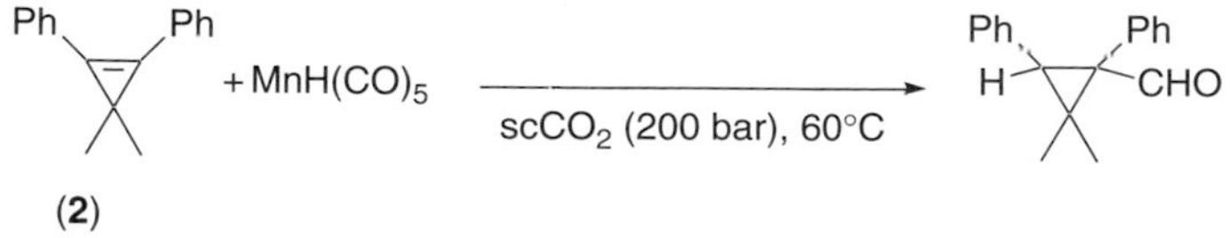

Scheme 2

hydroformylation of various olefins in $scCO_2$.[52,53] The rates in $scCO_2$ are considerably faster than in toluene or liquid CO_2. It was also reported that modified catalytic systems formed with perfluoroalkyl-substituted triarylphosphine and triaryl phosphite ligands exhibit high conversion and regioselectivity. For example, [Rh(hfacac)(cod)] with $P(C_6F_{13}C_2H_4\text{-}3\text{-}C_6H_4)_3$ catalyzes the hydroformylation of 1-octene to nonanal *n*-**3** and 2-methyloctanal *i*-**3** with a 92% conversion (*n* : *iso* ratio of 4.6) (Scheme 3), whereas the poorly soluble PPh_3 ligand gives a 26% conversion (*n* : *iso* ratio of 3.5).

The effect of various fluorinated phosphine ligands was investigated for the hydroformylation of 1-octene in $scCO_2$ using $[Rh(acac)(CO)_2]$.[54] The activity increased as the basicity of the phosphine decreased, that is, $[3,5\text{-}(CF_3)_2C_6H_3]_3P > [4\text{-}CF_3C_6H_4]_3P \sim [3\text{-}CF_3C_6H_4]_3P > [4\text{-}CF_3OC_6H_4]_3P > [4\text{-}F(CF_2)_4(CH_2)_3C_6H_4]_3P$. A perfluoroalkyl ponytail, $n\text{-}C_6F_{13}$, attached to PPh_3 produces a faster hydroformylation rate in $scCO_2$ than an ethylene-spaced one, $n\text{-}C_2H_4C_6F_{13}$, probably as a result of the stronger electron-withdrawing effect.[55] Arylphosphines with a long alkyl chain, such as $n\text{-}C_{10}H_{21}$ and $n\text{-}C_{16}H_{33}$, are not effective owing to their low solubility in $scCO_2$. On the other hand, PEt_3 is soluble in $scCO_2$ and is effective as a non-fluorinated phosphine ligand for rhodium-catalyzed hydroformylation in $scCO_2$.[56] Using $[Rh_2(OAc)_4]$ and PEt_3, 1-hexene is hydroformylated to C_7 aldehydes in 82% yield within 1 h at 100°C. Compared with the reaction in toluene, the reaction in $scCO_2$ has a similar turnover frequency (TOF, $57\,h^{-1}$), but a slightly improved *n* : *iso* ratio (2.4) (Scheme 4).

Rhodium-catalyzed asymmetric hydroformylation has been studied in $scCO_2$.[57–59] The hydroformylation of styrene in $scCO_2$ was first attempted using $[Rh(acac)(CO)_2]$ and (*R,S*)-BINAPHOS as chiral ligand. Although an appreciable enantiomeric excess (ee) of 66% was obtained at 0.48 g/mL (close to the critical density), the ee decreased dramatically as the CO_2 pressure increased, as a result of the low solubility of the chiral ligand in $scCO_2$. Modifying the chiral phosphine

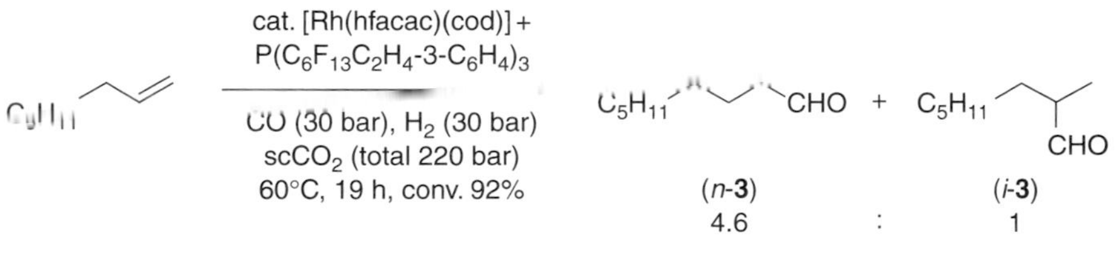

C_5H_{11}–CH=CH$_2$ → (cat. [Rh(hfacac)(cod)] + $P(C_6F_{13}C_2H_4\text{-}3\text{-}C_6H_4)_3$; CO (30 bar), H_2 (30 bar), $scCO_2$ (total 220 bar), 60°C, 19 h, conv. 92%) → C_5H_{11}–CH$_2$CH$_2$CHO (*n*-**3**) + C_5H_{11}–CH$_2$CH(CH$_3$)CHO (*i*-**3**)

4.6 : 1

Scheme 3

C_3H_7–CH=CH$_2$ → (cat. $[Rh_2(OAc)_4]$ + PEt_3; CO (20 bar), H_2 (20 bar), $scCO_2$ (65 bar at rt), 100°C, 1 h, 82%) → C_3H_7–CH$_2$CH$_2$CHO + C_3H_7–CH$_2$CH(CH$_3$)CHO

2.4 : 1

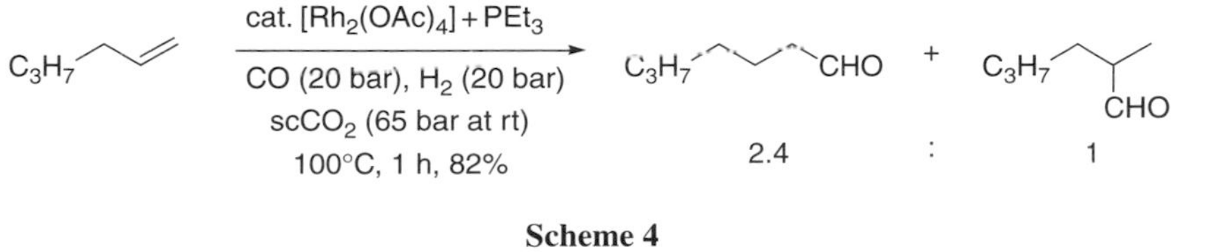

Scheme 4

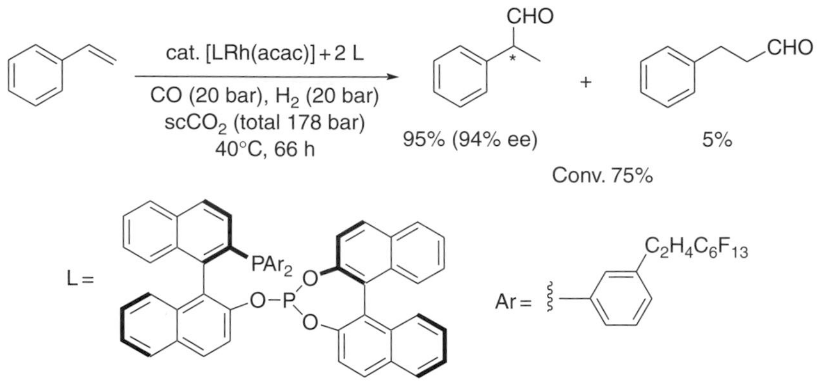

Scheme 5

with perfluoroalkyl substituents increased the solubility in $scCO_2$ and improved the enantioselectivity (up to 94%) (Scheme 5). Under similar conditions, the rate and selectivity in $scCO_2$ are comparable to those with the unmodified phosphine in benzene.

The rate of rhodium-catalyzed hydroformylation of alkyl acrylates is significantly accelerated using $scCO_2$ as a solvent.[60] At 80°C, the TOF for the hydroformylation of *t*-butyl acrylate by $[Rh(acac)(CO)_2]$ and $P(p\text{-}C_6H_4C_6F_{13})_3$ in $scCO_2$ to *t*-butyl-2-methyl-3-oxopropionate reaches $1800\,h^{-1}$, whereas the analogous reaction in toluene using bis(diphenylphosphino)butane, one of the best ligands for the reaction, is only $300\,h^{-1}$. A fluorous polymeric phosphine ligand is also effective for the hydroformylation of alkyl acrylates in $scCO_2$.[61] Interestingly, the hydroformylation of an equimolar mixture of 1-decene and ethyl acrylate gives only the aldehyde **4** arising from ethyl acrylate (branched : linear aldehyde ratio of 92) (Scheme 6).

For the hydroformylation of propylene in $scCO_2$, the critical point data (T_c and P_c) of various mixtures of the related components (CO_2, H_2, CO, propylene, *n*- and *iso*-butyraldehyde) have been measured.[62] These data will be very useful for designing reaction conditions in $scCO_2$ – for example, for avoiding phase separation during the reaction.

4.3.3 Hydrogenation

One of the seminal works on applying $scCO_2$ to organic synthetic processes was published in 1994, when the hydrogenation of CO_2 to form formic acid using a homogeneous ruthenium phosphine catalyst was reported, with CO_2 being used as both substrate and solvent. This work was later extended to produce both DMF and methyl formate from $scCO_2$ (see Section 4.3.10).

The first report made on asymmetric catalysis in $scCO_2$ was also for hydrogenation. In 1995, the hydrogenation of α-enamides to α-amino acid derivatives in $scCO_2$ was reported using cationic rhodium complexes with the chiral (*R*,*R*)-Et-DuPHOS

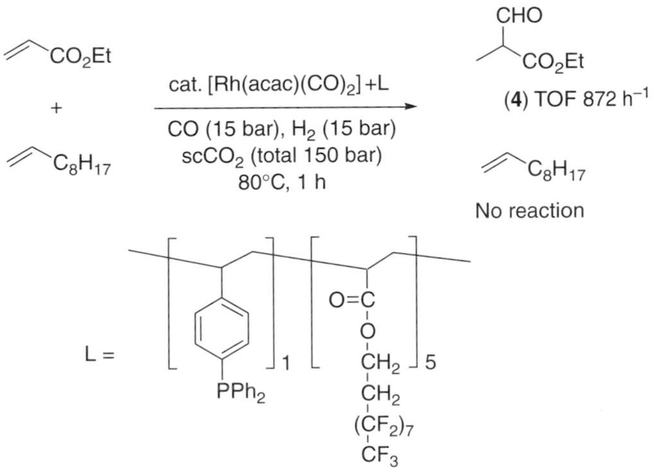

Scheme 6

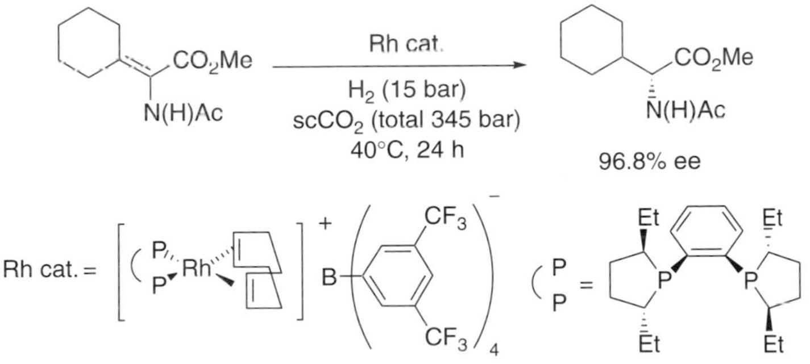

Scheme 7

ligand and either tetrakis-3,5-bis(trifluoromethyl)phenylborate (BARF) or triflate as counteranions (Scheme 7).[63] The hydrogenation of enantioselectively difficult β,β-disubstituted α-enamides in $scCO_2$ (345 bar, with 15 bar of H_2) produced 88.4–96.8% ee of valine derivatives, which are the highest levels of enantioselectivity for these substrates. The hydrogenation of tiglic acid in $scCO_2$ was reported to give 2-methylbutanoic acid **5** in 99% yield and 89% ee using a ruthenium complex with a chiral H_8-BINAP ligand and adding a fluorinated alcohol (Scheme 8).[64]

Cationic iridium(I) complexes with chiral phosphinodihydrooxazoles modified with perfluoroalkyl groups on the ligand and BARF counteranion catalyze the enantioselective hydrogenation of *N*-(1-phenylethylidene)aniline in $scCO_2$ (Scheme 9).[65] (*R*)-*N*-phenyl-1-phenylethylamine **6** is formed quantitatively with up to 81% ee, and the rate in $scCO_2$ is much faster than that in CH_2Cl_2. Furthermore, the extractive properties of $scCO_2$ enabled product separation and catalyst

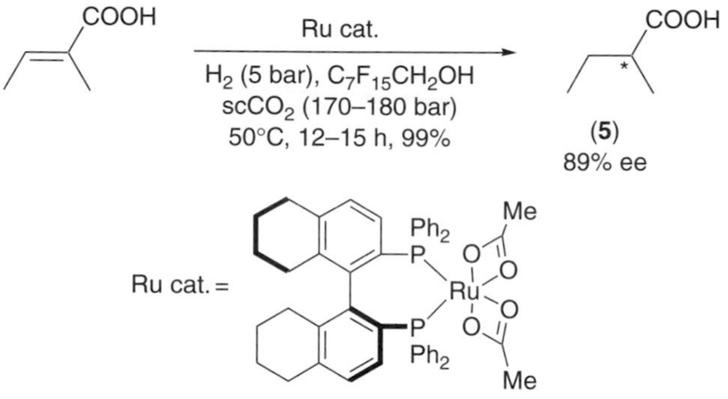

COOH

Ru cat.

H_2 (5 bar), $C_7F_{15}CH_2OH$
$scCO_2$ (170–180 bar)
50°C, 12–15 h, 99%

COOH

(**5**)
89% ee

Ru cat. =

Ph2 P Ru P Ph2 O O O O Me Me

Scheme 8

N Ph

Ph Me

Ir cat.

H_2 (30 bar), $scCO_2$ (d 0.75 g/mL)
40°C, 20 h, conv. 100%

N Ph

Ph Me

(*R*-**6**)
81% ee

Ir cat. =

$C_6F_{13}C_2H_4$ P N O Ir $C_6F_{13}C_2H_4$ B CF_3 CF_3 4

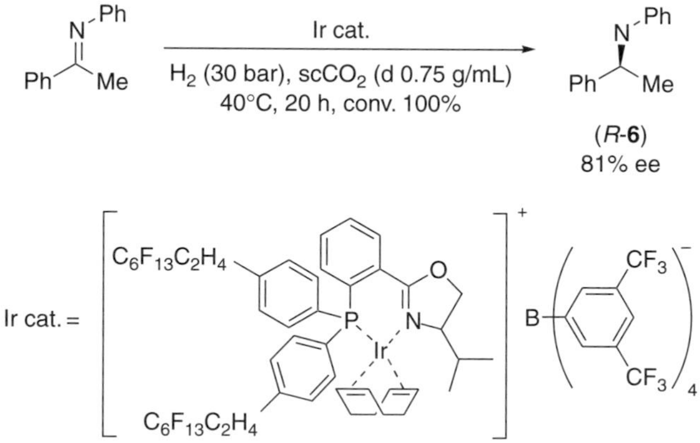

Scheme 9

recycling with almost identical levels of enantioselectivity. The continuous enantioselective hydrogenation of ethyl pyruvate in $scCO_2$ has been demonstrated using a cinchonidine-modified Pt/Al_2O_3 catalyst, although higher conversion and ee are obtained in supercritical ethane than in $scCO_2$.[66]

Metal nanoparticles have been used as highly active catalysts for hydrogenation in $scCO_2$. Polymer-supported colloidal Pd nanoparticles efficiently catalyze the hydrogenation of 1-hexyne in $scCO_2$ with a TOF in excess of $4\,000\,000\,h^{-1}$ at a low hydrogen pressure of 15 bar at 50°C.[67] Pd nanoparticle catalysts, which are stabilized by a water-in-CO_2 microemulsion and are uniformly dispersed in the $scCO_2$, rapidly and efficiently hydrogenate olefins.[68] For example, the hydrogenation of 4-methoxycinnamic acid to 4-methoxyhydrocinnamic acid **7** is completed (>99%) within 20 s at 50°C in $scCO_2$ (Scheme 10). Rh nanoparticle catalysts prepared in a similar manner are also effective for the hydrogenation of naphthalene and phenol in $scCO_2$.[69]

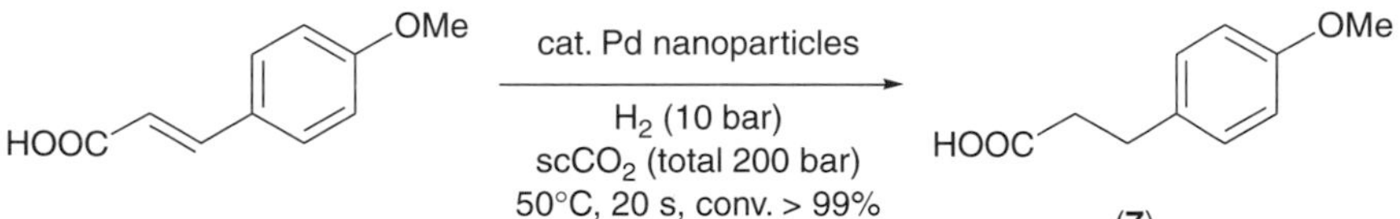

OMe
HOOC
cat. Pd nanoparticles
H_2 (10 bar)
$scCO_2$ (total 200 bar)
50°C, 20 s, conv. > 99%
OMe
HOOC
(**7**)

Scheme 10

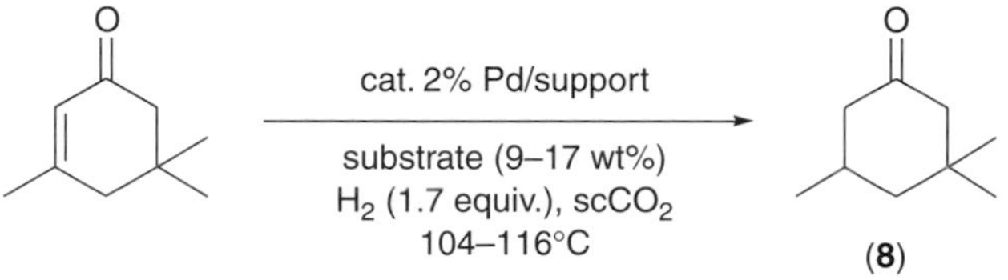

O
cat. 2% Pd/support
substrate (9–17 wt%)
H_2 (1.7 equiv.), $scCO_2$
104–116°C
O
(**8**)
1000 t/y

Scheme 11

Successful heterogeneous catalysis for hydrogenation in $scCO_2$ has been reported. Various organic compounds involving cyclohexene and acetophenone can be continuously hydrogenated with good throughput in $scCO_2$ using a flow reactor and supported Pd or Pt catalysts.[70] It is believed that the role of $scCO_2$ is not merely enhancing the concentration of hydrogen but rather reducing the viscosity and increasing the diffusion rate, which promotes transport to and from the catalytic surface, compared with liquid-phase reactions. Recently, the continuous hydrogenation of isophorone to trimethylcyclohexanone **8** in $scCO_2$ using a supported Pd catalyst was commercialized (Scheme 11).[71] As the reaction in $scCO_2$ is highly selective (i.e. overhydrogenated byproducts such as trimethylcyclohexanol are not formed), the process requires no purification steps after hydrogenation.

The selective hydrogenation of α,β-unsaturated aldehydes such as cinnamaldehyde **9** to unsaturated alcohols was investigated in $scCO_2$ using a Pt/Al_2O_3 catalyst (Scheme 12).[72] As the pressure increased, both the conversion and the selectivity increased. The solvent polarity and electronic state of platinum explained the preferential hydrogenation of the C=O bonds in $scCO_2$. The hydrogenation of dehydroisophytol to isophytol was carried out in $scCO_2$ using a palladium–silicon alloy catalyst and a flow reactor.[73] The catalytic performance was correlated with the phase behavior of the reaction system, and a high conversion and selectivity was obtained in a single-phase system. $ScCO_2$ is effective as a solvent for hydrogenation of low-vapor-pressure compounds.[74] Thus, using a Pd/Al_2O_3 catalyst, maleic anhydride is selectively hydrogenated in $scCO_2$ to γ-butyrolactone **10** in approximately 80% yield, whereas the yield is only 2% in polyethylene glycol or acetone (Scheme 13). The ability of $scCO_2$ to remove deactivating materials such as coke deposits from the surface of a Pd/Al_2O_3 catalyst has been shown for the hydrogenation of cyclododecatriene.[75]

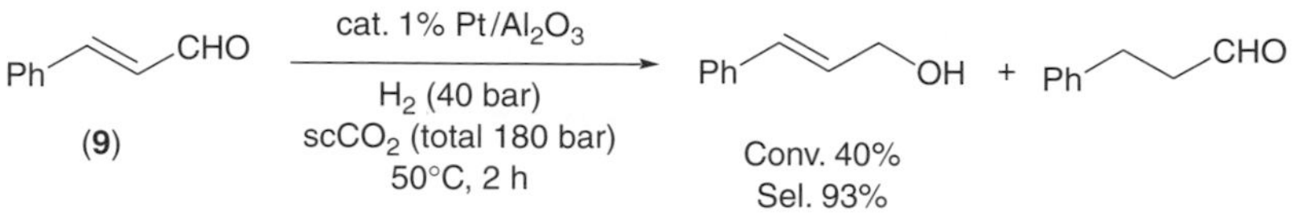

Ph–CH=CH–CHO (**9**) → [cat. 1% Pt/Al_2O_3; H_2 (40 bar), $scCO_2$ (total 180 bar), 50°C, 2 h] → Ph–CH=CH–CH$_2$OH + Ph–CH$_2$CH$_2$–CHO

Conv. 40%
Sel. 93%

Scheme 12

Maleic anhydride → [cat. 1% Pd/Al_2O_3; H_2 (2.1 MPa), $scCO_2$ (12 MPa), 200°C, 2 h, 80%] → (**10**)

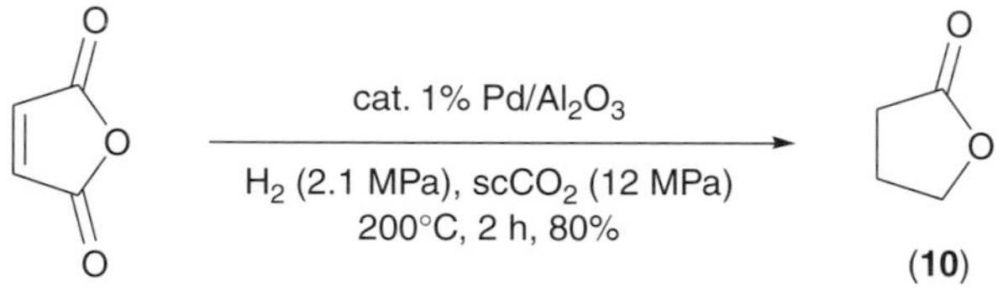

Scheme 13

4.3.4 Oxidation

Dense CO_2 is an ideal reaction medium for oxidation catalysis because its inertness with respect to oxidation offers safety and avoids side-products from solvent oxidation, and its complete miscibility with molecular oxygen provides high concentrations of the oxidant and eliminates mass transfer limitations. Furthermore, the excellent heat transport capacity of $scCO_2$ allows effective heat control in exothermic oxidation reactions. Recently, a review of catalytic oxidations in dense CO_2 has been published.[76]

One of the earliest studies of oxidation reactions in dense CO_2 is the oxidation of toluene with air using a CoO/Al_2O_3 catalyst.[77] In $scCO_2$, toluene was oxidized to primarily benzaldehyde at low rates and conversions. The characteristics of the reaction were similar to those of the catalytically assisted, free-radical, homogeneous oxidations in the liquid phase. The aerobic oxidation of cyclohexane in $scCO_2$ was investigated in the presence of acetaldehyde using iron(III)-5,10,15,20-tetrakis(pentafluorophenyl)porphyrin catalyst **11**.[78] The total yield of cyclohexanol and cyclohexanone was low ($<5\%$), but it depended strongly on the CO_2 pressure and showed a maximum near the critical pressure (60–80 bar).

The oxidation of cyclohexene with dioxygen in $scCO_2$ has also been reported.[79] The iron(III)-halogenated-porphyrin-catalyzed reaction produced epoxide and allylic oxidation products (Scheme 14). The turnover number (TON) reached up to 1530 for cyclohexene in 24 h at 80°C. The activity is lower in $scCO_2$ than in organic solvents, but the selectivity for epoxidation (up to 34%) is higher in $scCO_2$.

The cobalt–salen complex catalyzes the oxidation of substituted phenols such as 2,6-di-*tert*-butyl phenol with dioxygen in $scCO_2$ to produce the oxygenation **12** and coupling **13** products.[80] The reaction was also carried out in a mixed medium consisting of $scCO_2$ and acetonitrile.[81] Compared with the reaction in neat acetonitrile, the presence of CO_2 increases the oxygen solubility roughly 100-fold,

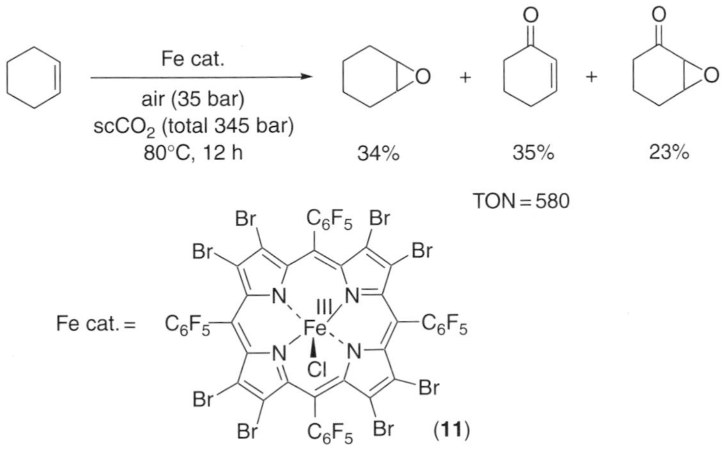

Fe cat.
air (35 bar)
$scCO_2$ (total 345 bar)
80°C, 12 h
34% 35% 23%
TON = 580

Fe cat. = (11)

Scheme 14

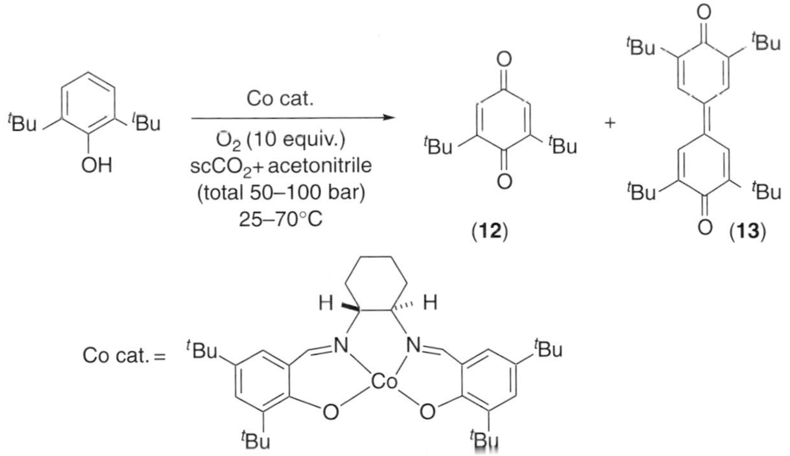

Co cat.
O_2 (10 equiv.)
$scCO_2$+ acetonitrile
(total 50–100 bar)
25–70°C
(12) (13)

Co cat. =

Scheme 15

and acetonitrile solubilizes the catalyst. Thus, the TOFs in the CO_2-expanded acetonitrile (the volume of acetonitrile is doubled) are between one and two orders of magnitude greater than those in $scCO_2$ and a few times greater than that in neat acetonitrile (Scheme 15, Table 4.2).

Some aerobic oxidation reactions progress effectively in $scCO_2$ without a catalyst. Aerobic oxidation of olefins (e.g. *cis*-cyclooctene and (*R*)-(+)-limonene) in the presence of aldehydes (e.g. 2-methyl-propionaldehyde) in $scCO_2$ ($d = 0.75$ g/mL) gave the corresponding epoxides without a catalyst.[82] It is speculated that the stainless steel from autoclave walls triggered the formation of acylperoxy radicals from the aldehyde and oxygen, and the reaction proceeded via a non-catalytic radical

Table 4.2

Solvent	Temperature (°C)	TOF (h^{-1})
$scCO_2$ + acetonitrile	25	3
$scCO_2$ + acetonitrile	50	25
$scCO_2$	50	1
Neat acetonitrile	28	1

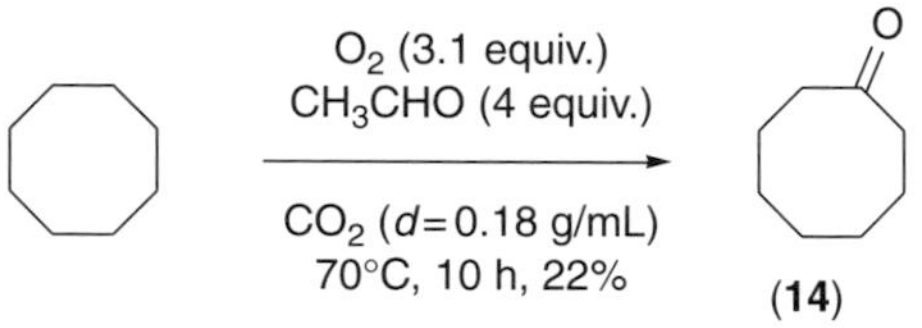

O_2 (3.1 equiv.)
CH_3CHO (4 equiv.)
CO_2 (d = 0.18 g/mL)
70°C, 10 h, 22%
(**14**)

Scheme 16

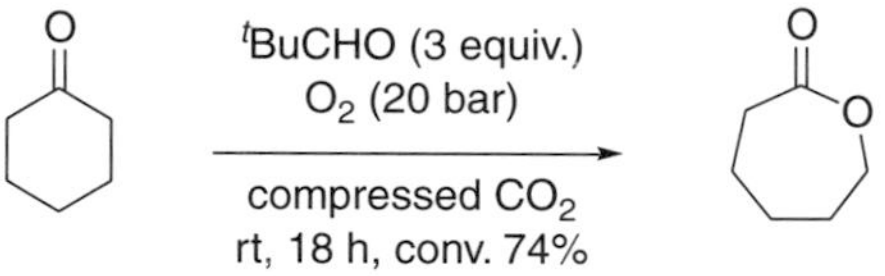

^tBuCHO (3 equiv.)
O_2 (20 bar)
compressed CO_2
rt, 18 h, conv. 74%

Scheme 17

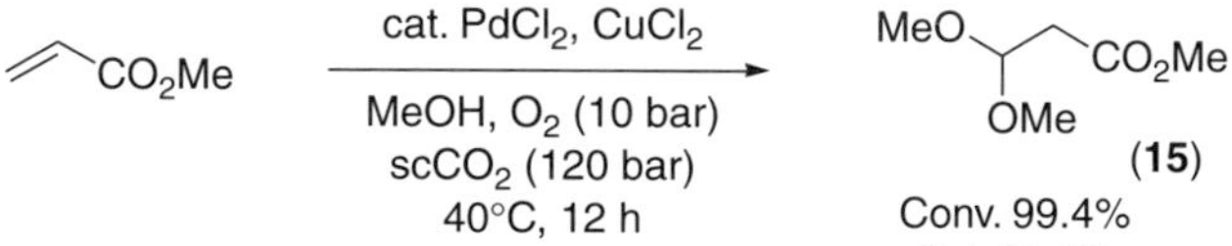

CO_2Me
cat. $PdCl_2$, $CuCl_2$
MeOH, O_2 (10 bar)
$scCO_2$ (120 bar)
40°C, 12 h
MeO, CO_2Me, OMe
(**15**)
Conv. 99.4%
Sel. 96.6%

Scheme 18

pathway. Similarly, the oxidation of cyclooctane with molecular oxygen in the presence of acetaldehyde and compressed CO_2 is promoted by the steel to produce cyclooctanone **14** in 22% yield (Scheme 16).[83]

The aerobic Baeyer–Villiger oxidation of both cyclic and acyclic ketones to the corresponding lactones and esters also proceeds efficiently in the presence of an aldehyde as co-reductant without a catalyst in compressed CO_2 (Scheme 17).[84]

Methyl acrylate is efficiently oxidized to the methyl 3,3-dimethoxypropanoate **15** in the presence of molecular oxygen (10 bar) and excess methanol using $PdCl_2$–$CuCl_2$ catalysts.[85] Excellent conversion (99.4%) and selectivity (96.6%) are reported at 40°C in $scCO_2$ (120 bar) (Scheme 18).

Water-insoluble primary and secondary alcohols are oxidized to the corresponding carbonyl compounds in $scCO_2$ with high reaction rates and yields using

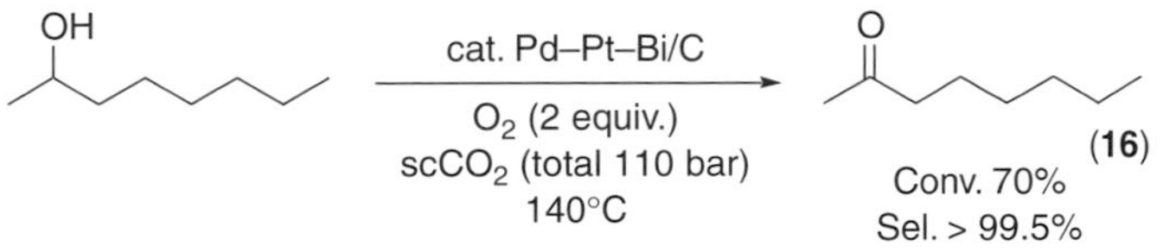

OH

cat. Pd–Pt–Bi/C

O_2 (2 equiv.)
$scCO_2$ (total 110 bar)
140°C

O

(**16**)

Conv. 70%
Sel. > 99.5%

Scheme 19

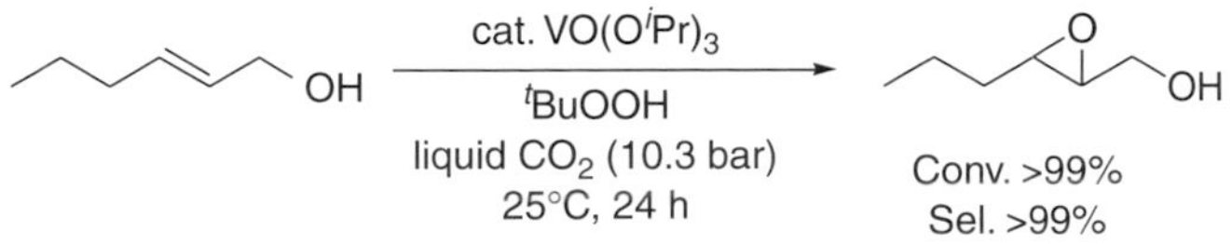

cat. $VO(O^iPr)_3$

OH

^tBuOOH
liquid CO_2 (10.3 bar)
25°C, 24 h

O

OH

Conv. >99%
Sel. >99%

Scheme 20

a solid catalyst and a continuous fixed-bed reactor.[86] For example, the oxidation of 2-octanol with oxygen in $scCO_2$ (110 bar) over a Pd–Pt–Bi/C catalyst gives 2-octanone **16** with a high selectivity (>99.5%) at approximately 70% conversion and a short residence time (17 s) (Scheme 19). The oxidation of benzyl alcohols in $scCO_2$ by Pt/C and Pd/C catalysts is reported to give high yields of aldehydes.[87] In terms of catalytic activity and life, using $scCO_2$ is advantageous.

Selective oxidation using peroxides as oxidants in liquid CO_2 or $scCO_2$ has also been demonstrated. Various allylic alcohols are oxidized with high conversions (>99%) and selectivities (85–100%) to the corresponding epoxides using *t*-BuOOH and $VO(O^iPr)_3$ as a catalyst in liquid CO_2 (10.3 bar) at 25°C (Scheme 20).[88,89] The reaction rates in liquid CO_2 are lower than those in CH_2Cl_2, but comparable to those in acetonitrile and toluene. For the $VO(acac)_2$-catalyzed epoxidation, the reactivity is enhanced using fluorinated acac-type ligands such as 1,1,1-trifluoro-2,4-pentanedione because the catalyst solubility is improved.

Using *t*-BuOOH and $Mo(CO)_6$, cyclohexene and cyclooctene are oxidized in quantitative yields to 1,2-cyclohexanediol and cyclooctane oxide, respectively, in $scCO_2$ at 95°C.[90] A vanadyl complex with a salen-type ligand is effective for the diastereoselective epoxidation of allylic alcohols with *t*-BuOOH in the presence of geraniol in $scCO_2$ (Scheme 21).[91] The epoxidation of cyclohexene proceeded in an $scCO_2$ and aqueous H_2O_2 system by adding $NaHCO_3$ and DMF as a co-solvent.[92] The process did not require a metallic catalyst. It was suggested that the reaction occurred within the aqueous phase, and that peroxycarbonic acid ($HOCO_3H$) – which was formed *in situ* – acted as an epoxidizing agent.

The sulfoxidation of chiral sulfides derived from cysteine proceeded efficiently at 40°C in $scCO_2$ using *t*-BuOOH with an Amberlyst™ 15 ion exchange resin catalyst.[93] Varying the pressure in $scCO_2$ dramatically enhanced the diastereoselectivity. The maximum selectivity (>95% de) was obtained at ~180 bar, whereas in toluene or CH_2Cl_2 no appreciable diastereoselectivity was observed (Scheme 22).

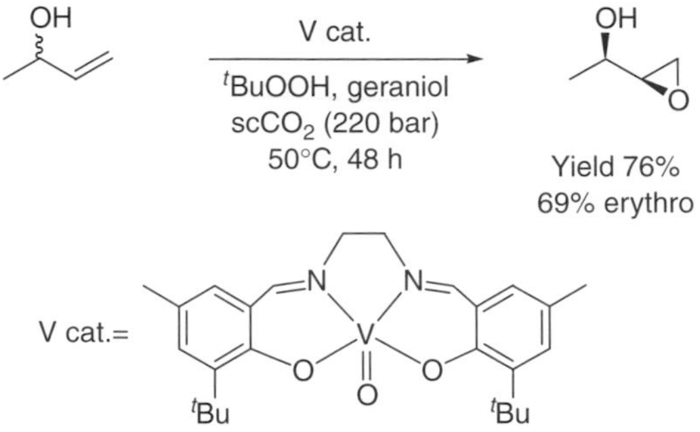

OH

V cat.

^{t}BuOOH, geraniol
$scCO_2$ (220 bar)
50°C, 48 h

OH

O

Yield 76%
69% erythro

V cat.=

N N V O O O ^{t}Bu ^{t}Bu

Scheme 21

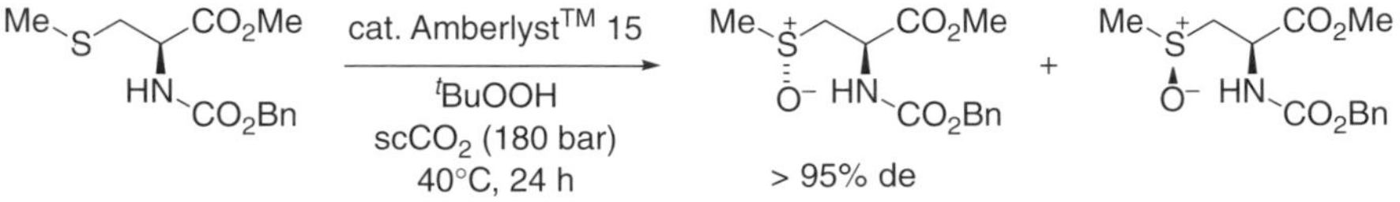

Me S CO_2Me HN CO_2Bn

cat. Amberlyst™ 15
^{t}BuOOH
$scCO_2$ (180 bar)
40°C, 24 h

Me S^+ O^- CO_2Me HN CO_2Bn
> 95% de

+

Me S^+ O^- CO_2Me HN CO_2Bn

Scheme 22

PhI + CO_2Me

cat. $Pd(OAc)_2$
$PPh(C_2H_4C_6F_{13})_2$
Et_3N, $scCO_2$
100°C, 64 h, 91%

Ph CO_2Me
(**17**)

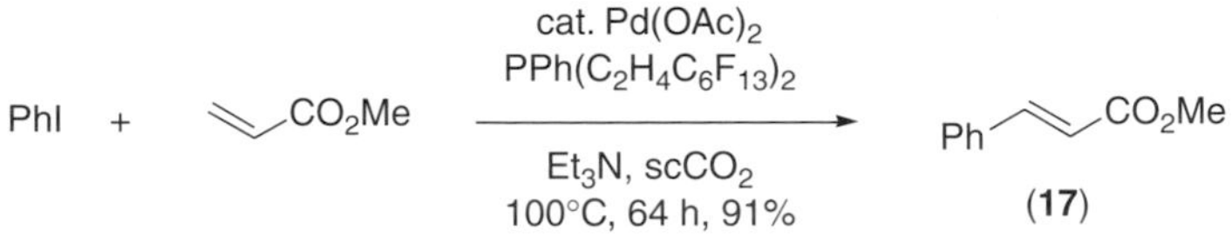

Scheme 23

4.3.5 Complex-catalyzed carbon–carbon bond formation

A variety of carbon–carbon bond-forming reactions have been conducted in $scCO_2$. In 1998, two groups independently reported palladium-catalyzed C–C coupling reactions in $scCO_2$ using fluorinated phosphine ligands.[94,95] $Pd(OAc)_2$ with fluorinated phosphines, $PPh_{(3-n)}(C_2H_4C_6F_{13})_n$ ($n = 1$ and 2), catalyzes various intermolecular and intramolecular Mizoroki–Heck reactions, and Suzuki–Miyaura and Sonogashira couplings in $scCO_2$.[94] For example, the Mizoroki–Heck reaction between iodobenzene and methyl acrylate in $scCO_2$ gives methyl cinnamate **17** in a superior yield (91%) to that reported for conventional solvents (Scheme 23).

For the Stille coupling of iodobenzene with vinyl(tributyl)tin in $scCO_2$, a quantitative conversion (>99%) is achieved using $Pd_2(dba)_3$ with a $P[3,5\text{-}(CF_3)_2C_6H_3]_3$ ligand, which is much higher than that achieved with PPh_3 and comparable to that achieved with tri-2-furylphosphine, a ligand noted for high palladium-coupling activity (Scheme 24).[95] It has been suggested that the high conversion of the fluorinated ligand arises from the enhanced solubility of the palladium complex in

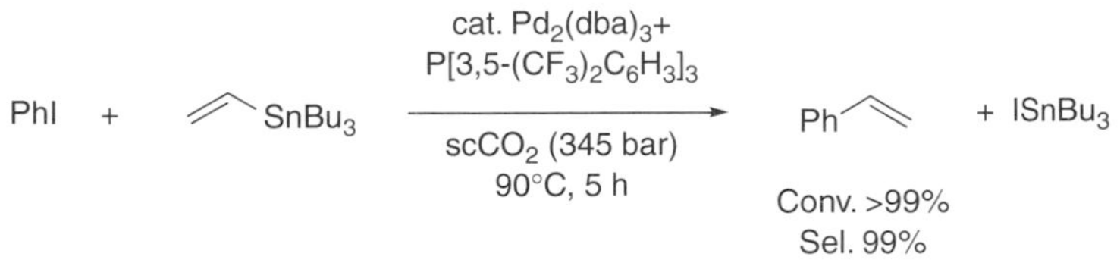

PhI + SnBu$_3$ → (cat. Pd$_2$(dba)$_3$+ P[3,5-(CF$_3$)$_2$C$_6$H$_3$]$_3$; scCO$_2$ (345 bar), 90°C, 5 h) Ph + ISnBu$_3$

Conv. >99%
Sel. 99%

Scheme 24

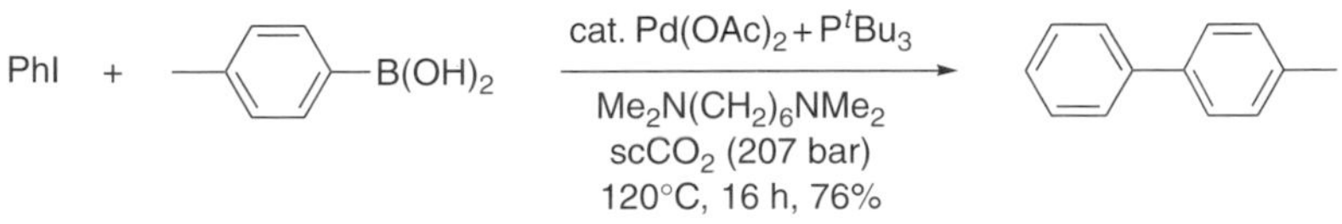

PhI + B(OH)$_2$ → (cat. Pd(OAc)$_2$ + P^{t}Bu$_3$; Me$_2$N(CH$_2$)$_6$NMe$_2$, scCO$_2$ (207 bar), 120°C, 16 h, 76%)

Scheme 25

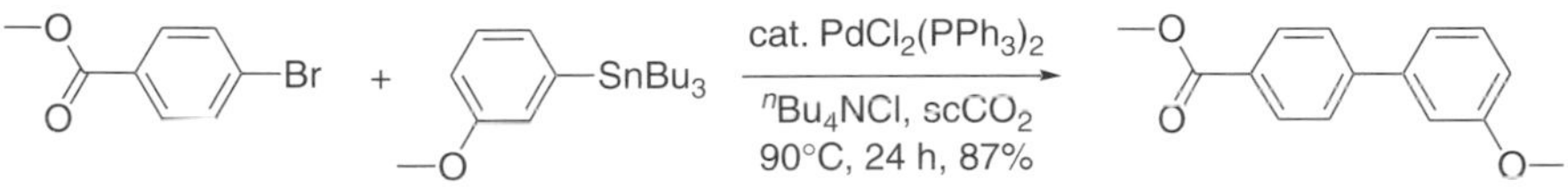

Br + SnBu$_3$ → (cat. PdCl$_2$(PPh$_3$)$_2$; ^{n}Bu$_4$NCl, scCO$_2$, 90°C, 24 h, 87%)

Scheme 26

scCO$_2$. A fluorinated phosphine ligand such as PPh(C$_6$F$_5$)$_2$ is also effective for the palladium-catalyzed Heck reaction in scCO$_2$.[96]

The choice of initial palladium precursors is also important for promoting coupling reactions in scCO$_2$. For the Mizoroki–Heck reaction of iodobenzene and methyl acrylate, for example, the use of fluorinated palladium sources such as Pd(OCOCF$_3$)$_2$ and Pd(hfacac)$_2$, instead of Pd(OAc)$_2$ and Pd$_2$(dba)$_3$, increases the reaction efficiency and allows the use of PCy$_3$ or PBu$_3$, which are usually regarded as poor ligands.[97] However, completely non-fluorous palladium catalytic systems have also been developed for coupling reactions in scCO$_2$. The combination of Pd(OAc)$_2$ and P(t-Bu)$_3$ is highly efficient for promoting both Mizoroki–Heck and Suzuki–Miyaura reactions in scCO$_2$ (Scheme 25).[98] Even PdCl$_2$(PPh$_3$)$_2$ effectively catalyzes the Stille coupling reactions in scCO$_2$, although the yields are slightly lower than those with perfluoro-tagged triphenylphosphine ligands (Scheme 26).[99]

Recently, metal carbene complexes that catalyze alkene metathesis to form new C=C double bonds have received a great deal of attention. A facile alkene metathesis reaction in scCO$_2$ has been reported (Scheme 27).[100] Both ring-closing and ring-opening reactions proceeded efficiently. For example, molybdenum–carbene complexes with perfluoroalkoxy ligands and ruthenium–carbene complexes with tricyclohexyl phosphine ligands work as efficient catalysts – both of them displaying similar catalytic activities in scCO$_2$ to the reaction in chlorinated solvents.

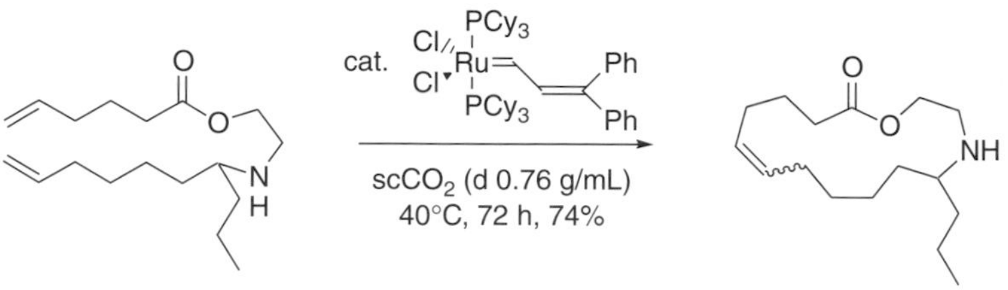

cat. $Cl_2(PCy_3)_2Ru{=}CH{-}CH{=}CPh_2$

$scCO_2$ (d 0.76 g/mL)
40°C, 72 h, 74%

Scheme 27

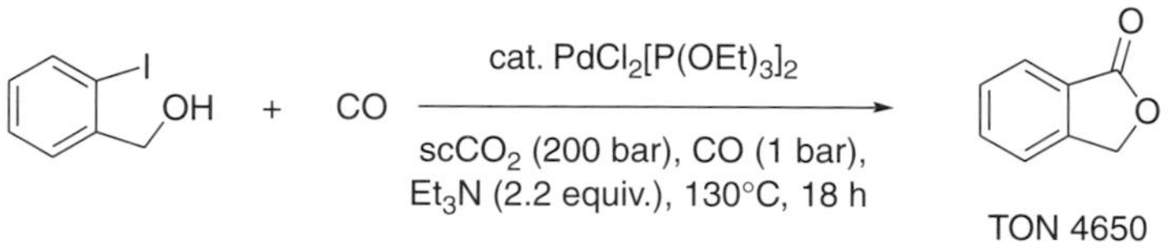

cat. $PdCl_2[P(OEt)_3]_2$

$scCO_2$ (200 bar), CO (1 bar),
Et_3N (2.2 equiv.), 130°C, 18 h

TON 4650

Scheme 28

The molybdenum complexes probably work as homogeneous catalysts, whereas ruthenium complexes seem to work as heterogeneous ones. The reaction in $scCO_2$ is different from that in organic solvents because it can be applied without difficulty to substrates with amino groups. It is presumed that the amino group is transformed to a carbamic acid moiety in $scCO_2$, with CO_2 acting as a protecting group. The metathesis of α,ω-diene gave macrocyclic compounds (intramolecular reaction) and oligomeric products (intermolecular compounds). The selectivity for the macrocycle increases in high density CO_2 (>0.7 g/mL).

Palladium complexes with phosphite ligands exhibit high catalytic activity for the intramolecular esterification of aromatic halides in $scCO_2$ (Scheme 28)[101], which is noteworthy because non-fluorous CO_2-philic ligands are used. Since $scCO_2$ is miscible with gaseous compounds, the carbonylation in $scCO_2$ proceeds remarkably faster than that in toluene, especially at a low CO pressure (e.g. 1 bar).

The addition of electrophilic alkenes to aldehydes in the presence of strong bases occurs under $scCO_2$ conditions (Morita–Baylis–Hillman reaction (Scheme 29).[102] The reaction proceeds faster in $scCO_2$ than in organic solvents. At relatively low CO_2 pressures (~80–100 bar), further intermolecular dehydration (dimerization) of the product leads to ethers **18**. Unsymmetrical ethers are synthesized using another alcohol in the etherification step.

Other C–C bond-forming reactions have been successfully developed using $scCO_2$ and liquid CO_2 as reaction media. Examples include the synthesis of cyclopentenones via cobalt-catalyzed cocyclizations of alkynes with alkenes and carbon monoxide (Pauson–Khand reaction) (Scheme 30)[103], enantioselective nickel-catalyzed hydrovinylation of styrenes (Scheme 31)[104], and the palladium-catalyzed hydroarylation of acyclic β-substituted-α,β-enones with aryl iodides (formal conjugate addition) (Scheme 32).[105]

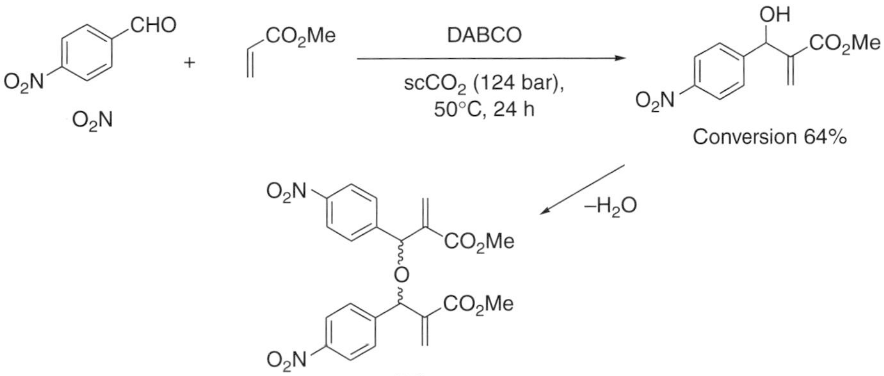

CHO
O_2N + CO_2Me
DABCO
$scCO_2$ (124 bar), 50°C, 24 h
OH
CO_2Me
O_2N
Conversion 64%
$-H_2O$
O_2N
CO_2Me
O
CO_2Me
O_2N
(**18**)

Scheme 29

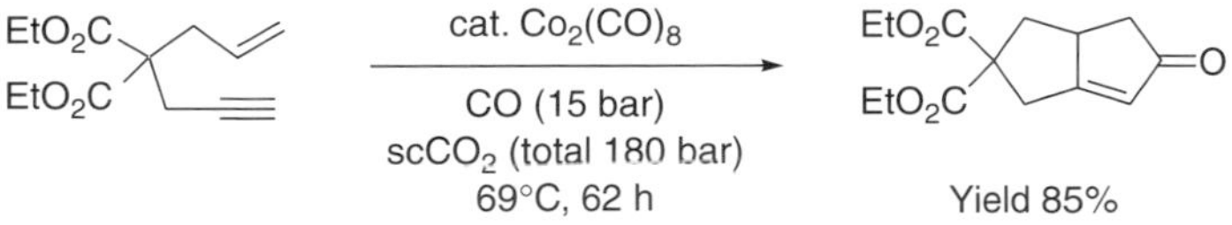

EtO_2C
EtO_2C
cat. $Co_2(CO)_8$
CO (15 bar)
$scCO_2$ (total 180 bar)
69°C, 62 h
EtO_2C
EtO_2C
O
Yield 85%

Scheme 30

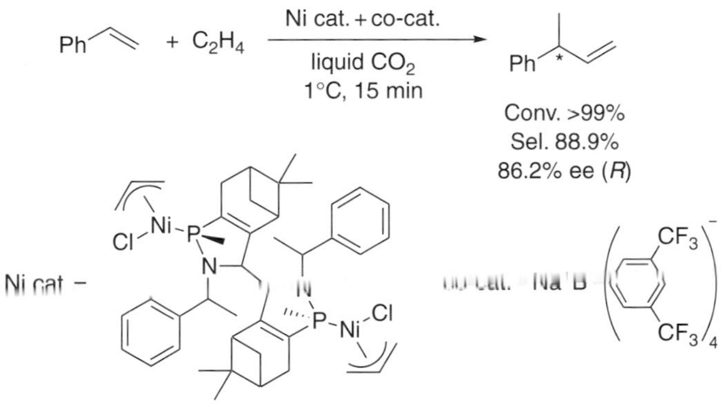

Ph + C_2H_4
Ni cat. + co-cat.
liquid CO_2
1°C, 15 min
Ph *
Conv. >99%
Sel. 88.9%
86.2% ee (*R*)

Ni cat =
Cl Ni P N
N P Ni Cl
co-cat. = Na^+B^- CF_3 CF_3 4

Scheme 31

4.3.6 Diels–Alder reactions

Stoichiometric and catalytic Diels–Alder reactions in $scCO_2$ have been studied extensively. The first report appeared in 1987.[43] The reaction of maleic anhydride and isoprene was conducted in $scCO_2$, and the effect of CO_2 pressure (80–430 bar) on the reaction rate was investigated (Scheme 33). A large increase in the rate was observed near the critical pressure at 35°C. At pressures of 200 bar and above, the

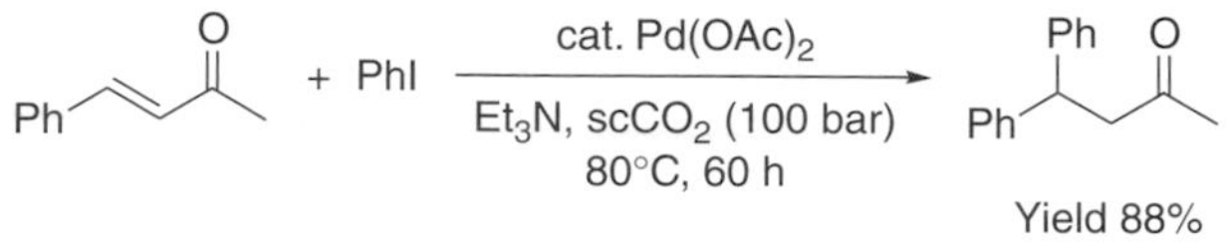

+ PhI
cat. $Pd(OAc)_2$
Et_3N, $scCO_2$ (100 bar)
80°C, 60 h
Yield 88%

Scheme 32

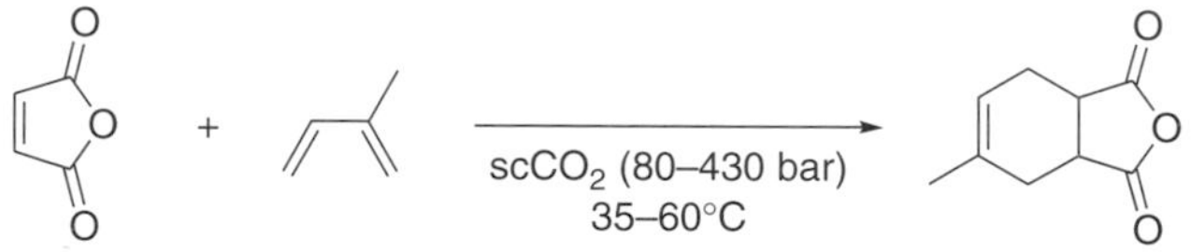

$scCO_2$ (80–430 bar)
35–60°C

Scheme 33

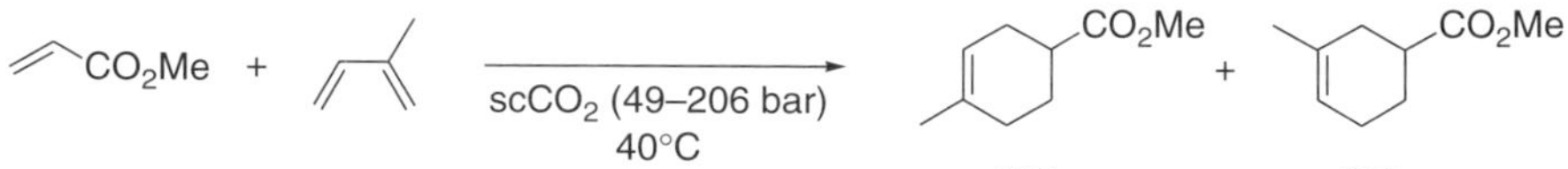

CO_2Me +
$scCO_2$ (49–206 bar)
40°C
(**19**) (**20**)

Scheme 34

rate was similar to that in ethyl acetate. On the other hand, it was reported that the reaction between cyclopentadiene and *p*-benzoquinone in CO_2 occurs throughout the liquid and supercritical ranges (25–40°C), without a discontinuity, at a constant density (0.8 g/mL) and is ~20% faster than the reaction in diethyl ether.[106]

For the Diels–Alder reaction of methyl acrylate with isoprene in $scCO_2$, the effect of CO_2 pressure on the distribution between 4-methyl-3-cyclohexene-1-carboxylate **19** and the corresponding 3-methyl **20** was investigated (Scheme 34).[107] The selectivity of 4-methyl increased from 67 to 86% as the CO_2 pressure increased from 49 to 206 bar, but an anomalous selectivity (39%) was obtained around the critical pressure, 74.5 bar. It was suggested that there were steric constraints of the reacting species caused by aggregation of the solvent molecules.[108] Such a dramatic reversal in regioselectivity is not observed, however, for several other Diels–Alder reactions, including the methyl acrylate–isoprene reaction.[109]

Mechanistic and kinetic studies have been made of Diels–Alder reactions in $scCO_2$. For the reaction of maleic anhydride and isoprene in the presence of aluminum chloride in $scCO_2$, a two-step mechanism in which a σ-bond is formed followed by cyclization was suggested.[110] For the reaction of ethyl acrylate and cyclopentadiene in $scCO_2$, good correlation was found between the density and the temperature-normalized rate constant.[111]

In the Diels–Alder reaction between cyclopentadiene and methyl acrylate in $scCO_2$, the ratio of *endo*-products to *exo*-products is maximal at a density higher than the critical density.[112] The effect is thought to be associated with interactions

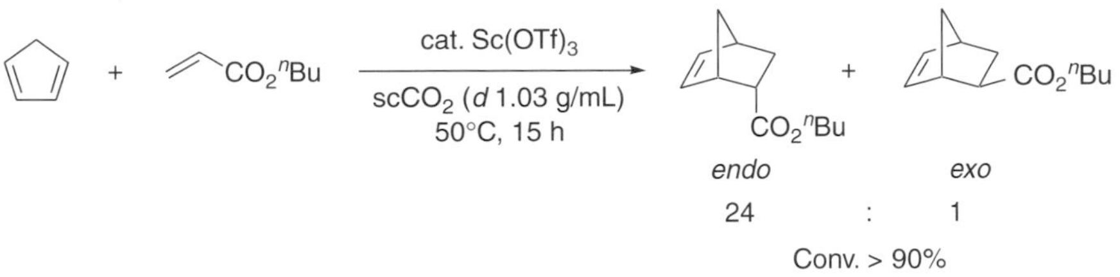

Scheme 35

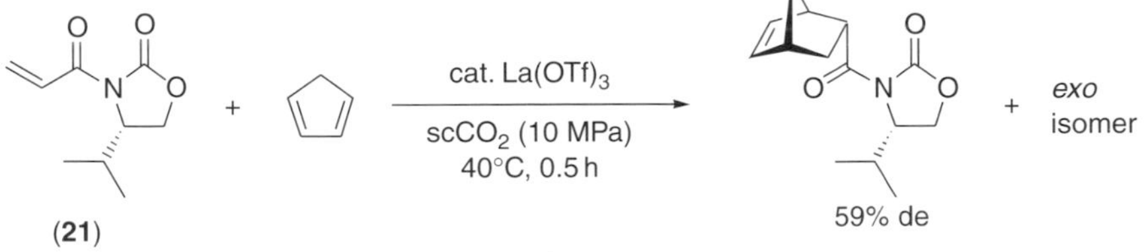

Scheme 36

between the solvent molecules and the transition states. The *endo* : *exo* stereoselectivity for similar reactions using a Lewis acid catalyst can also be enhanced by optimizing the CO_2 density.[113] For example, an *endo* : *exo* ratio of 24 : 1 for the reaction between *n*-butyl acrylate and cyclopentadiene catalyzed by scandium triflate is achieved with a CO_2 density of approximately 1.03 g/mL, whereas those in toluene and $CHCl_3$ were 10 : 1 and 11 : 1, respectively (Scheme 35). Diels–Alder reactions and aza-Diels–Alder reactions in $scCO_2$ using scandium perfluoroalkanesulfonates such as $Sc(OSO_2C_8F_{17})_3$ as a Lewis acid catalyst have also been successful.[114]

A diastereoselective Diels–Alder reaction of cyclopentadiene with a chiral dienophile, *N*,*N′*-fumaroyldi[(2*R*)-bornane-10,2-sultam], was carried out in $scCO_2$.[115] The highest diastereoselectivity (65% conversion and 93% de) is obtained around the critical point (74 bar) at 33°C. Lanthanide triflate catalyzes the reaction of cyclopentadiene and 3-acryloyl-(4*S*)-isopropyloxazolidin-2-one **21** in $scCO_2$ to give the *endo* adduct (*endo* : *exo* = 70 : 30) with a higher diastereoselectivity (59% de) than in CH_2Cl_2 (38–42% de) (Scheme 36).[116]

Amorphous fumed silica promotes several Diels–Alder reactions in gaseous CO_2 and $scCO_2$.[117] For the reaction of methyl vinyl ketone and penta-1,3-diene at 80°C, the yield decreases as the CO_2 pressure increases because the concentration of the reactant adsorbed on the silica surface decreases.

4.3.7 Acid catalysis

Continuous Friedel–Crafts alkylation in $scCO_2$ has been demonstrated using a fixed-bed flow reactor and Deloxan®, a polysiloxane-based solid acid catalyst.[118]

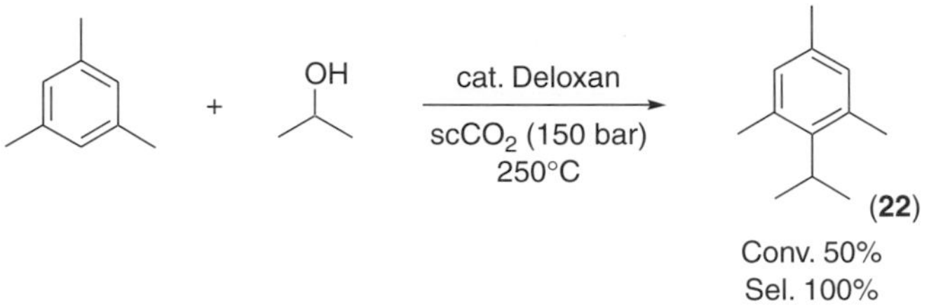

OH + cat. Deloxan, $scCO_2$ (150 bar), 250°C → (22)

Conv. 50%
Sel. 100%

Scheme 37

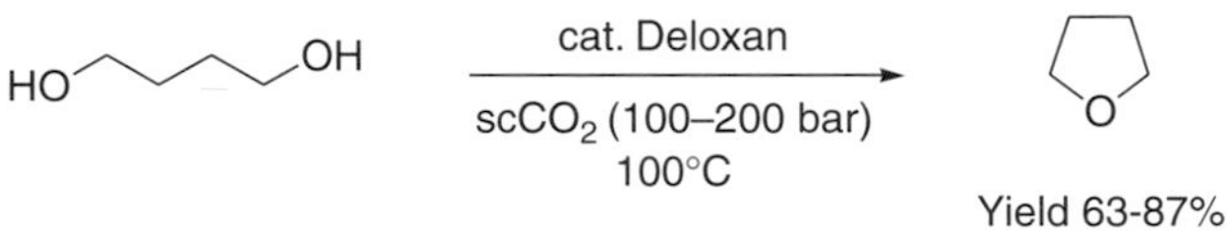

HO–(CH2)4–OH; cat. Deloxan, $scCO_2$ (100–200 bar), 100°C → O

Yield 63-87%

Scheme 38

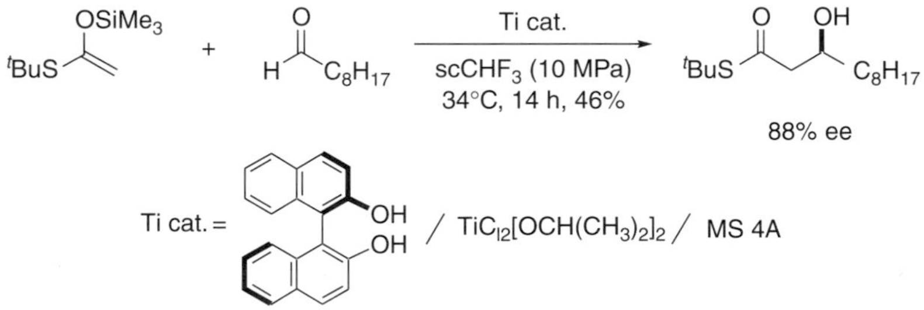

OSiMe3, ^tBuS + O, H, C_8H_{17}; Ti cat., $scCHF_3$ (10 MPa), 34°C, 14 h, 46% → ^tBuS, O, OH, C_8H_{17}

88% ee

Ti cat. = OH, OH / $TiCl_2[OCH(CH_3)_2]_2$ / MS 4A

Scheme 39

Alkylation of mesitylene by 2-propanol gives 100% selectivity for 1-isopropyl-2,4,6-trimethylbenzene **22** at a conversion of 50% (Scheme 37). Compared with liquid-phase or gas-phase reactions, $scCO_2$ reduces the mass transport restrictions at the catalytic surface, which increases the residence time of the substrate and reduces coking of the catalysts. This reaction method is also successful in the formation of cyclic ethers, acetals, linear alkyl ethers and aryl ethers.[119] For example, when synthesizing THF from 1,4-butanediol in $scCO_2$ at 100°C, increasing the pressure from 100 to 200 bar increases the yield from 63 to 87% (Scheme 38).

A binaphthol-derived chiral titanium(IV) complex effectively catalyzes the Mukaiyama aldol reaction in $scCHF_3$ or $scCO_2$.[120] It was found that the chemical yield and enantioselectivity of the reaction in supercritical fluids could be tuned by changing the supercritical fluids ($scCHF_3$ vs. $scCO_2$) and adjusting the matched polarities by varying the pressure of CHF_3 (Scheme 39).

The addition of poly(ethylene glycol) (PEG) is effective for Lewis acid-catalyzed reactions in $scCO_2$.[121] PEG works as a surfactant in $scCO_2$ to form emulsions in which the catalyst and substrates are packed, thus leading to an acceleration of the reactions. The CO_2–PEG system has been successfully applied to an

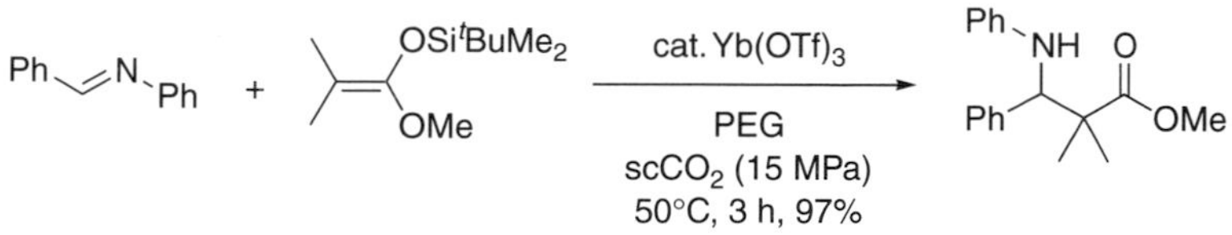

Ph–CH=N–Ph + $(CH_3)_2C=C(OSi^tBuMe_2)(OMe)$ → (cat. Yb(OTf)$_3$; PEG; $scCO_2$ (15 MPa); 50°C, 3 h, 97%) → PhNH–CH(Ph)–C(CH$_3$)$_2$–CO–OMe

Scheme 40

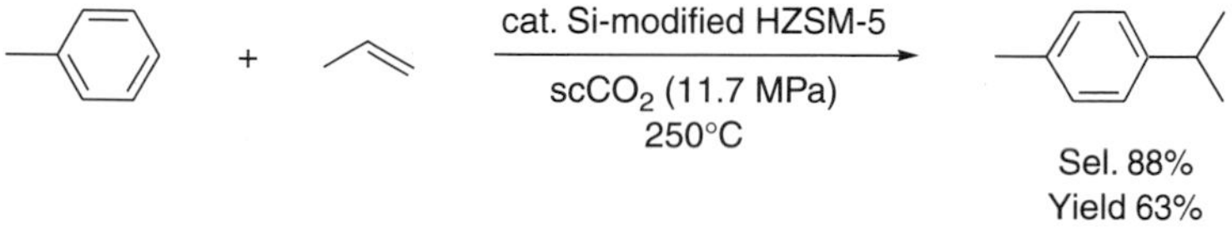

toluene + propylene → (cat. Si-modified HZSM-5; $scCO_2$ (11.7 MPa); 250°C) → cymene

Sel. 88%
Yield 63%

Scheme 41

ytterbium-catalyzed Mannich reaction of silyl enolates with imines and scandium-catalyzed aldol reactions of silyl enolates with aldehydes (Scheme 40). Lewis acid-catalyzed Diels–Alder reactions in $scCO_2$ have been described in Section 4.3.6.

The alkylation of isobutane with 1-butene was carried out over an ultrastable Y-type zeolite (USY) catalyst in $scCO_2$.[122] Using supercritical CO_2 rather than isobutene/1-butene as the reaction medium allowed the reaction temperature to be lowered. As a result, the C_8 alkylate selectivity increased. The effectiveness of $scCO_2$ in terms of alkylate selectivity and coke precursor removal has also been reported.[123] On the other hand, for the USY-catalyzed alkylation of isobutane with *trans*-2-butene at high levels of conversion (100%), adding $scCO_2$ decreased the catalytic longevity and product selectivity.[124] The alkylation of toluene with propylene over silicon-modified HZSM-5 zeolite using $scCO_2$ increased the yield of cymene and reduced the cracking of propylene compared with the reaction under atmospheric pressure (Scheme 41).[125]

4.3.8 Polymer synthesis

In 1992, it was demonstrated that $scCO_2$ works as an alternative solvent to chlorofluorocarbons (CFCs) for the homogeneous free-radical polymerization of highly fluorinated monomers.[45] The homogeneous polymerization of 1,1-dihydroperfluorooctyl acrylate **23** using azobisisobutyronitrile (AIBN) in $scCO_2$ (59.4°C, 207 bar) gave perfluoropolymer in a 65% yield with a molecular weight of 270 000 (Scheme 42). In 1994, the first example of free-radical dispersion polymerization using amphiphilic polymers as a stabilizer in $scCO_2$ was reported.[37] $ScCO_2$ has also been successfully employed in cationic polymerizations. One example is the polymerization of isobutyl vinyl ether (IBVE) **24** using an adduct of acetic acid and IBVE as the initiator, ethylaluminum dichloride as a Lewis acid and ethyl acetate

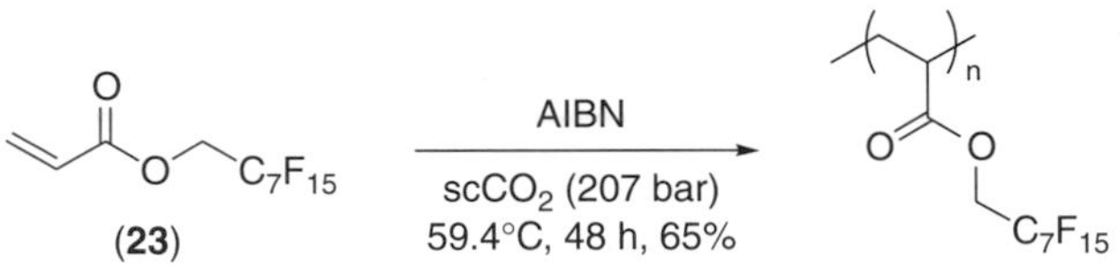

(**23**)

AIBN

$scCO_2$ (207 bar)
59.4°C, 48 h, 65%

C_7F_{15}

Scheme 42

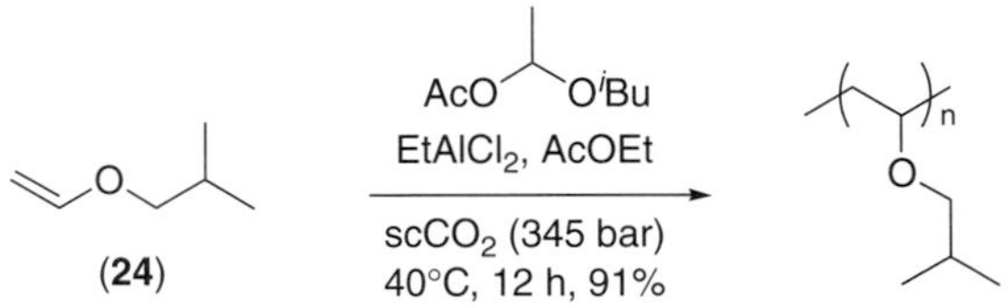

AcO O^{i}Bu
$EtAlCl_2$, AcOEt

$scCO_2$ (345 bar)
40°C, 12 h, 91%

(**24**)

Scheme 43

cat. $Cl_2(PCy_3)_2Ru{=}CHPh$

$scCO_2$ (92 bar)
46°C, 1 h, 97%

(**25**)

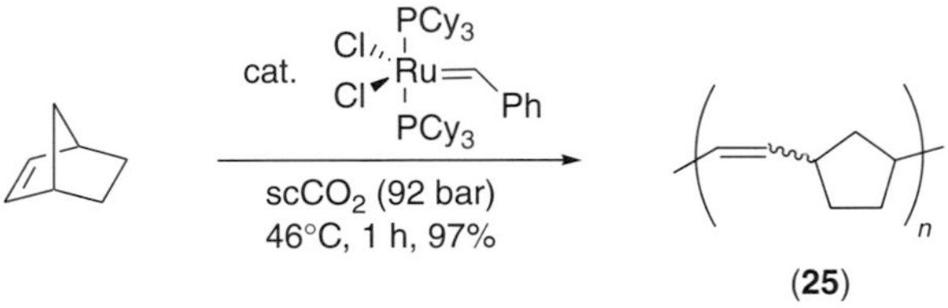

Scheme 44

as a Lewis base deactivator. The reaction proceeded via a heterogeneous precipitation process in $scCO_2$ (40°C, 345 bar) to form poly(IBVE) in a 91% yield with a molecular weight distribution of 1.8 (Scheme 43).[126] There are several reviews covering the history and recent developments of homogeneous and heterogeneous polymerizations in $scCO_2$.[127–129]

This section focuses on transition-metal-catalyzed polymerizations in $scCO_2$, which are related to the other homogeneous catalysis reactions in $scCO_2$. A variety of metal-catalyzed polymerizations in $scCO_2$ have been demonstrated.

$Ru(H_2O)_6(O_3SC_6H_4\text{-}p\text{-}Me)_2$ catalyzes a ring-opening metathesis polymerization of norbornene in $scCO_2$ (60–350 bar).[130,131] Although the catalyst is insoluble in $scCO_2$, adding methanol (up to 16 wt%) solubilizes the catalyst and increases the yield of polynorbornene **25**. The resulting polymers have a higher *trans*-vinylene content (>70%) than that achieved in pure $scCO_2$ (17%), suggesting the favorable formation of the *trans*-propagating species at the metal center in methanol/$scCO_2$. The polymerization of norbornene in $scCO_2$ is more efficiently catalyzed by a ruthenium carbene complex.[100] Using a small amount of catalyst (a monomer : catalyst ratio of 625), polynorbornene is produced in a 97% yield within 1 h at 46°C in $scCO_2$

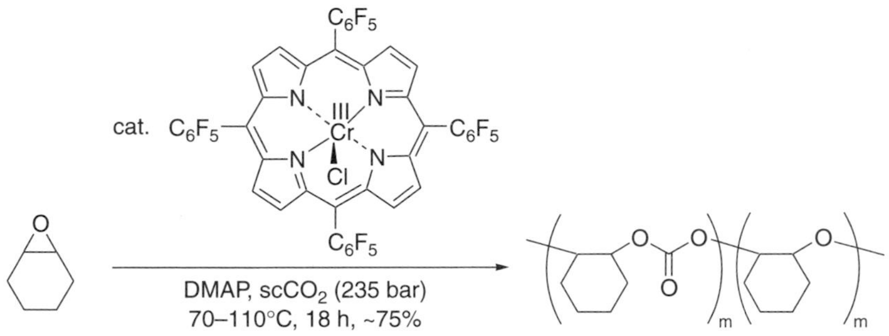

DMAP, $scCO_2$ (235 bar)
70–110°C, 18 h, ~75%

(**26**)

Scheme 45

(92 bar) (Scheme 44). The molecular weight and microstructure of the polymer obtained in $scCO_2$ are comparable to those for polymers prepared in CH_2Cl_2.

Copolymerization of CO_2 and epoxides has been investigated in compressed CO_2. The copolymerization of CO_2 with propylene oxide was attempted using a heterogeneous zinc glutarate catalyst.[132] The use of $scCO_2$ (>77 bar) results in a high proportion of polycarbonate linkage (84–98%) in the polymer, which contains both polycarbonate and polypropylene oxide linkages, but propylene carbonate is formed as a byproduct in 8–14%. A CO_2-soluble zinc fluoroalkyl catalyst copolymerizes cyclohexene oxide and CO_2 in $scCO_2$.[133] Polymers containing >90% polycarbonate linkages with molecular weights (Mw) ranging from 50 000 to 180 000 are obtained. The copolymerization of cyclohexene oxide and CO_2 is efficiently catalyzed by a chromium(III)-fluorinated porphyrin complex in the presence of (dimethylamino)pyridine (DMAP) in $scCO_2$ to produce polycarbonate **26** in good yields (up to 75%) (Scheme 45).[134] The copolymers have a high degree of incorporated CO_2 (90–97%) and narrow molecular weight distributions (Mw/Mn < 1.4).

Oxidative coupling polymerization in $scCO_2$ has been demonstrated for the synthesis of poly(2,6-dimethylphenylene oxide) **27**.[135] The polymerization of 2,6-dimethylphenol in the presence of oxygen is conducted using CuBr as the catalyst, pyridine and a block copolymer of styrene and 1,1-dihydroperfluorooctyl acrylate (PS-*b*-PFOA) as a stabilizer in $scCO_2$ (350 bar) at 40°C. The polymerization occurred via a dispersion process and produced the polymer in a 74% yield with high molecular weight (17 000) (Scheme 46).

The rhodium-catalyzed polymerization of phenylacetylene in $scCO_2$ has been reported using [Rh(acac)(nbd)] (nbd = norbornadiene) and quinuclidine as a base and yielded 70% polyphenylacetylene with a molecular weight of 42 000 (Mw/Mn = 9.5) in $scCO_2$ (132 bar) (Scheme 47).[136] Although the catalyst was insoluble in $scCO_2$, the rate was much faster than that in THF or hexane. The preferential formation of a *cis*-cisoid polymer **28** (*cis*-cisoid : *cis*-transoid ratio 81 : 19) in $scCO_2$ contrasts with polymerization in THF (ratio 1 : 99). The use of $scCO_2$ as a solvent for metal-catalyzed polymerization has also been reported for the palladium-catalyzed

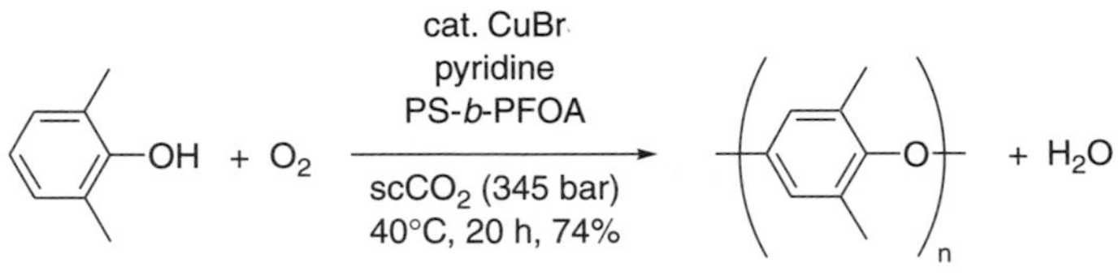

cat. CuBr
pyridine
PS-*b*-PFOA
—OH + O_2 → + H_2O
$scCO_2$ (345 bar)
40°C, 20 h, 74%
(**27**)

Scheme 46

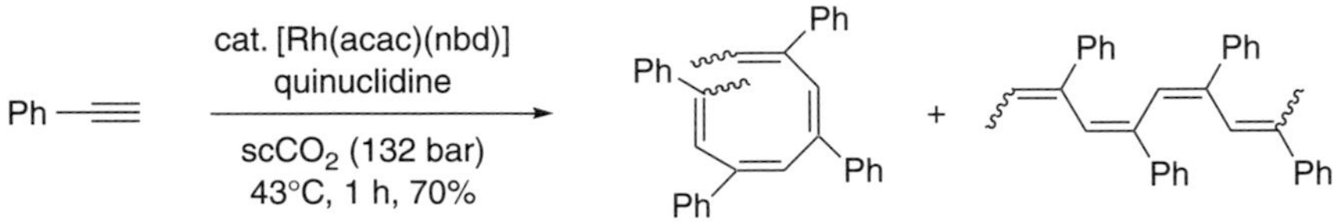

cat. [Rh(acac)(nbd)]
quinuclidine
Ph—≡ →
$scCO_2$ (132 bar)
43°C, 1 h, 70%

cis-cisoid-**28** : *cis*-transoid-**28**
81 : 19

Scheme 47

polymerization of α-olefins[137] and nickel-catalyzed copolymerization of ethylene and carbon monoxide to form polyketone.[138]

4.3.9 Biocatalysis

Enzymatic reactions are usually conducted in aqueous media. It was discovered in the mid-1980s that enzymes are active in supercritical CO_2.[139] Enzymatic reactions in $scCO_2$ are expected to have several merits over those in water: (1) because enzymes are usually insoluble in $scCO_2$, filtering easily separates the catalysts; (2) as $scCO_2$ has very effective mass transfer ability, the reaction in an $scCO_2$/solid system is faster than that in an ordinary liquid/solid system; (3) the products are easily separated from $scCO_2$ whereas extraction from aqueous media is often troublesome; (4) adjusting the pressure and temperature continuously tunes the physical properties. It is noteworthy that a small amount of water (i.e. buffer solution) is often present in the reaction mixture and that the reaction may actually be occurring in the aqueous phase.

Most reported enzymatic reactions are dehydration, *trans*-glycosilation and *trans*-esterification using hydrolytic enzymes such as glycosidase and lipase. The glycosilation of xylan **29** is used in the synthesis of biosurfactants. The reaction is faster in $scCO_2$ (150 bar) than without CO_2 pressure (Scheme 48).[140] On the other hand, lipid-coated enzymes are soluble in organic solvents and in $scCO_2$. *Trans*-galactosilation by a lipid-coated galactosidase **30** proceeds in $scCO_2$ and is 15 times faster than in organic solvents (e.g. isopropyl ether) (Scheme 49).[141] The initial rate of transgalactosilation jumps around the critical temperature and pressure.

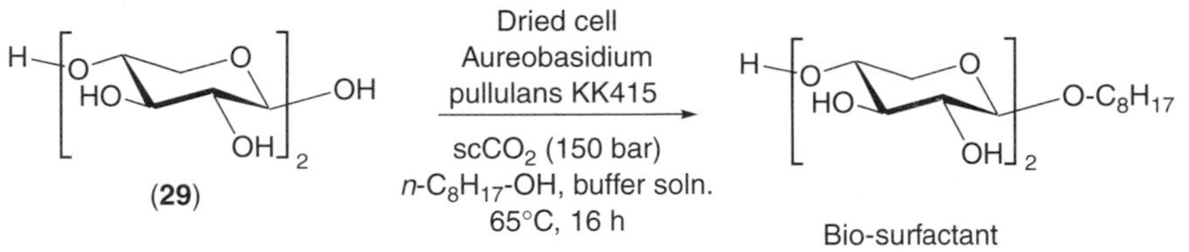

Bio-surfactant

CO_2 (bar)	Yield (mg/g xylan)
0	18
150	52

Scheme 48

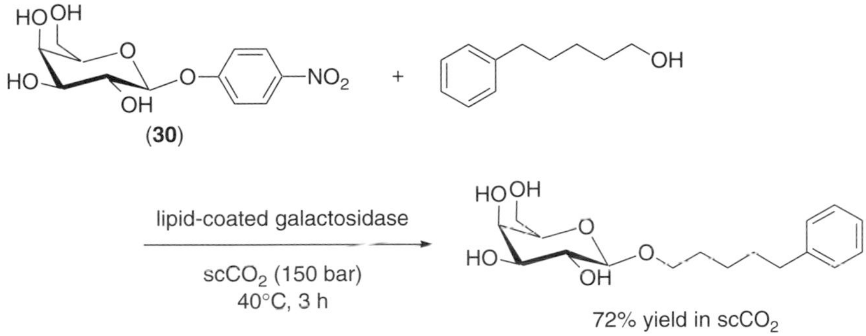

Scheme 49

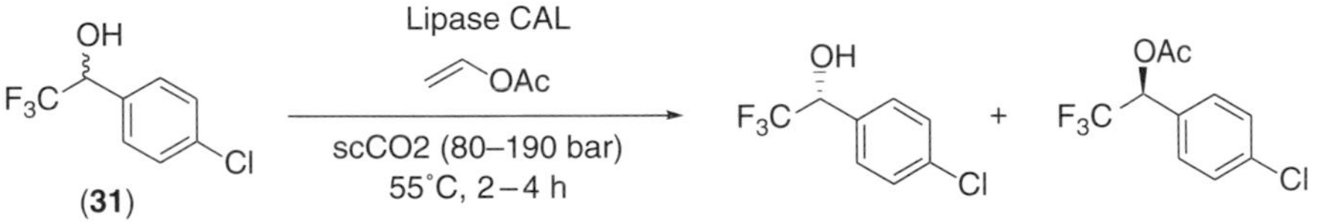

Scheme 50

Using the *Candida antarctica* lipase (CAL) in $scCO_2$ kinetically resolves racemic secondary alcohol **31** (Scheme 50).[142] The optical purity of the product is continuously tunable by changing the CO_2 pressure. The enantiomeric ratio decreases from 50 to 10 as the CO_2 pressure varies from 80 to 190 bar.

The utilization of biocatalysts other than hydrolytic enzymes has also been investigated. Alcohol dehydrogenase is used in the asymmetric reduction of ketones to yield optically active secondary alcohols (Scheme 51).[143] Although the productivity (substrate concentration) is low, high yields and excellent enantiomeric excess are obtained. On the other hand, carboxylation of pyrrole is efficient in $scCO_2$ (Scheme 52). The reaction under supercritical conditions is 12 times faster than that at atmospheric pressure, although 30 bar of CO_2 are as effective as pressure above P_c.[144] This last example also introduces the use of scCO2 as a carbon resource, which will be treated in the next section.

4.3.10 Utilization as a carbon resource

Supercritical CO_2 is usually regarded as an inactive reaction medium. However, supercritical conditions can enhance the reactivity of CO_2, allowing its utilization as a carbon resource. The $RuCl_2(PMe_3)_4$-catalyzed CO_2 hydrogenation (synthesis of formic acid) is remarkably accelerated under supercritical conditions (Scheme 53).[48] $RuCl_2(PMe_3)_4$ also efficiently catalyzes the formation of formic acid derivatives such as *N*,*N*-dimethylformamide (DMF)[145] and methyl formate[31] in the presence of dimethyl amine and methyl alcohol, respectively (Schemes 54 and 55). These findings inspired new research into homogeneous catalysis under supercritical CO_2.

It is commonly believed that for a reaction to progress smoothly, the catalyst must be soluble in $scCO_2$; this is consistent with the observation that common triphenylphosphine analogues are insoluble and thus inactive. In these reactions, it is essential to understand the phase behavior. Although the reaction mixture is initially

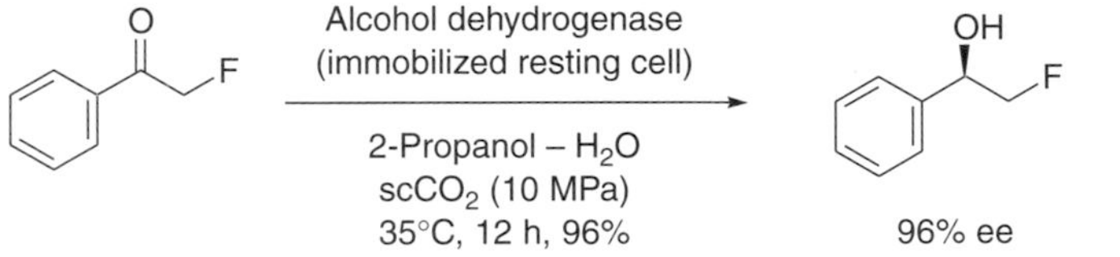

Alcohol dehydrogenase (immobilized resting cell)

2-Propanol – H_2O
$scCO_2$ (10 MPa)
35°C, 12 h, 96%

96% ee

Scheme 51

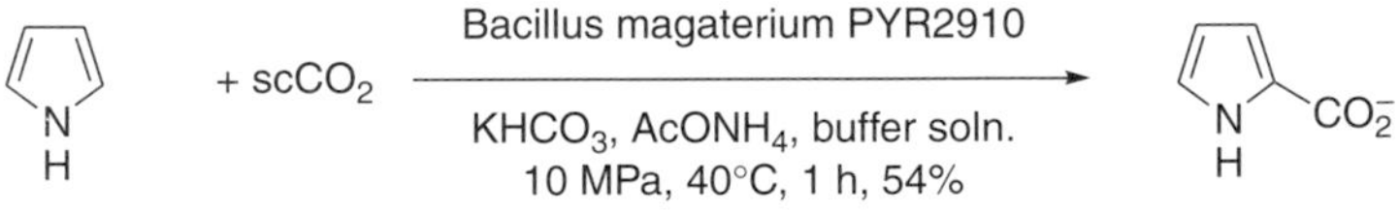

+ $scCO_2$

Bacillus magaterium PYR2910

$KHCO_3$, $AcONH_4$, buffer soln.
10 MPa, 40°C, 1 h, 54%

Scheme 52

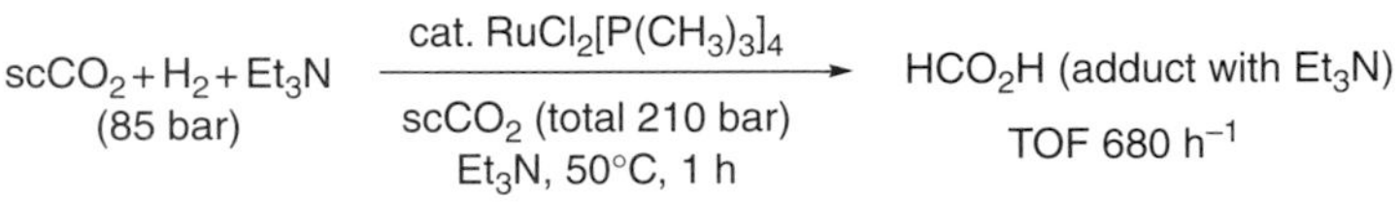

$scCO_2 + H_2 + Et_3N$ (85 bar) → cat. $RuCl_2[P(CH_3)_3]_4$; $scCO_2$ (total 210 bar), Et_3N, 50°C, 1 h → HCO_2H (adduct with Et_3N), TOF 680 h^{-1}

Scheme 53

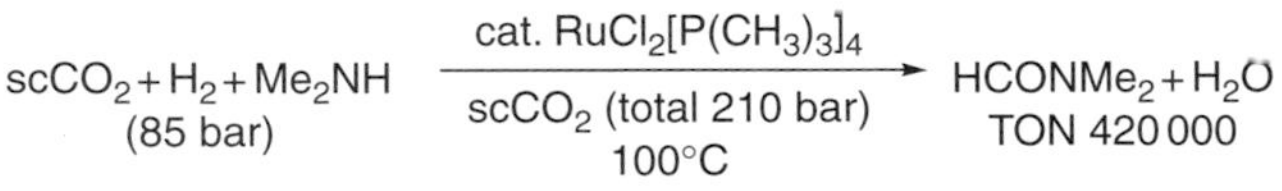

$scCO_2 + H_2 + Me_2NH$ (85 bar) → cat. $RuCl_2[P(CH_3)_3]_4$; $scCO_2$ (total 210 bar), 100°C → $HCONMe_2 + H_2O$, TON 420 000

Scheme 54

composed of a single phase, polar phases soon appear. In formic acid synthesis, an ammonium formate phase appears, and an aqueous phase appears in the syntheses of DMF and methyl formate. Phase separation often decreases the reaction rate because of catalyst/reactant separation. In the DMF and methyl formate syntheses, water-soluble phosphine complexes improve the catalytic performance.[146]

Cyclic carbonate synthesis from epoxide and CO_2 is one of the few industrial processes that use CO_2 as a feedstock. The reaction proceeds in the presence of catalysts that are typically metal halides or tetraalkylammonium halides. The products (e.g. ethylene carbonate, which is an important intermediate of dimethyl carbonate) are utilized for polycarbonate synthesis via a non-phosgene, melt polymerization process.

Recently, reactions under supercritical conditions have been investigated. For example, styrene oxide is converted to styrene carbonate **32** at 150°C in the presence of dimethyl formamide without a catalyst (Scheme 56)[147], but substrates are rather limited and propylene oxide does not react under similar conditions. On the other hand, solid catalysts on a fixed-bed flow reactor are being investigated as a replacement for batch processes.[148,149] The reactivity of solid catalysts is usually lower than that of homogeneous catalysts at the same temperature. It is expected that the reaction under supercritical conditions will compensate for the lack of catalytic activity (Scheme 57). In addition, $scCO_2$ can be easily recycled without depressurization and cooling (see below) since the resulting cyclic carbonates are not soluble in supercritical CO_2. Some metal complexes, such as aluminum salen complexes, have also been proposed as active homogeneous catalysts under supercritical conditions.[150,151]

Reactions that require phosgene are dangerous, and dimethyl carbonate (DMC) **33** is the most promising phosgene substitute. Current industrial processes for DMC synthesis are oxidative carbonylation of methanol and transesterification of ethylene carbonate with methanol. Hence, the direct reaction of CO_2 and methanol (Scheme 58) is regarded as an attractive, next-generation process, but the limitation

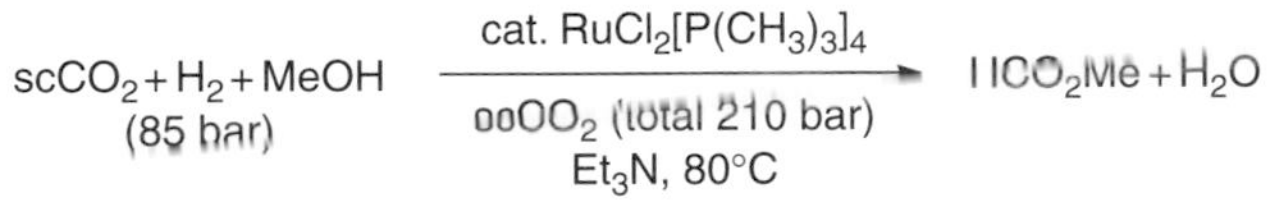

$scCO_2 + H_2 + MeOH$ (85 bar) —[cat. $RuCl_2[P(CH_3)_3]_4$ / $scCO_2$ (total 210 bar), Et_3N, 80°C]→ $HCO_2Me + H_2O$

Scheme 55

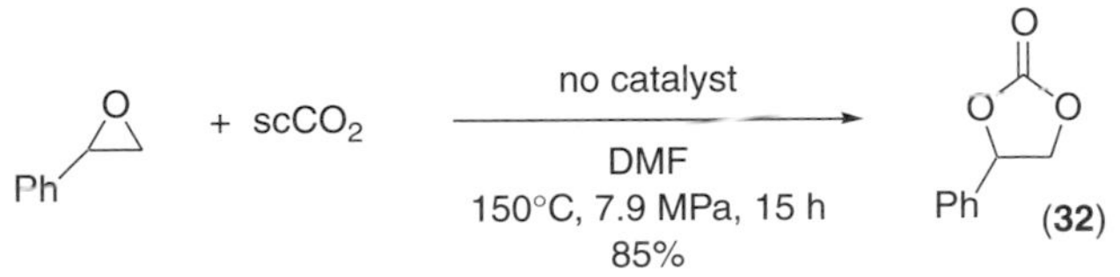

Ph-epoxide + $scCO_2$ —[no catalyst / DMF, 150°C, 7.9 MPa, 15 h, 85%]→ (**32**)

Scheme 56

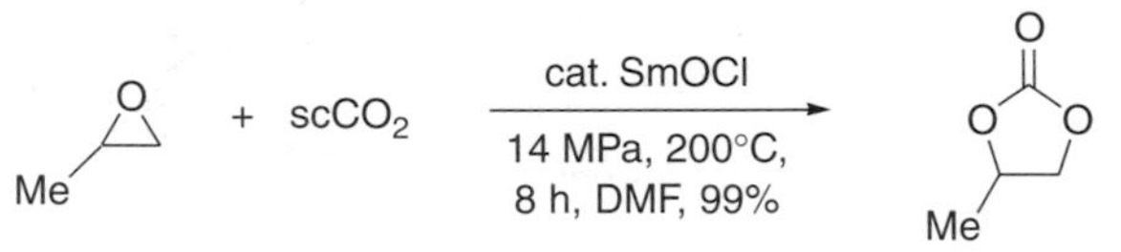

Me-epoxide + $scCO_2$ → (cat. SmOCl; 14 MPa, 200°C, 8 h, DMF, 99%) → propylene carbonate (Me)

Scheme 57

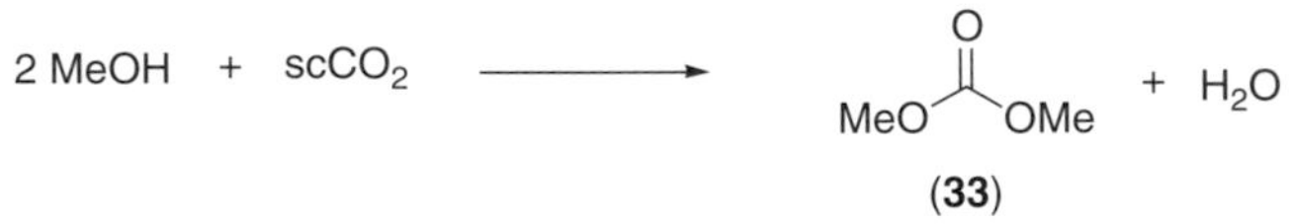

2 MeOH + $scCO_2$ → MeO(C=O)OMe (33) + H_2O

Scheme 58

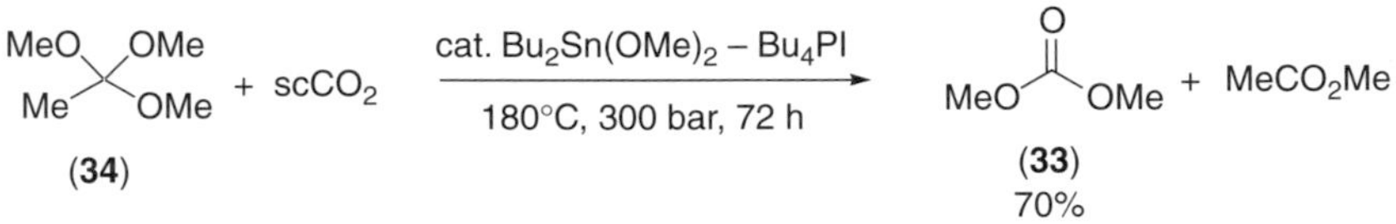

MeC(OMe)$_3$ (34) + $scCO_2$ → (cat. $Bu_2Sn(OMe)_2 – Bu_4PI$; 180°C, 300 bar, 72 h) → MeO(C=O)OMe (33) 70% + $MeCO_2Me$

Scheme 59

+ MeOH

$Me_2C(OMe)_2$ (35) + $scCO_2$ → (cat. $Bu_2Sn(OMe)_2$; MeOH, 180°C, 2000 bar, 24 h) → MeO(C=O)OMe (33) + acetone

cat.	Yield (% / acetal)
$Bu_2Sn(OMe)_2$	88
$Bu_2Sn{=}O$	79

Scheme 60

is that the reverse reaction, hydrolysis of DMC, also occurs. In other words, thermodynamics limits the conversion to a few percent. In order to overcome this problem, reactions of dehydrated derivatives of methanol (i.e. trimethyl orthoacetate **34**[152] and dimethyl acetal **35**[153]) have been proposed (Schemes 59 and 60). If the acetone in the acetal reaction is recycled, then the overall reaction is DMC synthesis from methanol and CO_2. In order to overcome the low methanol conversion in Scheme 58, an efficient dehydration process must be developed. A high methanol conversion (~50%) is possible using zeolite to continuously dehydrate the reaction mixture (Scheme 61).[154,155] Catalytic activity (reaction rate) and dehydration methods must

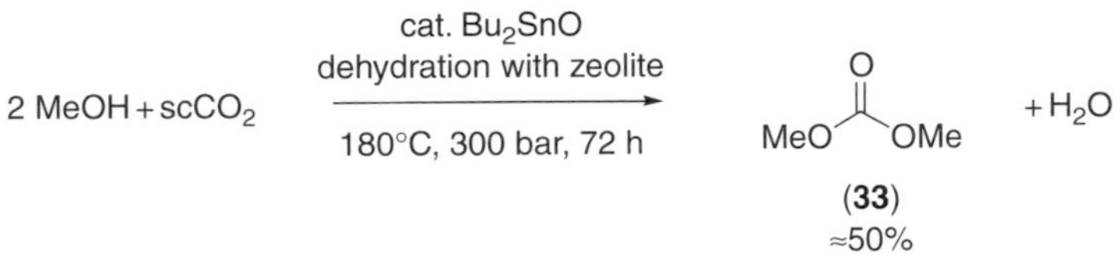

2 MeOH + $scCO_2$ —(cat. Bu_2SnO, dehydration with zeolite; 180°C, 300 bar, 72 h)→ MeO(C=O)OMe (**33**) ≈50% + H_2O

Scheme 61

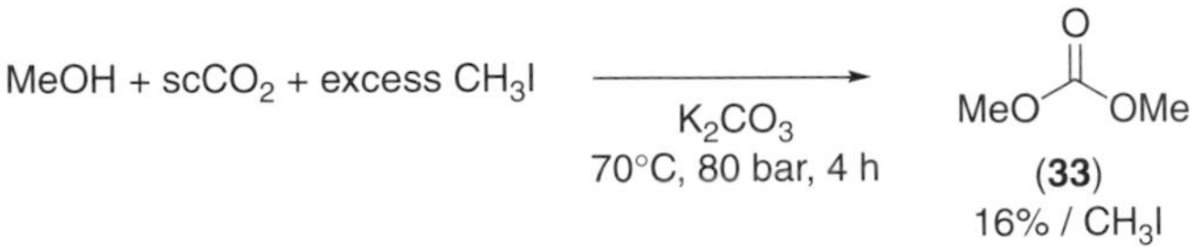

MeOH + $scCO_2$ + excess CH_3I —(K_2CO_3; 70°C, 80 bar, 4 h)→ MeO(C=O)OMe (**33**) 16% / CH_3I

Scheme 62

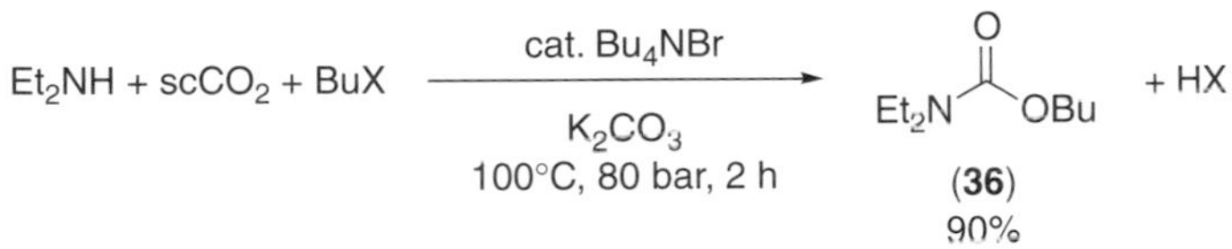

Et_2NH + $scCO_2$ + BuX —(cat. Bu_4NBr; K_2CO_3, 100°C, 80 bar, 2 h)→ Et_2N(C=O)OBu (**36**) 90% + HX

Scheme 63

be further improved for practical use. As a catalyst, nickel acetate also shows a low activity.[156]

The synthesis of DMC by reacting methanol and supercritical CO_2 in the presence of methyl iodide has also been proposed.[157] This reaction requires a stoichiometric amount of potassium carbonate and excess methyl iodide to yield 16% of DMC based on methyl iodide (Scheme 62). The yield is maximized near the critical point of CO_2.

Urethane **36** can be prepared by reacting an amine, $scCO_2$ and an alkyl halide (Scheme 63).[158] Adding a base such as potassium carbonate is necessary to neutralize the hydrogen halide. In a way similar to DMC synthesis, utilizing acetals as a dehydrating agent is effective for the synthesis of urethane **37** from an amine, $scCO_2$ and an alcohol (Scheme 64).[159] The selectivity of the reaction is highly dependent on the reaction pressure. Under high CO_2 pressure, amines are converted to carbamic acid, which prevents the *N*-alkylation, imine formation and urea formation. Nickel phenanthroline complexes exhibit similar catalytic activities to the tin complex.[160]

Vinylcarbamate **38** can be prepared using a ruthenium-catalyzed reaction between diethylamine and phenylacetylene in $scCO_2$ (Scheme 65). This reaction is three times faster in $scCO_2$ than in toluene.[161] In addition, aziridines[162] and

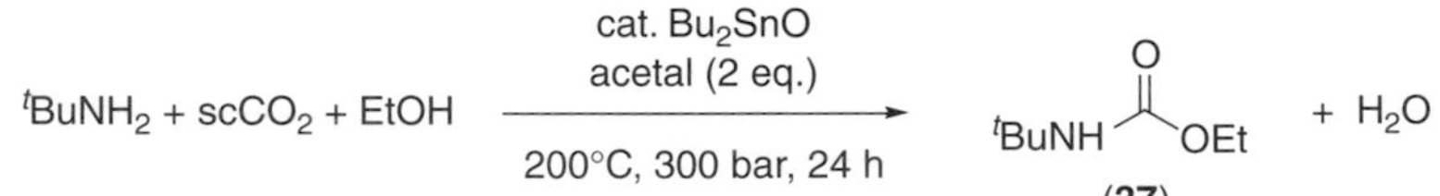

$^tBuNH_2 + scCO_2 + EtOH \xrightarrow[\text{200°C, 300 bar, 24 h}]{\text{cat. } Bu_2SnO\text{, acetal (2 eq.)}} {}^tBuNHC(O)OEt$ (**37**) 84% $+ H_2O$

Scheme 64

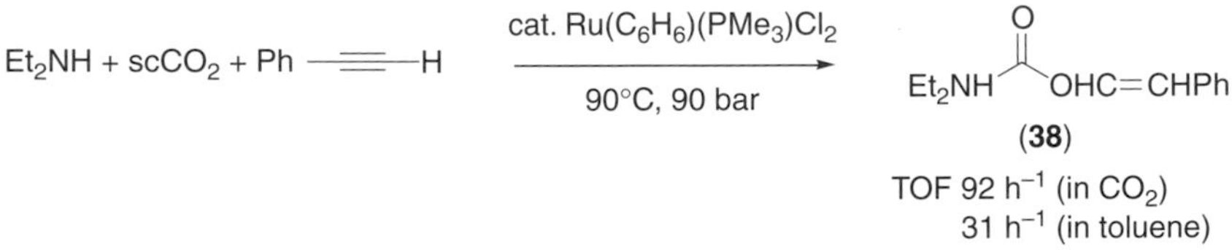

$Et_2NH + scCO_2 + Ph{-}C{\equiv}C{-}H \xrightarrow[\text{90°C, 90 bar}]{\text{cat. } Ru(C_6H_6)(PMe_3)Cl_2} Et_2NC(O)OHC{=}CHPh$ (**38**)

TOF 92 h^{-1} (in CO_2)
31 h^{-1} (in toluene)

Scheme 65

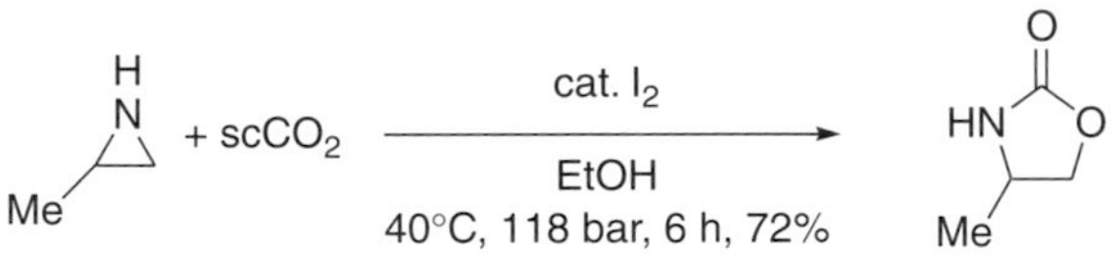

2-methylaziridine $+ scCO_2 \xrightarrow[\text{EtOH, 40°C, 118 bar, 6 h, 72\%}]{\text{cat. } I_2}$ 4-methyl-2-oxazolidinone (**39**)

Scheme 66

amino alcohols[163] react in $scCO_2$ to give 2-oxazolidinone **39** (Scheme 66). Again, the yield is maximized near the critical point.

4.3.11 Miscellaneous reactions

A key issue for synthetic chemists is the direct and selective functionalization of alkanes under mild conditions. A major problem in C–H bond activation by molecular catalysis is the lack of a suitable reaction medium, because most organic solvents are not inert under alkane activation conditions and therefore prevent the desired reactions. In this context, dense carbon dioxide seems to be a promising reaction medium as it is miscible with organics, including organometallics, and potentially stable under alkane activation conditions. Indeed, methane carbonylation and alkane dehydrogenation by molecular catalysis have been reported using dense carbon dioxide as the reaction medium (Scheme 67).[164–166]

Hydrometallations of alkenes and acetylenes such as hydroboration, hydrosilation, hydroalumination, hydrostannation and hydrozirconation are useful synthetic reactions. Rhodium–phosphine complexes catalyze hydroboration of substituted styrenes with a catecolborane (Scheme 68).[167] An extensive survey of the phosphine

cat. $RhCl(PMe_3)_3$ – hν; $scCO_2$ (150 bar), 50°C, 48 h; + H_2; TON 602

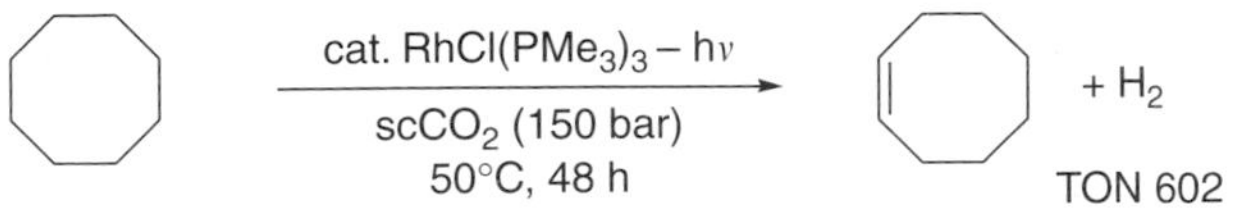

Scheme 67

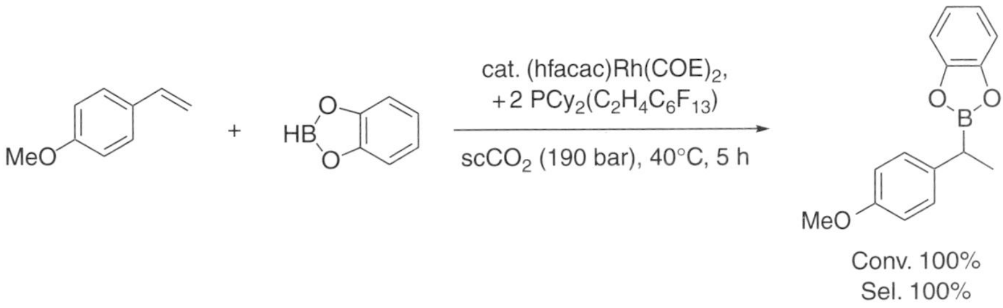

MeO ... + HB(O...O) ; cat. (hfacac)Rh(COE)$_2$, +2 $PCy_2(C_2H_4C_6F_{13})$; $scCO_2$ (190 bar), 40°C, 5 h; MeO; Conv. 100%; Sel. 100%

Scheme 68

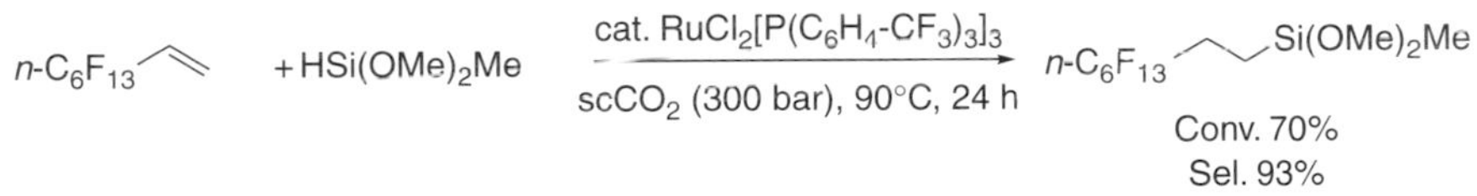

n-C_6F_{13} + $HSi(OMe)_2Me$; cat. $RuCl_2[P(C_6H_4\text{-}CF_3)_3]_3$; $scCO_2$ (300 bar), 90°C, 24 h; n-C_6F_{13} $Si(OMe)_2Me$; Conv. 70%; Sel. 93%

Scheme 69

effect on the yields and selectivities revealed that the use of $PCy_2(C_2H_4C_6F_{13})$ as ligand leads to high performance.

Hydrosilation of perfluoroalkenes with $HSi(OMe)_2Me$ occurs efficiently in supercritical CO_2 (Scheme 69).[168] In order to achieve high yields in organic solvents, halogenated compounds such as CH_2Cl_2 are necessary. The key to high yields in $scCO_2$ is the solubilization of the catalyst: a ruthenium complex with $P(C_6H_4\text{-}p\text{-}CF_3)_3$ ligand can be used, but the scope of the reaction is rather limited. When ordinary alkenes are used instead of fluorous ones, rapid isomerization of the double bond to generate the internal alkenes is observed. On the other hand, using more electron-rich hydrosilanes results in CO_2 reduction and formic acid derivatives.

The effect of the solvent on the product selectivity for the rhodium-catalyzed cyclization of an allylamine under a CO/H_2 atmosphere (Scheme 70)[169] is such that carbonylation in conventional organic solvents forms a cyclic amide **41**, whereas the reaction in $scCO_2$ preferentially produces reduced five-membered cyclic amine **40**. In $scCO_2$, the amino group of the starting material is transformed to the carbamic

cat. Rh(hfacac)(COD) + $P(C_6H_4\text{-}C_2H_4C_6F_{13})_3$, $scCO_2$ (d 0.78 g/mL), CO / H_2 (44 bar), 80°C, 20 h

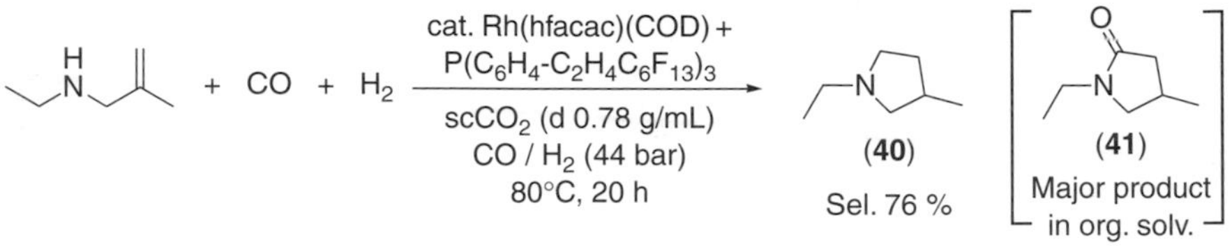

(**40**) Sel. 76 %

(**41**) Major product in org. solv.

Scheme 70

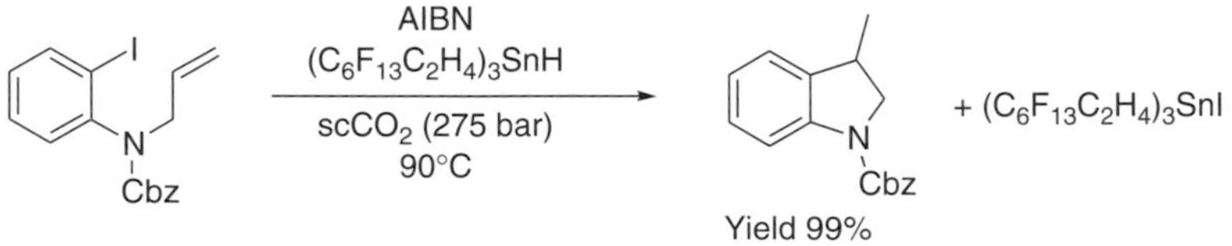

AIBN, $(C_6F_{13}C_2H_4)_3SnH$, $scCO_2$ (275 bar), 90°C

Cbz; Yield 99%; + $(C_6F_{13}C_2H_4)_3SnI$

Scheme 71

$C_6H_{17}OSO_3Me$ + KI → C_6H_{17}–I

cat. Bu_4NI or cat. silica-supported onium salts; 50°C, $scCO_2$ 130 bar, 180–300 min

Bu_4NI	conv. 7%
Silica-supported cat.	conv. 60%

$P^+Bu_3\ I^-$ (**42**)

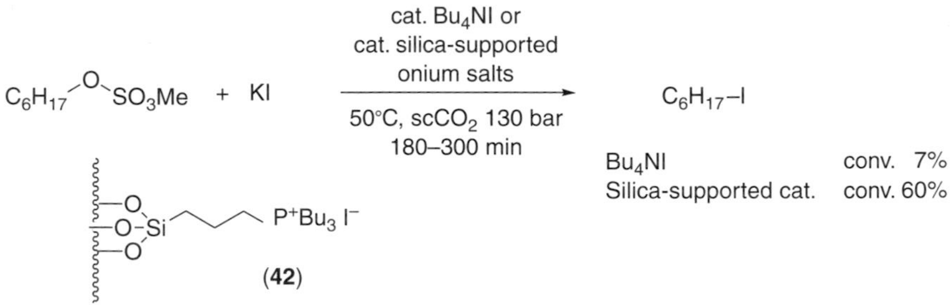

Scheme 72

acid moiety. Thus, the intermolecular hydrogenolysis of the acyl-rhodium intermediate is faster than the intramolecular attack of the amino group: the subsequent condensation/hydrogenation sequence leads to the saturated amine **40**.

Since C–H bonds can be reactive toward radicals, synthetic reactions proceeding via a radical mechanism often require the use of halogenated solvents. Because of its inertness, $scCO_2$ was used as alternative medium for the halogenation using bromine or *N*-bromosuccinimide (NBS).[46] A radical reaction (reduction or addition to alkenes) in $scCO_2$ using hydrostannanes is shown in Scheme 71.[170] Radical polymerization in $scCO_2$ is discussed in Section 4.3.8.

S_N2 nucleophilic substitutions take place in $scCO_2$.[171] The presence of silica-supported onium salts **42** as a phase-transfer catalyst enhances the reaction (Scheme 72). Esters formation from carboxylic acids and alcohols via dehydration also proceeds in $scCO_2$.[172] The conversion increases as the CO_2 pressure increases. A single homogeneous phase is obtained around the critical point, at which the conversion is maximized.

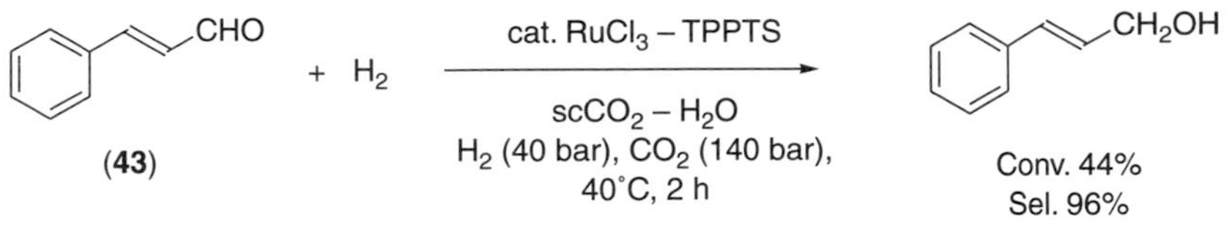

CHO + H_2 → cat. $RuCl_3$ – TPPTS; $scCO_2$ – H_2O; H_2 (40 bar), CO_2 (140 bar), 40°C, 2 h → CH_2OH

(**43**)

Conv. 44%
Sel. 96%

Scheme 73

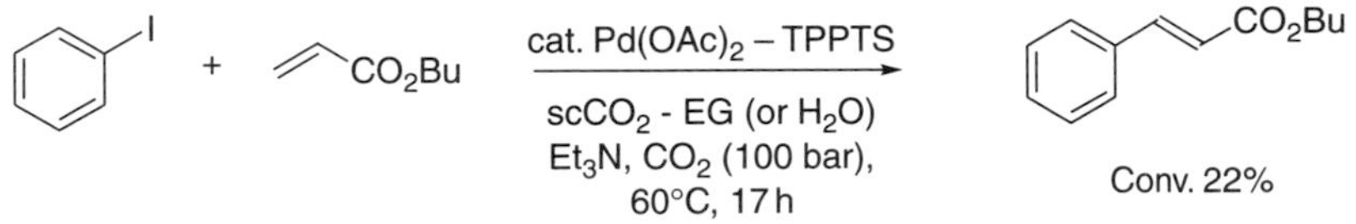

I + CO_2Bu → cat. $Pd(OAc)_2$ – TPPTS; $scCO_2$ - EG (or H_2O); Et_3N, CO_2 (100 bar), 60°C, 17 h → CO_2Bu

Conv. 22%

Scheme 74

4.4 Synthesis and separation

Several multiphase reactions can be designed by combining four different fluids (water, ionic liquids, supercritical CO_2 and organics) in addition to gases and solids. Indeed, a large number of biphasic systems containing $scCO_2$ have recently been proposed, and most aim at the efficient separation of homogeneous catalysts. Examples are $scCO_2/H_2O$, $scCO_2$/IL, $scCO_2$/liquid organics and $scCO_2$/solid systems. The high-density phase (the lower phase) usually contains the catalyst.

4.4.1 CO_2/water two-phase system

Water is an attractive reaction medium because it is abundant, non-toxic, non-flammable and of course economical. The most widely used water-soluble complexes are phosphine complexes bearing sulfonic acid groups. For example, a ruthenium complex catalyzes carbonyl-selective hydrogenation of cinnamaldehyde **43** in an $scCO_2/H_2O$ mixture using a TPPTS ($P(C_6H_4SO_3Na)_3$) ligand, which easily separates the catalyst (Scheme 73).[173] Compared with the reaction in a toluene/H_2O biphasic media, the $scCO_2/H_2O$ medium has a higher conversion. TPPTS complexes are also applicable to the Heck reaction. Although the conversion is not very high (22%), the palladium-catalyzed reaction between phenyl iodide and butyl acrylate proceeds in an $scCO_2/H_2O$ or $scCO_2$/EG biphasic system (Scheme 74).[174]

The addition of surfactants accelerates $scCO_2/H_2O$ biphasic reactions significantly. For example, hydrogenation of styrene in an $scCO_2/H_2O$ emulsion is much faster than that in a conventional biphasic system (Scheme 75).[175] Although reactions in emulsions are usually troublesome during the product separation step, $scCO_2/H_2O$ emulsions are easily broken down into simple biphasic systems by depressurizing, which allows product separation.

It is also possible to maintain catalysts in the upper $scCO_2$ phase using a CO_2-philic phosphine as a ligand. Hydroformylation of water-soluble alkenes is catalyzed

cat. RhCl $[P(C_6H_3(SO_3H)_2)_3]_3$

+ H_2 → $scCO_2 - H_2O$, surfactant, 275 bar total, 40°C

	TOF [h^{-1}]
toluene - H_2O	4
CO_2 - H_2O	26
CO_2 - H_2O emulsions	300

Scheme 75

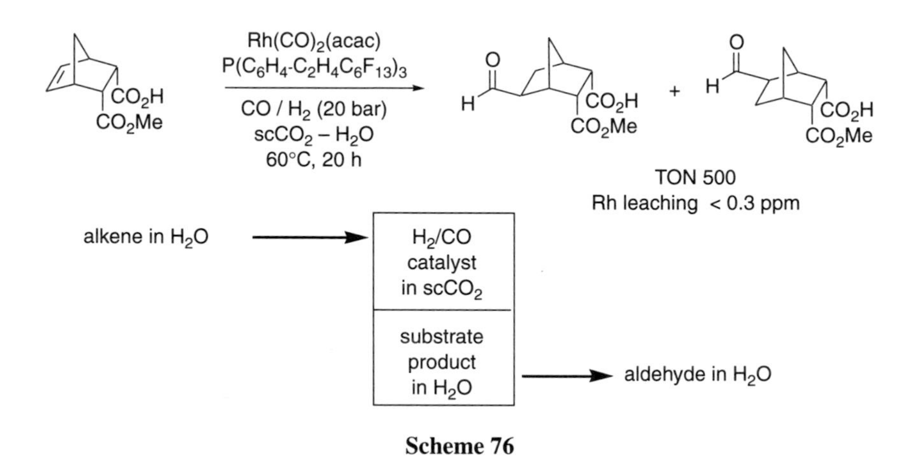

Scheme 76

by a rhodium complex with fluorous phosphine ligands (Scheme 76).[176] The starting alkene is fed in as an aqueous solution and the resulting aldehyde is removed from the bottom of the reactor.

A similar inverse biphasic system directly produced hydrogen peroxide from hydrogen and oxygen (Scheme 77).[177] A palladium complex with a fluorous phosphine ligand solubilized in the $scCO_2$ phase promoted the reaction between H_2 and O_2 to produce H_2O_2, which was extracted into the lower, water phase. The merits of this process are a high reactant concentration, a reduced explosion risk, a highly stable reaction medium with respect to oxidation and easy separation of the product from homogeneous catalysts.

The asymmetric reduction of ketones is carried out with a flow reactor using an alcohol dehydrogenase immobilized on a water-absorbing polymer.[178] Although the water phase is small, it is presumed that the reaction occurs in the aqueous phase. Hence, the reaction can be regarded as an $scCO_2/H_2O$ biphasic system.

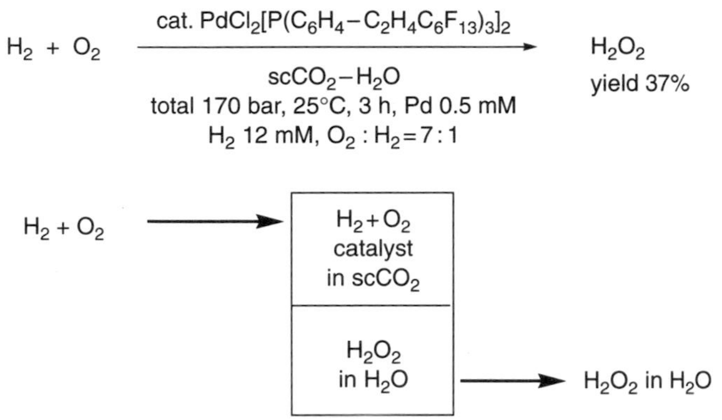

$H_2 + O_2$ —— cat. $PdCl_2[P(C_6H_4-C_2H_4C_6F_{13})_3]_2$ / $scCO_2-H_2O$ / total 170 bar, 25°C, 3 h, Pd 0.5 mM / H_2 12 mM, $O_2 : H_2 = 7 : 1$ ——→ H_2O_2 yield 37%

$H_2 + O_2$ ——→ [$H_2 + O_2$ catalyst in $scCO_2$ / H_2O_2 in H_2O] ——→ H_2O_2 in H_2O

Scheme 77

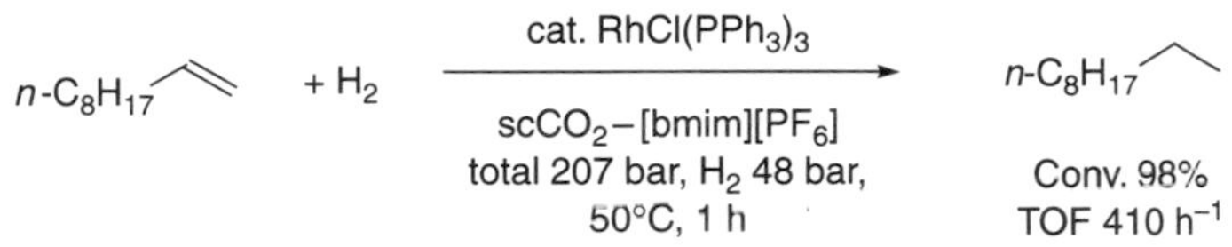

n-C_8H_{17}-CH=CH$_2$ + H_2 —— cat. $RhCl(PPh_3)_3$ / $scCO_2$–[bmim][PF_6] / total 207 bar, H_2 48 bar, 50°C, 1 h ——→ n-C_8H_{17}-CH$_2$CH$_3$ Conv. 98% TOF 410 h^{-1}

Scheme 78

4.4.2 CO_2/ionic liquid two-phase system

Reactions using ionic liquids as a substitute for organic solvents have recently attracted a great deal of interest: they can solubilize a large variety of organic, organometallic and inorganic materials and have no vapor pressure. Furthermore, ionic liquids have tunable physical properties. For example, BF_4 salts are hydrophilic, whereas PF_6 salts are usually insoluble in water (see Section 2.2). Biphasic systems with supercritical CO_2 are especially promising: ionic liquids are polar, can be used at high temperature and are immiscible with $scCO_2$. One of the most frequently used ionic liquids, [bmim][PF_6] (bmim: 1-butyl-3-methylimmidazolium), displays these typical properties. Supercritical CO_2 is used to extract products from ionic liquids without using organic solvents.

Unlike water, ionic liquids have high solubilizing power for organic compounds. Hence, homogeneous catalysts are usually soluble in ionic liquids without special modifications. Many reactions have been successfully tested in ionic liquids (see Section 2.3); the products are usually isolated by extraction with an organic solvent or direct distillation, when possible. The use of supercritical CO_2 represents an attractive alternative for the isolation of the product. For example, the hydrogenation of long-chain alkenes by Wilkinson's catalyst ($RhCl(PPh_3)_3$) as well as DMF synthesis from carbon dioxide using $RuCl_2$(dppe) (dppe: 1,2-bisdiphenylphosphinoethane) proceed in an $scCO_2$/[bmim][PF_6] mixture (Schemes 78 and 79).[179]

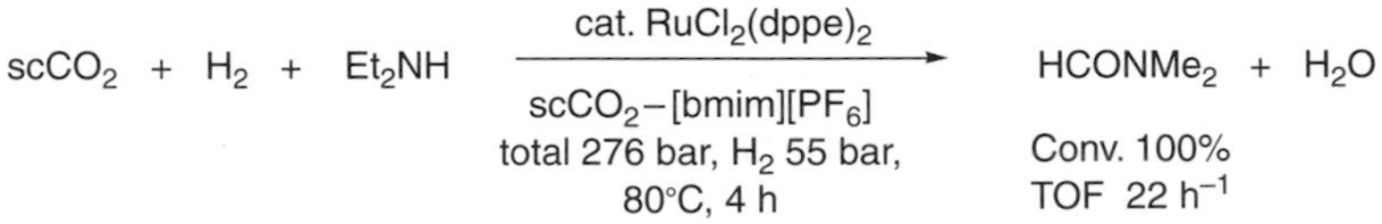

$$scCO_2 + H_2 + Et_2NH \xrightarrow[\text{scCO}_2\text{–[bmim][PF}_6\text{], total 276 bar, H}_2\text{ 55 bar, 80°C, 4 h}]{\text{cat. RuCl}_2\text{(dppe)}_2} HCONMe_2 + H_2O$$

Conv. 100%
TOF 22 h^{-1}

Scheme 79

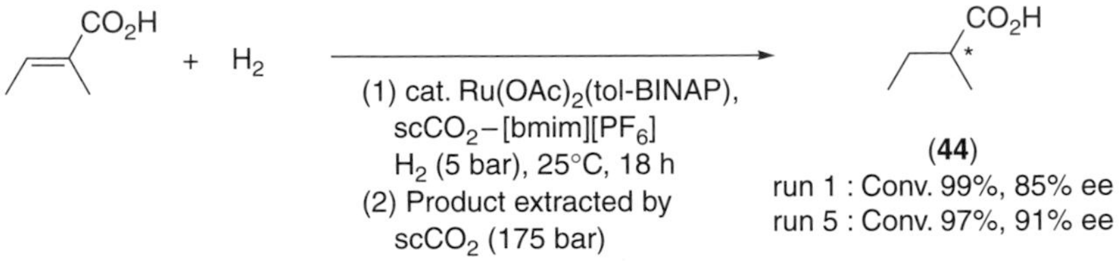

CO_2H + H_2 → CO_2H (44)

(1) cat. $Ru(OAc)_2$(tol-BINAP), $scCO_2$–[bmim][PF_6] H_2 (5 bar), 25°C, 18 h
(2) Product extracted by $scCO_2$ (175 bar)

(**44**)
run 1 : Conv. 99%, 85% ee
run 5 : Conv. 97%, 91% ee

Scheme 80

To limit the amount of waste material – and to achieve cost-efficient processes – the recycling of the homogeneous catalyst is crucial. Ru-BINAP and its derivatives catalyze the asymmetric hydrogenation of tiglic acid in [bmim][PF_6]. The 2-methylbutanoic acid **44** obtained can be extracted from the ionic liquid with $scCO_2$, allowing recovery and reuse of the ionic liquid phase containing the catalyst (Scheme 80, no loss of catalytic activity or enantioselectivity for the fifth run).[180]

The hydroformylation of long-chain alkenes was investigated in an $scCO_2$/[bmim][PF_6] biphasic system.[181] A rhodium complex with $P(OPh)_3$ as the ligand was first tested as the catalyst for the hydroformylation of terminal alkenes in a batch reactor. The product was extracted using $scCO_2$, and the catalyst remained in the ionic liquid phase. However, catalyst recycling resulted in the partial elution and hydrolysis of the ligand, which decreased the activity and selectivity. With the catalytic ligand changed to $PPh_2(C_6H_4SO_3)$, gradual oxidation of the ligand occurred. Ultimately, a continuous flow process was developed using $scCO_2$ as a mobile phase (Scheme 81). In this process, the starting alkene is introduced into the reactor with $scCO_2$, and the resulting aldehyde **45** is extracted from the ionic liquid phase, which also contains the rhodium complex. The catalyst is prepared *in situ* from $Rh_2(OAc)_4$ and $[PPh(C_6H_4SO_3)_2]$[1-propyl-3-methylimidazolium]$._2$ The reaction rate is constant over 72 h and shows high catalyst stability under these reaction conditions. The regioselectivity is also unchanged during the reaction (linear : branched ratio 3.1). In addition, rhodium leaching into the product is negligible (<1 ppm).

The $scCO_2$/IL flow process developed for hydroformylation is applicable to other homogeneous catalysis reactions, including enzymatic reactions. For instance, two groups independently reported the kinetic resolution of racemic secondary alcohols via transesterification by lipase in an $scCO_2$/[bmim][NTf_2] system (Scheme 82).[182,183] The reaction is advantageous with respect to the conventional process because of the ease of the product/enzyme and product/solvent separations.

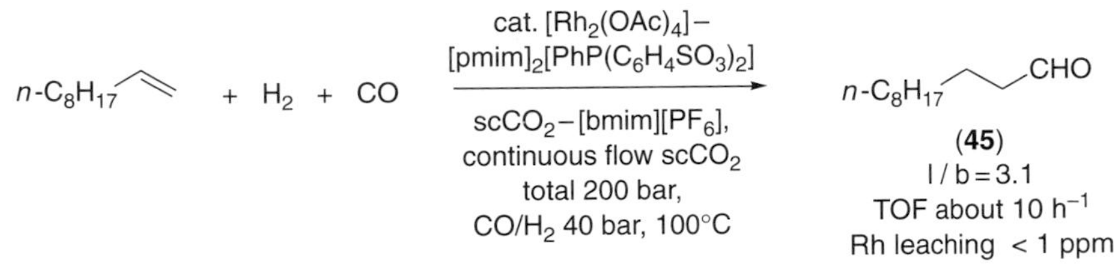

n-C_8H_{17} + H_2 + CO → (cat. $[Rh_2(OAc)_4]$–$[pmim]_2[PhP(C_6H_4SO_3)_2]$; $scCO_2$–[bmim][PF_6], continuous flow $scCO_2$, total 200 bar, CO/H_2 40 bar, 100°C) → n-C_8H_{17}–CHO (**45**)
l/b = 3.1
TOF about 10 h^{-1}
Rh leaching < 1 ppm

Scheme 81

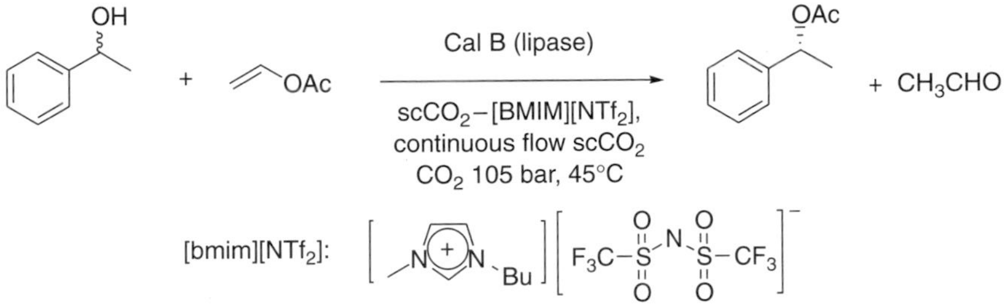

OH + OAc → (Cal B (lipase); $scCO_2$–[BMIM][NTf_2], continuous flow $scCO_2$, CO_2 105 bar, 45°C) → OAc + CH_3CHO

[bmim][NTf_2]: N+N Bu; F_3C–$S(O_2)$–N–$S(O_2)$–CF_3 ⁻

Scheme 82

O + BuOH → (Cal B (lipase); $scCO_2$–[bmim][NTf_2], 40°C, 150 bar) → O Bu + AcH

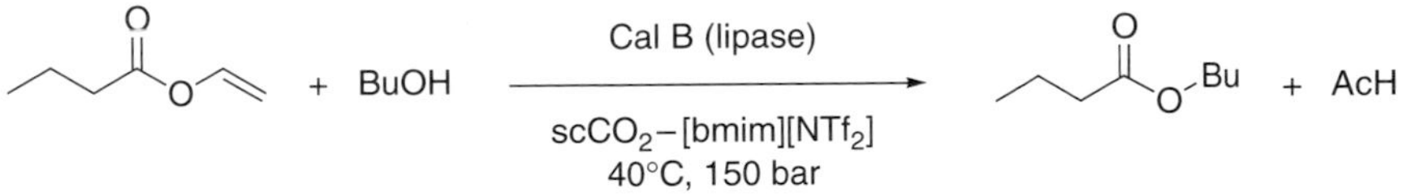

Scheme 83

Moreover, no substantial loss of enzyme activity is observed during recycling of the enzyme at 40°C. A related enzymatic transesterification was reported using $scCO_2$/[bmim][NTf_2] and $scCO_2$/[bmim][PF_6] biphasic systems (Scheme 83).[184]

4.4.3 *Recoverable catalysts*

When the catalyst is either solid or liquid and insoluble in $scCO_2$, simplification of catalyst separation is expected. The merits of the procedure compared with liquid/solid or liquid/liquid biphasic systems are the relatively high mass transfer ability, no need for solvent removal and less catalyst leaching owing to the relatively low solubilizing power of $scCO_2$.

Rhodium-catalyzed hydroformylation of 1-octene using a continuous flow reactor was conducted with a phosphine ligand immobilized on silica (Scheme 84).[185] A linear/branched ratio >30 was achieved, although the conversion was rather low. In addition, no activity loss was observed over several days.

On the other hand, a rhodium–phosphite complex formed a CO_2-insoluble liquid phase in the hydroformylation of 1-nonene (Scheme 85).[186] No leaching of rhodium was observed during the reaction. An activated carbon-supported rhodium catalyst was also utilized for hydroformylation by a flow system using $scCO_2$ as a mobile phase.[187]

Polyurea-encapsulated palladium catalysts promote the phosphine-free Mizoroki–Heck reaction, which results in a high yield of cinnamates **46** (Scheme 86).[188] The catalyst, which is easily recovered by filtration, is also applicable to the Suzuki–Miyaura and Stille coupling reactions.

Copolymerization of cyclohexene oxide and CO_2 proceeds with an immobilized chromium phorphyrin complex[189], although the molecular weight of the resulting copolymer **47** is limited to the oligomer level (Scheme 87).

Supported catalysts involving palladium on carbon[190] and dendrimer-encapsulated palladium[191] and a polymer-supported phosphine palladium catalyst[192] have facilitated C–C coupling reactions in $scCO_2$. Polymer-tethered substrates[98] or amine bases[192] have also been successfully used for the Mizoroki–Heck and Suzuki–Miyaura reactions in $scCO_2$. For example, REM resin underwent a Mizoroki–Heck reaction with iodobenzene to yield, after cleavage, (*E*)-methyl cinnamate **48** (74%) (Scheme 88).[98] It is assumed that $scCO_2$ acts as a good solvent that swells the polymers and exposes reactive sites.

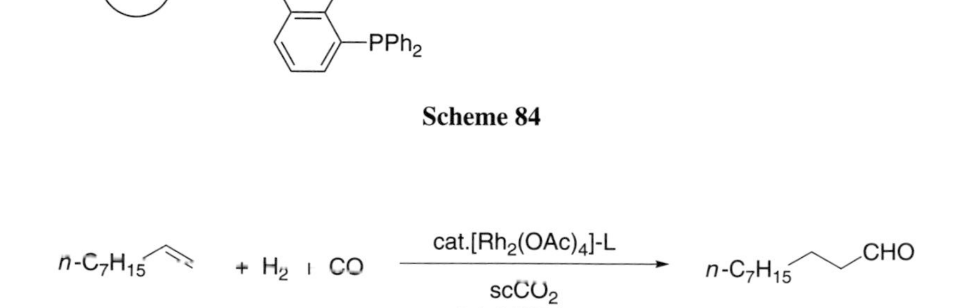

n-C_6H_{13} + H_2 + CO → (Rh immobilized on silica; continuous flow $scCO_2$, CO_2 120 bar, CO/H_2 50 bar, 90°C) n-C_6H_{13}–CHO

Conv. 14%
l/b = 33

immobilized catalyst: silica–N(phenoxazine)(PPh$_2$)$_2$ + Rh(acac)(CO)$_2$

Scheme 84

n-C_7H_{15} + H_2 + CO → (cat.[Rh$_2$(OAc)$_4$]-L; $scCO_2$, total 220 bar, CO/H_2 40 bar, 100°C, 1 h) n-C_7H_{15}–CHO

L: P–[O–C$_6$H$_4$–C$_9$H$_{19}$]$_3$

Yield 42–85%
l/b = 2.5–5.5
no Rh leaching

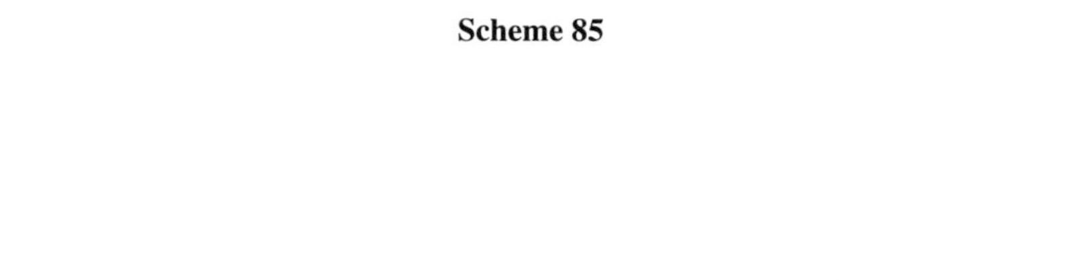

Scheme 85

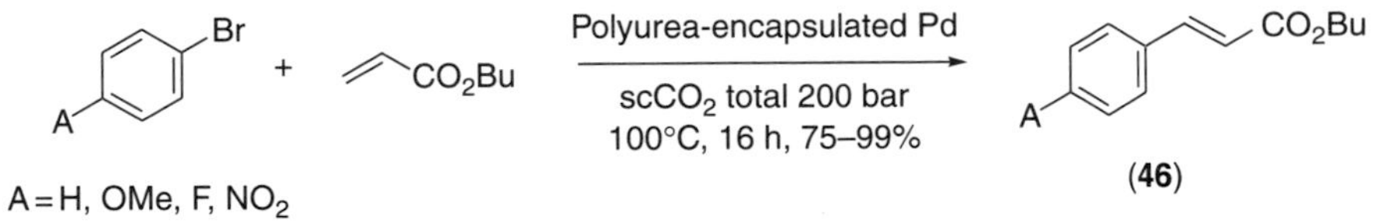

Br + CO_2Bu — Polyurea-encapsulated Pd; $scCO_2$ total 200 bar, 100°C, 16 h, 75–99% → CO_2Bu (**46**)

A = H, OMe, F, NO_2

Scheme 86

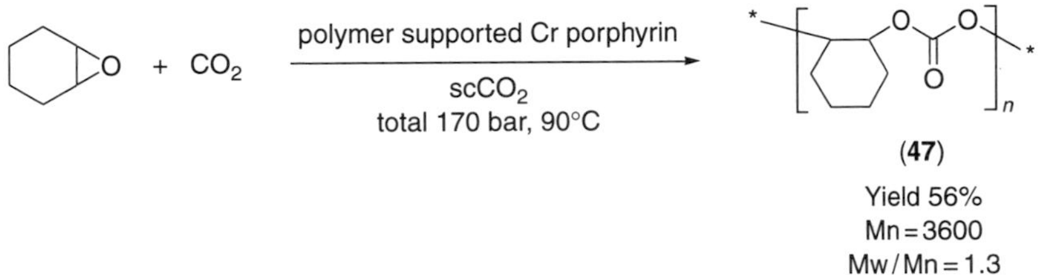

O + CO_2 — polymer supported Cr porphyrin; $scCO_2$ total 170 bar, 90°C → (**47**) Yield 56%, Mn = 3600, Mw/Mn = 1.3

Scheme 87

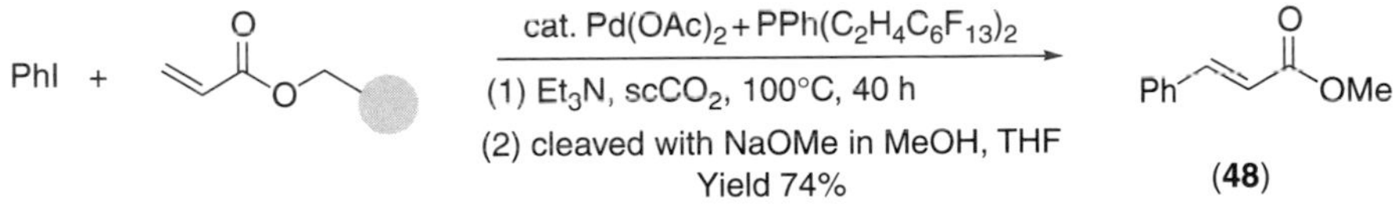

PhI + — cat. $Pd(OAc)_2 + PPh(C_2H_4C_6F_{13})_2$; (1) Et_3N, $scCO_2$, 100°C, 40 h; (2) cleaved with NaOMe in MeOH, THF; Yield 74% → Ph OMe (**48**)

Scheme 88

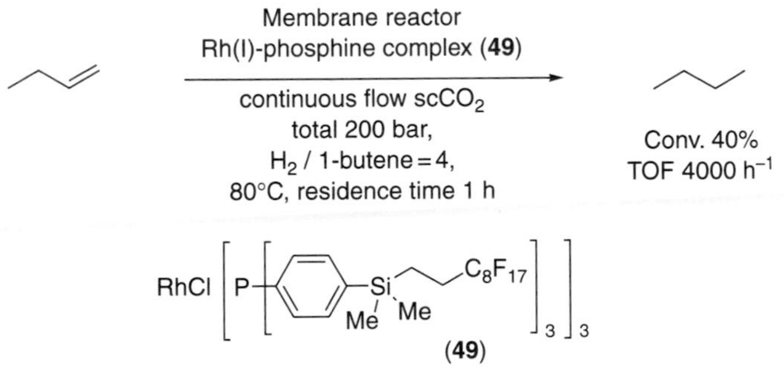

Membrane reactor, Rh(I)-phosphine complex (**49**); continuous flow $scCO_2$ total 200 bar, H_2 / 1-butene = 4, 80°C, residence time 1 h → Conv. 40%, TOF 4000 h^{-1}

$RhCl[P[C_6H_4-Si(Me)(Me)CH_2CH_2C_8F_{17}]_3]_3$ (**49**)

Scheme 89

4.4.4 *Other separation processes*

A new separation procedure for homogeneous catalysts that combines a membrane reactor and $scCO_2$ was recently developed (Scheme 89 and Figure 4.6).[193] A mixture of a rhodium complex with a fluorous phosphine ligand **49**, 1-butene, hydrogen and $scCO_2$ is introduced into a membrane reactor with a pore size of 0.6 nm. Since

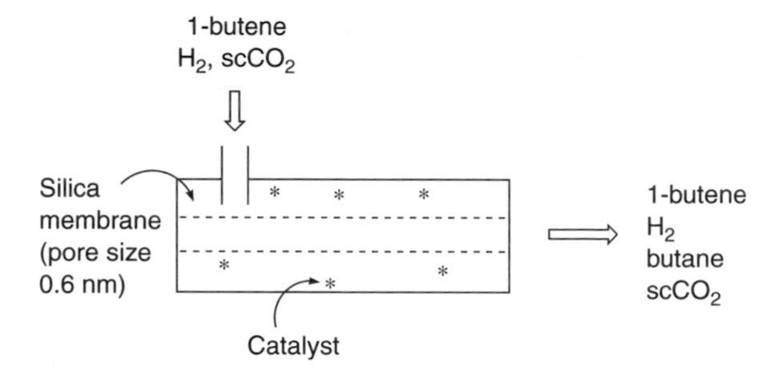

Figure 4.6

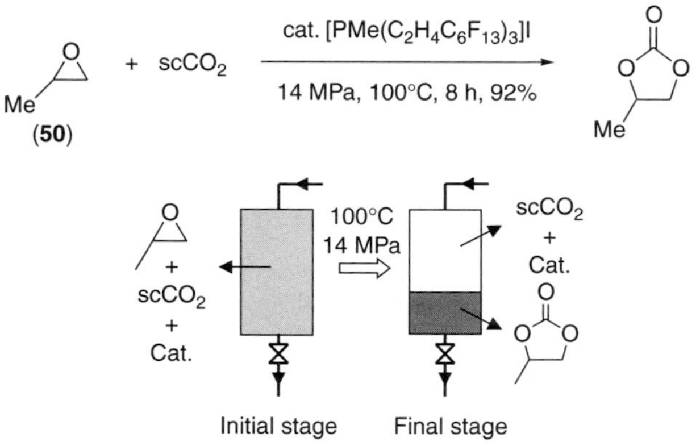

Me (**50**) + $scCO_2$ → cat. $[PMe(C_2H_4C_6F_{13})_3]I$; 14 MPa, 100°C, 8 h, 92%

+ $scCO_2$ + Cat.
100°C
14 MPa
$scCO_2$ + Cat.
Initial stage
Final stage

Scheme 90

the catalyst cannot permeate the membrane, butane is continuously obtained in a catalyst-free form with a conversion of 40% and a TOF of 4000 h^{-1}.

Alkylene carbonate synthesis in $scCO_2$ displays ideal phase behavior for $scCO_2$/product and product/catalyst separation. Initially, starting epoxides such as propylene oxide **50** form a homogeneous phase with $scCO_2$. As the reaction proceeds, the resulting alkylene carbonate spontaneously separates from the supercritical CO_2 phase as a lower liquid phase. The product is recovered from the bottom of the reactor, while maintaining high CO_2 pressure and temperature inside the reactor. In addition, CO_2-philic polyfluoroalkyl phosphonium iodides (e.g. $(C_6F_{13}C_2H_4)_3CH_3PI$) can be employed as catalysts that can, as a result of their high solubility in supercritical CO_2, be used repeatedly (Scheme 90).[194]

Hydrogenation of long-chain alkenes is conducted in $scCO_2$ using a rhodium phosphine complex immobilized on a fluorous polymer (Scheme 91).[195] The polymer is soluble in $scCO_2$ under the reaction conditions and can be filtered off after depressurization.

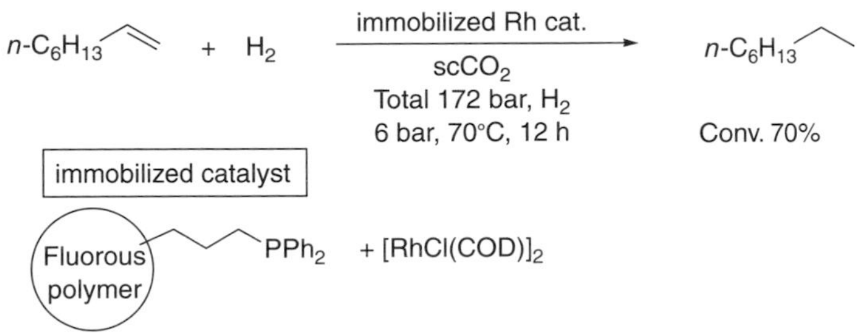

$n\text{-}C_6H_{13}$–CH=CH$_2$ + H_2 → (immobilized Rh cat., $scCO_2$, Total 172 bar, H_2 6 bar, 70°C, 12 h) → $n\text{-}C_6H_{13}$–CH$_2$CH$_3$, Conv. 70%

immobilized catalyst: Fluorous polymer–(CH$_2$)$_n$–PPh$_2$ + $[RhCl(COD)]_2$

Scheme 91

Table 4.3

Ligands	Authors	Ref.
PMe_3	Noyori *et al.*	48
PEt_3	Cole-Hamilton *et al.*	56
$P(2\text{-furyl})_3$	Tumas, Rayner *et al.*	95, 97
$P(OEt)_3$	Ikariya *et al.*	101
$P(C_6H_{11})_2(R_f)$	Tumas *et al.*	167
$P(R_f)_3$	Horváth *et al.*	164, 168, 196
$P(C_6H_4\text{-}R_f)_3$	Leitner *et al.*	52
$P(O\text{-}C_6H_4\text{-}R_f)_3$	Leitner *et al.*	53
$P(C_6H_4\text{-O-fluoroalkyl})_3$	Erkey *et al.*	54, 197
$P(C_6H_4\text{-}CF_3)_3$	Erkey *et al.*	54

$R_f = C_2H_4C_6F_{13}$.

4.4.5 New CO_2-philes

The best-known methodology for solubilizing substrates and catalysts in $scCO_2$ is the introduction of multiple fluorine atoms. In addition, sterically small molecules and compounds with a high oxygen content usually exhibit high CO_2-philicity. Table 4.3 summarizes typical phosphine ligands used for preparing homogeneous catalysts with high solubility in $scCO_2$. As $scCO_2$ technology expands, variations in solubilizing methods, which are environmentally friendlier than introducing fluorous substituents, are becoming important.

Copolymers obtained from epoxides and CO_2 **51** have recently been found to be highly soluble in $scCO_2$ (Figure 4.7).[39] Also, new surfactants have been proposed based on an Anionic Surfactant Aerosol OT (AOT)-like structure with high degrees of chain branching **52**[198] and a siloxane-based CO_2-philic group.[199] In addition, acetylated sugars have been reported to have high CO_2-philicity.[40] In general, ester and ether groups that have structures similar to CO_2 effectively promote CO_2-philicity. For example, phosphines with furyl groups (see Section 4.3.5) or alkoxycarbonyl substituents[200] enhance CO_2-philicity.

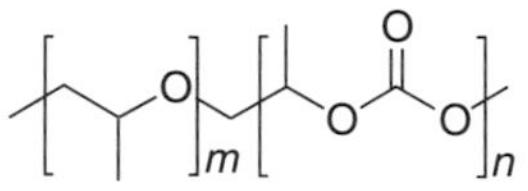

(51)

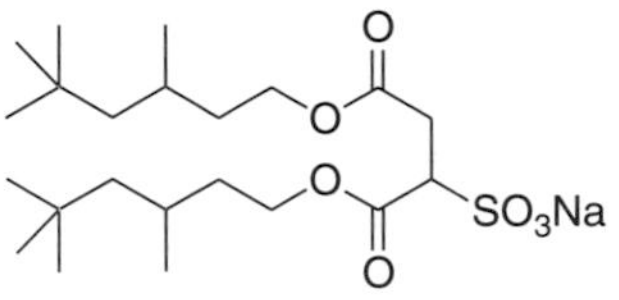

(52)

Figure 4.7

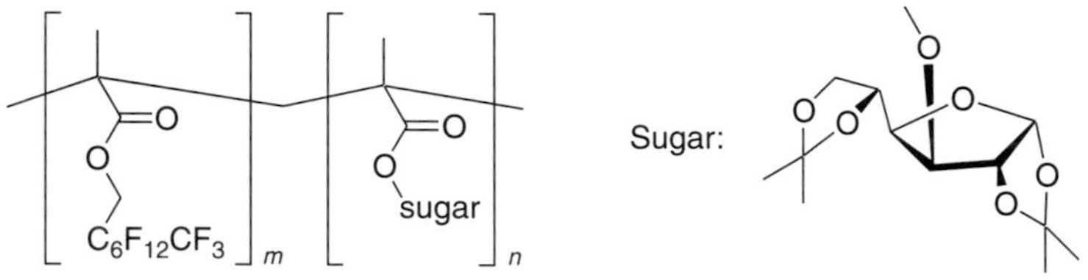

(53)

Figure 4.8

Recently, a combination of fluorous substituents and a sugar-derived structure allowed the preparation of the $scCO_2$-soluble copolymer **53** as a novel phase-transfer catalyst (Figure 4.8).[201] Dendrimers with fluorous substituents were also prepared for the same use.[202] They are soluble in dense carbon dioxide and can solubilize otherwise CO_2-insoluble compounds such as Pd-nanoparticles (Scheme 92).[191] The resulting dendrimer-encapsulated Pd catalyzes the hydrogenation of styrene and the Heck reaction of phenyl iodide.

4.5 Experimental methods

4.5.1 Hydroformylation in $scCO_2$

(Ref. 56) A rhodium catalyst solution was prepared by dissolving $[Rh_2(OAc)_4]$ (0.052 g, 0.1176 mmol) in PEt_3 (0.1 mL, 0.67 mmol) under a nitrogen atmosphere. To the red solution was added degassed 1-hexene (2 mL, 16 mmol). The mixed solution was transferred into a 36 mL Hastelloy autoclave equipped with a mechanical stirrer and a sapphire base for visual observation of the contents. The autoclave was then pressurized with CO (20 bar) and H_2 (20 bar) and the mixture was stirred at room temperature for 1 h. Liquid CO_2 was introduced into the autoclave using a cooled-head high performance liquid chromatography (HPLC) pump until the total pressure reached 65 bar. The autoclave was then heated to 100°C and stirred for 1 h. The pressure and temperature were monitored throughout the reaction. Visual

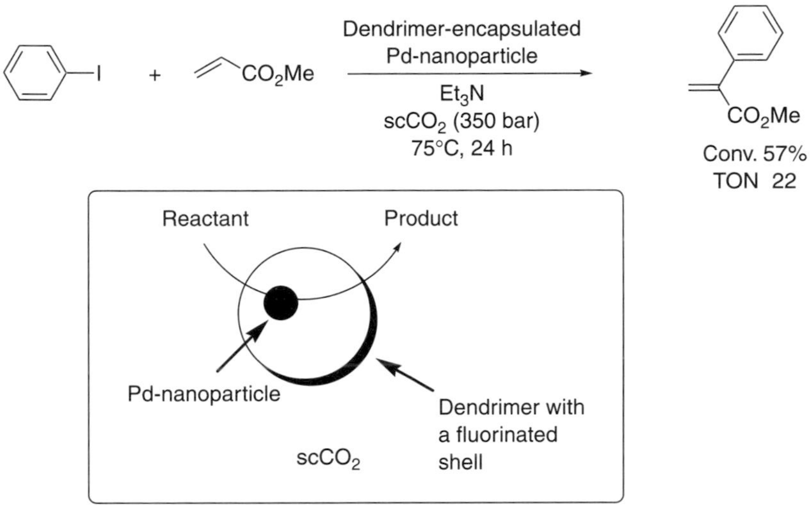

Dendrimer-encapsulated Pd-nanoparticle; Et_3N; $scCO_2$ (350 bar); 75°C, 24 h

CO_2Me

Conv. 57%
TON 22

Reactant
Product
Pd-nanoparticle
Dendrimer with a fluorinated shell
$scCO_2$

Scheme 92

observation through the sapphire window showed that the reaction proceeded homogeneously in a monophasic pale yellow solution. The autoclave was cooled to -50°C using dry ice, and the CO_2 was vented. An aliquot of 1.8 mL of the liquid product was collected and analyzed using gas liquid chromatography (GLC) (quantitative analysis) and gas chromatography–mass spectrometry (GC-MS) (identification of products). C_7 aldehydes with an $n : i$ ratio of 2.4 were produced in 82% yield with heptanol in 2.3%, and the TOF was 57 h^{-1}. $[RhH(CO)_2(PEt_3)_2]$ was found by ^{31}P-NMR (nuclear magnetic resonance) to be a major species formed in the final liquid.

4.5.2 *Polymer synthesis in $scCO_2$*

(Ref. 45) A 10 mL high-pressure reaction view cell containing a magnetic stirring bar was charged with 1,1-dihydroperfluorooctyl acrylate (5.0 g, purified using column chromatography) and recrystallized azobisisobutyronitrile (50 mg), purged with argon for $\sim$10 min, and then filled with CO_2 to $<$70 bar. The cell was heated to 59.4 $\pm$ 0.1°C and the pressure was adjusted to 207 $\pm$ 0.5 bar by the addition of CO_2 (the total amount of CO_2 added was $\sim$6.6 g). The polymerization proceeded homogeneously in an optically clear system. When CO_2 was vented after 48 h, the polymer precipitated because it is insoluble in CO_2 at lower pressures. The polymer and any unreacted monomer were dissolved in 1,1,2-trifluorotrichloroethane (Freon-113) and quantitatively removed from the reaction vessel. The polymer was precipitated into a large excess of methanol washed several times with methanol and then dried *in vacuo* overnight. A transparent viscous polymer was obtained in

65% yield (3.25 g). The molecular weight of the polymer determined using gel permeation chromatography (GPC) was 270 000 g/mol. The polymer showed identical IR and ^{1}H-NMR spectra to the polymer synthesized in CFCs.

4.5.3 Suzuki–Miyaura reaction in $scCO_2$

(Ref. 98) Palladium(II) acetate (11 mg, 0.05 mmol) and tolylboronic acid (204 mg, 1.5 mmol) were placed in a 10 mL stainless steel cell. The cell was transferred into a glove box and tri-*tert*-butylphosphine (20 mg, 0.1 mmol) was added under a nitrogen atmosphere. After the cell was sealed and taken out of the glove box, iodobenzene (0.204 g, 1 mmol) and N,N,N',N'-tetramethylhexane-1,6-diamine (0.172 g, 1 mmol) were injected through the inlet port. The cell was then charged with CO_2 (~5 mL of liquid CO_2) to ~60 bar and heated to 120°C. At 120°C, the pressure was adjusted to 207 bar by the addition of CO_2 and the reaction was started. After 16 h, the cell was cooled to room temperature and gaseous contents were vented into ethyl acetate (100 mL) until an atmospheric pressure was reached. The cell was then opened and washed out with ethyl acetate (20 mL). Both ethyl acetate solutions were combined and concentrated *in vacuo* to give the crude product. This product was adsorbed onto a flash silica column using CH_2Cl_2 and eluted with hexane. A white crystalline solid of 4-phenyltoluene was obtained in 76% yield (128 mg).

4.5.4 Continuous etherification in $scCO_2$

(Ref. 119) The continuous etherification of 1,4-butanediol to tetrahydrofuran in $scCO_2$ was carried out using a fixed-bed flow reaction system. Figure 4.9 shows a schematic view of the apparatus, which consisted of a CO_2 cylinder (Cyl), excess flow cutout valves (EF), a pneumatic pump (20–200 bar, module PM 101, NWA GmbH) (PP), a regulator that determined the system pressure (S), an HPLC pump (0.3–20.0 mL/min, Gilson model 303) (P), a mixer (M), a reactor made from 316

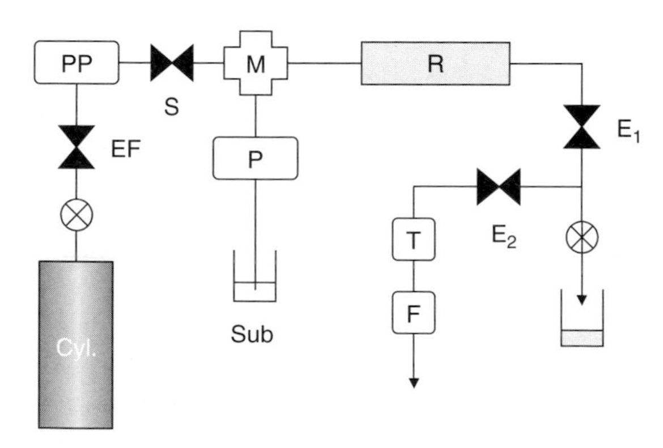

Figure 4.9

stainless steel tubing (9 mm i.d., length 152 mm, 10 mL volume) containing the solid catalyst (R), expansion valves for pressure reduction and flow control (module PE 103, NWA GmbH) (E_1 and E_2), a trap for any droplets inadvertently carried through with the CO_2 (T) and a flow meter (F). Sub stands for 1,4-butanediol. The reactor was heated by three cartridge-heaters in an aluminum block, which encased the reactor. A thermocouple reading the temperature of the catalyst bed thermostatically controlled the heaters. Additional thermocouples monitored the temperatures of the Al heating block and the fluid at the end of the reactor. CO_2 was pressurized using PP and then reduced to system pressure through S. The CO_2 was mixed with 1,4-butanediol in M and the mixture was passed through R containing a DELOXAN ASP catalyst (Degussa-Huls AG, Frankfurt, Germany). After passing through R, the fluid was depressurized stepwise using E_1 and the products were separated from CO_2. Adjusting E_2 set the flow rate of CO_2 to 0.65 L/min at 1 bar and 20°C (1.06 g of CO_2/min), which was measured using F. The recovered products were analyzed using NMR, GC or GC-MS without further workup. In a typical experiment, when the reaction was run with a flow rate of 1,4-butanediol of 0.5 mL/min at a total pressure of 100 bar at 150°C, 1,4-butanediol was fully converted and the yield of tetrahydrofuran reached 100%.

4.5.5 Enzymatic reaction in scCO₂

(Ref. 143) A freshly prepared cell of *Geotrichum candidum* IFO 5767 (0.25 g wet weight) was suspended in H_2O (0.75 mL) and 2-propanol (0.05 mL), and immobilized on water-absorbing polymer (BL-100®, 0.125 g). A 10 mL high-pressure-resistant stainless steel vessel was charged with the immobilized cell, a magnetic stirring bar and fluoromethyl phenyl ketone (0.017 mmol). The ketone was placed in a glass tube to prevent it from coming into contact with the biocatalyst before achieving supercritical conditions. The vessel was then warmed to 35°C and charged with preheated (35°C) CO_2 to 100 bar. After the mixture was stirred at 35°C for 12 h, the CO_2 was liquefied at −10°C and the gas pressure was released. The resulting residue was dissolved in ether, and the mixture was put on Extrelut and quickly eluted with ether. The yield and ee of the alcohol produced were measured using a chiral GC column (Chirasil-DEX CB, 25 m, He 2 mL/min). The absolute configuration was determined by comparing the GC retention time with those of authentic samples. Thus, (*R*)-1-phenyl-2-fluoroethanol was obtained in a 96% yield with 96% ee.

4.5.6 Photochemical reaction in scCO₂

(Ref. 164) Figure 4.10 shows a schematic view of apparatus for the rhodium-catalyzed carbonylation of methane. A 20 mL autoclave with sapphire windows was charged with $RhCl(PMe_3)_3$ (0.014 mmol), benzaldehyde (18 μL), carbon monoxide (3 bar) and methane (110 bar). The autoclave was then charged with CO_2 up to a total pressure of 300 bar at 4°C. The reaction mixture was irradiated through the

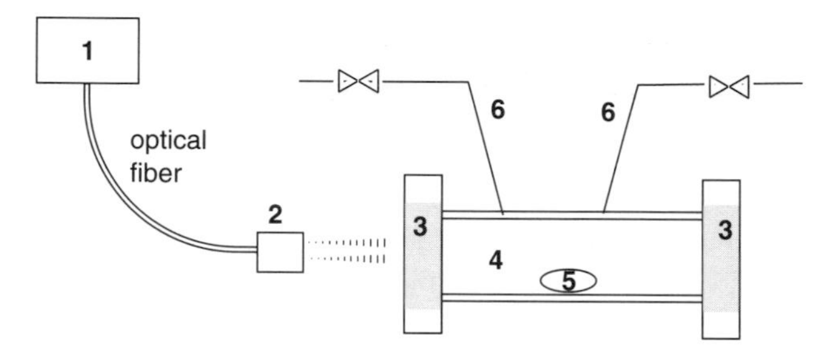

1 High-pressure mercury lamp with an optical fiber; **2** outlet for light;
3 sapphire windows; **4** high-pressure reactor (diameter 20 mm, inner volume 20 mL);
5 magnetic bar for stirring; **6** CO_2 inlet and outlet.

Figure 4.10

sapphire windows with a 250 W high-pressure mercury lamp (USHIO SP3-250) for 16 h. The turnover for acetaldehyde was 77.

4.5.7 *Hydrogenation of styrene in an scCO2/H2O emulsion*

(Ref. 175) Three different surfactants formed water in CO_2 (w/c) or CO_2 in water (c/w) emulsions: (1) anionic perfluoropolyether ammonium carboxylate (PFPE $COO^-NH_4^+$, Mw = 740 and 2500 g/mol), (2) cationic Lodyne 106A ($C_6F_{13}(CH_2)_2SCH_2CH(OH)CH_2N^+(CH_3)_3Cl^-$, Mw = 531.5 g/mol) and (3) non-ionic poly(butylene oxide)-*b*-poly(ethylene oxide) (PBO-PEO, MW) 860-*b*-660 g/mol). PFPE was synthesized according to the literature[203]. Lodyne 106A and PBO-*b*-PEO were purchased from CIBA and PPG Industries, respectively. These surfactants formed emulsions with droplet diameters of 0.5–15 μm and surface areas up to 105 m^2/L, and could easily be broken simply by decreasing the pressure. Owing to the low interfacial tension, γ, between water and CO_2 (17 $mN\,m^{-1}$ at pressures above 70 bar), emulsions with only 0.1–2.0 wt% surfactant were formed. These emulsions dissolved both hydrophobic and hydrophilic substrates. Conductivity measurements showed that the hydrophilic PBO-PEO (400 μS/cm) and Lodyne 106A (800 μS/cm) formed c/w emulsions, whereas the water-insoluble PFPE surfactant formed c/w emulsions when the molecular weight was 740 (100 μS/cm) and w/c emulsions when the molecular weight was 2500 (0.1 μS/cm). Optical microscopic measurements revealed that the PBO PEO and PFPE emulsion droplets were 3–5 μm while the Lodyne 106A droplets were 10–15 μm. The pH of the aqueous phase for water/CO_2 mixtures was ~3 as a result of the formation and dissociation of carbonic acid.

To demonstrate the advantages of surfactant-stabilized emulsions for biphasic homogeneous catalysis, the hydrogenation of styrene using the water-soluble catalyst $RhCl(tppds)_3$ (tppds = tris(3,5-disulfonatophenyl)phosphine) was investigated.

The reactions were conducted in custom-designed variable-volume view cells[204] equipped with a six-port high-pressure sampling valve for reagent addition and sample analysis via GC. The system was stirred with a magnetic bar kept at a steady rotation speed for all reactions. Each reaction was run under the following conditions: 50/50 wt% water/CO_2 (4.75 g each), 1.5 wt% surfactant, 80 mmol/L styrene, 1 mol% catalyst to substrate, Rh/L = 1/6, 40°C, 275 bar. The TOF thus measured in an H_2O/CO_2 emulsion system was 300 h^{-1} at 50% conversion, and the TOFs in biphasic H_2O/toluene and H_2O/CO_2 systems were 4 and 26 h^{-1}, respectively.

4.5.8 *Hydroformylation in scCO$_2$/H$_2$O*

(Ref. 176) $[Rh(CO)_2(acac)]$ (0.004 mmol) and $P(C_6H_4\text{-}p\text{-}C_2H_4C_6F_{13})_3$ (0.04 mmol) were placed in a 100 mL autoclave under an argon atmosphere. The mixture was pressurized with CO/H_2 (1 : 1) (20 bar), and then CO_2 (50 g) was introduced using a compressor. *Endo*-5-carboxy-6-carbomethoxy-2-norbornen (8 mmol) was dissolved in distilled and deoxygenated water (20 mL) and the aqueous solution was added to the autoclave using an HPLC pump. When the reaction mixture was heated to 60°C and stirred at 2000 rpm, a white emulsion was formed. After 20 h, the heating and stirring were stopped and a rapid phase separation occurred. The lower water layer was then taken out using a capillary needle valve at the bottom of the autoclave and collected in a Schlenk flask. The substrate/product mixture of more than 90 vol% was recovered from the autoclave. A 2 mL sample was worked up by removing the water *in vacuo*. The conversion, determined using ^{1}H-NMR spectroscopy, was 86% and the turnover number was 1720. The rhodium leaching, determined using atomic absorption spectroscopy, was 0.05 ppm. Adding a new aqueous solution of the olefin into the autoclave using the HPLC pump meant that the run could be repeated.

4.5.9 *Hydroformylation in an scCO$_2$/IL biphasic system using a continuous flow reactor*

(Ref. 181) Hydroformylation of oct-1-ene was carried out using a continuous flow reactor assembled as shown in Figure 4.11. The oct-1-ene, CO/H_2 and CO_2 joined above the reactor and passed through the mixture of $scCO_2$ and $[bmim][PF_6]$ containing the catalyst. The reaction mixture was then removed from the reactor and underwent a two-stage decompression. The products recovered were analyzed using GC. The reaction conditions were a continuous flow of oct-1-ene (0.03 cm^3/min, 1.91 × 10^{-4} mol/min), $[Rh2(OAc)_4]$ (10 mg, 4.5 × 10^{-5} mol Rh) and $[pmim][PhP(C_6H_4SO_3)_2]$ (0.47 g, 7.1 × 10^{-4} mol) as catalyst precursors, $[bmim][PF_6]$ (6 cm^3) as a reaction medium, a continuous flow of CO/H_2 (0.45 dm^3/min), 200 bar total pressure, 100°C. Aldehydes could be produced at a constant rate and selectivity (l : b = 3.8) over 72 h.

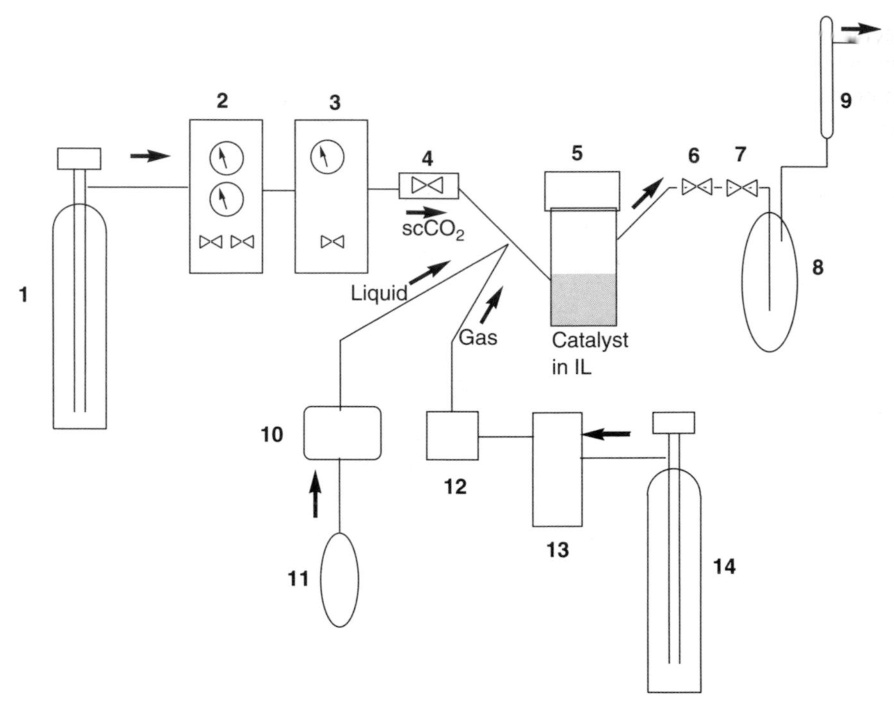

1 CO_2 supply; **2** liquid CO_2 pump; **3** ressure regulator; **4** non-return valve, **5** autoclave with magnetically driven paddle stirrer, **6** first expansion valve; **7** second expansion valve, **8** collection vessel; **9** flow metre; **10** liquid substrate pump; **11** liquid substrate supply; **12** CO/H_2 dosemeter; **13** gas booster; **14** CO/H_2 supply.

Figure 4.11

Table 4.4 Characteristics of the membrane reactor.

Thickness of selective silica layer	200 nm
Length of membrane	0.30 m
Outer diameter of ceramic membrane	0.014 m
Volume of membrane reactor	35.0 mL
CO_2 permeance[a] (200 bar, 353 K)	3.0×10^{-3} mol/m^2 s bar

[a] Flux divided by the pressure difference across the membrane.

4.5.10 Hydrogenation of 1-butene in scCO$_2$ using a membrane reactor

(Ref. 192) The setup of the continuous flow membrane reactor is shown in Figure 4.6. A microporous silica membrane with an average pore size of 0.6 nm[205] was used in the reactor. The characteristics of the reactor are summarized in Table 4.4. First, the active catalyst was generated *in situ* in a high-pressure reactor from $[RhCl(COD)]_2$ and $P(C_6H_4$-*p*-$SiMe_2CH_2CH_2C_8F_{17})_3$, which were introduced into the system by gently flushing CO_2 through the high-pressure reactor. Then, the reaction was started by adding the 1-butene and hydrogen to the membrane module with $scCO_2$. The membrane was pressurized up to 200 bar with the reaction mixture using an LKB HPLC pump and was monitored using a Meyvis 802-C pressure module. The permeate needle valve was opened to vary the transmembrane pressure between 0.5 and 10 bar. The permeate was analyzed using GC and GC-MS. During the continuous reaction experiments, the concentration of the rhodium complex was 1.0×10^{-6} mol/L, and 1-butene and hydrogen were present in a concentration of 0.02 and 0.08 mol/L, respectively. In a typical experiment, the TOF based on the initial amount of catalyst added was 4000 h^{-1} at a pressure difference of 3 bar (a residence time of 62 min).

4.5.11 *Catalyst/product separation for cyclic carbonate synthesis from* $scCO_2$

(Ref. 194) A $(C_6F_{13}C_2H_4)_3MePI$ catalyst was prepared as follows: $(C_6F_{13}C_2H_4)_3P$ was synthesized according to the literature.[206] A solution of $(C_6F_{13}C_2H_4)_3P$ (8.58 g, 8.00 mmol) and methyl iodide (0.50 mL, 8.00 mmol) in acetone (20 mL) was refluxed for 3 days under an argon atmosphere. After evaporation of the acetone under reduced pressure, the crude residue was washed several times with dry ether and dried *in vacuo* (yield 5.83 g, 60%).

A typical procedure for catalyst recycling is as follows. The reactions were carried out using a stainless steel reactor (20 cm^3 inner volume) equipped with a mechanical stirrer, sapphire windows and a valve at the bottom of the reactor for recovering the product. The mixture of propylene oxide (57.2 mmol) and CO_2 was heated at 100°C under a total pressure of 140 bar in the presence of $(C_6F_{13}C_2H_4)_3MePI$ (0.572 mmol, 1 mol%) and biphenyl (internal standard for GC analysis). Visual inspection through the sapphire windows revealed that all the constituents were miscible and formed a homogeneous phase at the beginning of the reaction, confirming the homogeneous catalysis of $(C_6F_{13}C_2H_4)_3MePI$. As the reaction proceeded, the resulting propylene carbonate separated from the uniform phase and gradually accumulated to form a lower phase in the reactor; finally, the volume ratio of the upper supercritical phase to the lower phase became ~3 (Figure 4.12). After 24 h, the lower phase

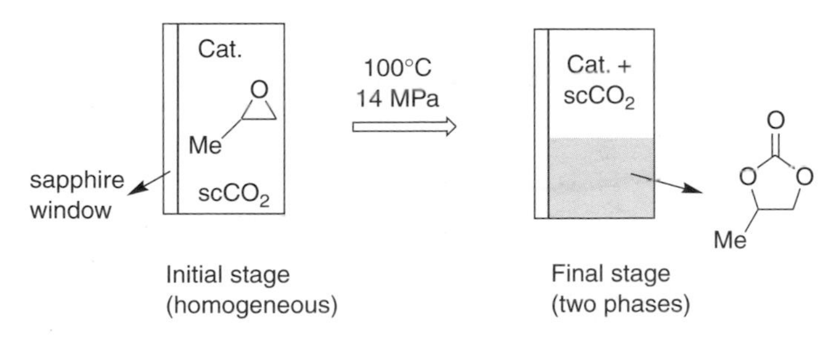

Figure 4.12

was taken out of the reactor by slowly opening the valve at the bottom. During this procedure the pressure decreased from 140 to 110 bar. The reaction was then repeated by supplying propylene oxide (57.2 mmol) containing biphenyl (internal standard for GC analysis) to the reactor at 110 bar followed by readjusting the pressure to 140 bar upon the introduction of CO_2. Propylene carbonate was produced in the second run with almost the same yield as the first run (>90%), showing that the $(C_6F_{13}C_2H_4)_3$MePI remained in the upper supercritical phase.

References

[1] Cagniard de LaTour, C. *Ann. Chim. Phys.* **1823**, *22*, 410.
[2] Thurston, R. H. *A History of the Growth of the Steam Engine*; Kegan, Paul, Trench, Trubner & Co. Ltd.: London, 1895.
[3] Jessop, P. G.; Leitner, W., Eds. *Chemical Synthesis Using Supercritical Fluids*; Wiley-VCH: Weinheim, 1999.
[4] Zosel, K. *Angew. Chem., Int. Ed. Engl.* **1978**, *17*, 702.
[5] Williams, J. R.; Clifford, A. A., Eds. *Supercritical Fluid Methods and Protocols*; Humana Press: Totowa, NJ, 2000.
[6] Canelas, D. A.; Betts, D. E.; DeSimone, J. M.; Yates, M. Z.; Johnson, K. P. *Macromolecules* **1998**, *31*, 6794.
[7] Anon. *Chem. Br.* **1997**, *33(12)*, 16.
[8] McCoy, M. *Chem. Eng. News* **1999**, 11.
[9] McCoy, M. *Chem. Eng. News* **1999**, 13.
[10] Donohue, M. D.; Geiger, J. L.; Kiamos, A. A.; Nielson, K. A. In *ACS Symposium Series*; American Chemical Society, 1996; Vol. 626, pp. 152–167.
[11] Canelas, D. A.; DeSimone, J. M. *Chem. Br.* **1998**, *34(8)*, 38.
[12] Henon, F.; Camaiti, M.; Burke, A.; Carbonell, R.; DeSimone, J. M.; Paicenti, F. *J. Supercrit. Fluids* **1999**, *15*, 173.
[13] Wells, S. L.; DeSimone, J. *Angew. Chem., Int. Ed. Engl.* **2001**, *40*, 518.
[14] Ozcan, A. S.; Clifford, A. A.; Bartle, K. D. *Dyes Pigm.* **1998**, *36*, 103.
[15] Ozcan, A. S.; Clifford, A. A.; Bartle, K. D.; Lewis, D. M. *Dyes Pigm.* **1998**, *36*, 103.
[16] Montero, G.; Hinks, D.; Hooker, J. *J. Supercrit. Fluids* **2003**, *26*, 47.
[17] Maeda, S.; Hongyou, S.; Kunitou, K.; Mishima, K. *Text. Res. J.* **2002**, *72*, 240.
[18] Hendrix, W. A. *J. Ind. Text.* **2001**, *31*, 43.
[19] Han, X.; Koelling, K. W.; Tomasko, D. L.; Lee, L. J. *Polym. Eng. Sci.* **2002**, *42*, 2094.
[20] Liang, M.-T.; Wang, C.-M. *Ind. Eng. Chem. Res.* **2000**, *39*, 4622.
[21] Stewart, I. H.; Derouane, E. G. *Curr. Top. Catal.* **1999**, *2*, 17.
[22] Jessop, P. G.; Ikariya, T.; Noyori, R. *Chem. Rev.* **1999**, *99*, 475.
[23] Leitner, W. *Acc. Chem. Res.* **2002**, *35*, 746.
[24] Baiker, A. *Chem. Rev.* **1999**, *99*, 453.
[25] Tester, J. W.; Danheiser, R. L.; Weinstein, R. D.; Renslo, A.; Taylor, J. D.; Steinfeld, I. J. In *ACS Symposium Series*; American Chemical Society: Washington, 2000; Vol. 767, p. 270.
[26] Oakes, R. S.; Clifford, A. A.; Rayner, C. M. *J. Chem. Soc., Perkin Trans. 1* **2001**, 917.
[27] Clifford, A. A. *Fundamentals of Supercritical Fluids*; Oxford University Press: Oxford, 1998.
[28] Clifford, A. A.; Bartle, K. D.; Rayner, C. M. *University of Leeds.*
[29] Mesiano, A. J.; Beckman, E. J.; Russell, A. J. *Chem. Rev.* **1999**, *99*, 623.
[30] Johns, A. I.; Raschid, S.; Watson, J. T. R.; Clifford, A. A. *J. Chem. Soc., Faraday Trans. 1* **1986**, *82*, 2235.
[31] Jessop, P. G.; Hsiao, Y.; Ikariya, T.; Noyori, R. *J. Am. Chem. Soc.* **1996**, *118*, 344.
[32] Jessop, P. G.; Ikariya, T.; Noyori, R. *Science* **1995**, *269*, 1065.
[33] Brennecke, J. F.; Chateauneuf, J. E. *Chem. Rev.* **1999**, *99*, 433.

[34] Tucker, S. C. *Chem. Rev.* **1999**, *99*, 391.
[35] Diep, P.; Jordan, K. D.; Johnson, J. K.; Beckman, E. J. *J. Phys. Chem. A* **1998**, *102*, 2231.
[36] Dardin, A.; DeSimone, J. M.; Samulski, E. T. *J. Phys. Chem. B* **1998**, *102*, 1775.
[37] DeSimone, J. M.; Murray, E. E.; Menceloglu, Y. Z.; McClain, J. B.; Romack, T. J.; Combes, J. R. *Science* **1994**, *265*, 356.
[38] Lin, Y.; Smart, N. G.; Wai, C. M. *Trends Anal. Chem.* **1995**, *14*, 123.
[39] Sarbu, T.; Styranec, T.; Beckman, E. J. *Nature* **2000**, *405*, 165.
[40] Raveendran, P.; Wallen, S. L. *J. Am. Chem. Soc.* **2002**, *124*, 7274.
[41] Potluri, V. K.; Xu, J. H.; Enick, R.; Beckman, E. J.; Hamilton, A. D. *Org. Lett.* **2002**, *4*, 2333.
[42] Randolph, T. W.; Blanch, H. W.; Prausintz, J. M.; Wilke, C. R. *Biotechnol. Lett.* **1985**, *7*, 325.
[43] Paulaitis, M. E.; Alexander, G. C. *Pure Appl. Chem.* **1987**, *59*, 61.
[44] Aida, T.; Squires, T. G. In *ACS Symposium Series*; American Chemical Society: Washington, 1987; Vol. 329, p. 58.
[45] DeSimone, J. M.; Guan, Z.; Elsbernd, C. S. *Science* **1992**, *257*, 945.
[46] Tanko, J. M.; Blackert, J. F. *Science* **1994**, *263*, 203.
[47] Rathke, J. W.; Klingler, R. J.; Krause, T. R. *Organometallics* **1991**, *10*, 1350.
[48] Jessop, P. G.; Ikariya, T.; Noyori, R. *Nature* **1994**, *368*, 231.
[49] Morgensten, D. A.; LeLacheur, R. M.; Morita, D. K.; Borkovsky, S. L.; Feng, S.; Brown, G. H.; Luan, L.; Gross, M. F.; Burk, M. J.; Tumas, W. *ACS Symposium Series*; American Chemical Society: Washington, 1996, Vol. 626, p. 132.
[50] Guo, Y.; Akgerman, A. *Ind. Eng. Chem. Res.* **1997**, *36*, 4581.
[51] Jessop, P. G.; Ikariya, T.; Noyori, R. *Organometallics* **1995**, *14*, 1510.
[52] Kainz, S.; Koch, D.; Baumann, W.; Leitner, W. *Angew. Chem. Int. Ed. Engl.* **1997**, *36*, 1628.
[53] Koch, D.; Leitner, W. *J. Am. Chem. Soc.* **1998**, *120*, 13398.
[54] Palo, D. R.; Erkey, C. *Organometallics* **2000**, *19*, 81.
[55] Banet Osuna, A. M.; Chen, W.; Hope, E. G.; Kemmitt, R. D. W.; Paige, D. R.; Stuart, A. M.; Xiao, J.; Xu, L. *J. Chem. Soc., Dalton Trans.* **2000**, 4052.
[56] Bach, I.; Cole-Hamilton, D. J. *Chem. Commun.* **1998**, 1463.
[57] Kainz, S.; Leitner, W. *Catal. Lett.* **1998**, *55*, 223.
[58] Franciò, G.; Leitner, W. *Chem. Commun.* **1999**, 1663.
[59] Franciò, G.; Wittmann, K.; Leitner, W. *J. Organometal. Chem.* **2001**, *621*, 130.
[60] Hu, Y.; Chen, W.; Banet Osuna, A. M.; Stuart, A. M.; Hope, E. G.; Xiao, J. *Chem. Commun.* **2001**, 725.
[61] Hu, Y.; Chen, W.; Banet Osuna, A. M.; Iggo, J. A.; Xiao, J. *Chem. Commun.* **2002**, 788.
[62] Ke, J.; Han, B.; George, M. W.; Yan, H.; Poliakoff, M. *J. Am. Chem. Soc.* **2001**, *123*, 3661.
[63] Burk, M. J.; Feng, S.; Gross, M. F.; Tumas, W. *J. Am. Chem. Soc.* **1995**, *117*, 8277.
[64] Xiao, J.; Nefkens, S. C. A.; Jessop, P. G.; Ikariya, T.; Noyori, R. *Tetrahedron Lett.* **1996**, *37*, 2813.
[65] Kainz, S.; Brinkmann, A.; Leitner, W.; Pfaltz, A. *J. Am. Chem. Soc.* **1999**, *121*, 6421.
[66] Wandeler, R.; Künzle, N.; Schneider, M. S.; Mallat, T.; Baiker, A. *Chem. Commun.* **2001**, 673.
[67] Niessen, H. G.; Eichhorn, A.; Woelk, K.; Bargon, J. *J. Mol. Catal. A* **2002**, *182–183*, 463.
[68] Ohde, H.; Wai, C. M.; Kim, H.; Kim, J.; Ohde, M. *J. Am. Chem. Soc.* **2002**, *124*, 4540.
[69] Ohde, M.; Ohde, H.; Wai, C. M. *Chem. Commun.* **2002**, 2388.
[70] Hitzler, M. G.; Poliakoff, M. *Chem. Commun.* **1997**, 1667.
[71] Licence, P.; Ke, J.; Sokolova, M.; Ross, S. K.; Poliakoff, M. *Green Chem.* **2003**, *5*, 99.
[72] Bhanage, B. M.; Ikushima, Y.; Shirai, M.; Arai, M. *Catal. Lett.* **1999**, *62*, 175.
[73] Tschan, R.; Wandeler, R.; Schneider, M. S.; Burgener, M.; Schubert, M. M.; Baiker, A. *Appl. Catal. A* **2002**, *223*, 173.
[74] Pillai, U. R.; Sahle-Demessie, E. *Chem. Commun.* **2002**, 422.
[75] Trabelsi, F.; Stüber, F.; Abaroudi, K.; Larrayoz, M. A.; Recasens, F.; Sueiras, J. E. *Ind. Eng. Chem. Res.* **2000**, *39*, 3666.
[76] Musie, G.; Wei, M.; Subramaniam, B.; Busch, D. H. *Coord. Chem. Rev.* **2001**, *219–221*, 789.
[77] Dooley, K. M.; Knopf, F. C. *Ind. Eng. Chem. Res.* **1987**, *26*, 1910.
[78] Wu, X.-W.; Oshima, Y.; Koda, S. *Chem. Lett.* **2001**, 1045.
[79] Birnbaum, E. R.; Le Lacheur, R. M.; Horton, A. C.; Tumas, W. *J. Mol. Catal. A* **1999**, *139*, 11.

[80] Musie, G. T.; Wei, M.; Subramaniam, B.; Busch, D. H. *Inorg. Chem.* **2001**, *40*, 3336.
[81] Wei, M.; Musie, G. T.; Busch, D. H.; Subramaniam, B. *J. Am. Chem. Soc.* **2002**, *124*, 2513.
[82] Loeker, F.; Leitner, W. *Chem. Eur. J.* **2000**, *6*, 2011.
[83] Theyssen, N.; Leitner, W. *Chem. Commun.* **2002**, 410.
[84] Bolm, C.; Palazzi, C.; Franciò, G.; Leitner, W. *Chem. Commun.* **2002**, 1588.
[85] Jia, L.; Jiang, H.; Li, J. *Chem. Commun.* **1999**, 985.
[86] Jenzer, G.; Sueur, D.; Mallat, T.; Baiker, A. *Chem. Commun.* **2000**, 2247.
[87] Steele, A. M.; Zhu, J.; Tsang, S. C. *Catal. Lett.* **2001**, *73*, 9.
[88] Pesiri, D. R.; Morita, D. K.; Glaze, W.; Tumas, W. *Chem. Commun.* **1998**, 1015.
[89] Pesiri, D. R.; Morita, D. K.; Walker, T.; Tumas, W. *Organometallics* **1999**, *18*, 4916.
[90] Haas, G. R.; Kolis, J. W. *Organometallics* **1998**, *17*, 4454.
[91] Haas, G. R.; Kolis, J. W. *Tetrahedron Lett.* **1998**, *39*, 5923.
[92] Nolen, S. A.; Lu, J.; Brown, J. S.; Pollet, P.; Eason, B. C.; Griffith, K. N.; Gläser, R.; Bush, D.; Lamb, D. R.; Liotta, C. L.; Eckert, C. A.; Thiele, G. F.; Bartels, K. A. *Ind. Eng. Chem. Res.* **2002**, *41*, 316.
[93] Oakes, R. S.; Clifford, A. A.; Bartle, K. D.; Pett, M. T.; Rayner, C. M. *Chem. Commun.* **1999**, 247.
[94] Carroll, M. A.; Holmes, A. B. *Chem. Commun.* **1998**, 1395.
[95] Morita, D. K.; Pesiri, D. R.; David, S. A.; Glaze, W. H.; Tumas, W. *Chem. Commun.* **1998**, 1397.
[96] Fujita, S.; Yuzawa, K.; Bhanage, B. M.; Ikushima, Y.; Arai, M. *J. Mol. Catal. A* **2002**, *180*, 35.
[97] Shezad, N.; Oakes, R. S.; Clifford, A. A.; Rayner, C. M. *Tetrahedron Lett.* **1999**, *40*, 2221.
[98] Early, T. R.; Gordon, R. S.; Carroll, M. A.; Holmes, A. B.; Shute, R. E.; McConvey, I. F. *Chem. Commun.* **2001**, 1966.
[99] Osswald, T.; Schneider, S.; Wang, S.; Bannwarth, W. *Tetrahedron Lett.* **2001**, *42*, 2965.
[100] Fürstner, A.; Ackermann, L.; Beck, K.; Hori, H.; Koch, D.; Langemann, K.; Liebl, M.; Six, C.; Leitner, W. *J. Am. Chem. Soc.* **2001**, *123*, 9000.
[101] Kayaki, Y.; Noguchi, Y.; Iwasawa, S.; Ikariya, T.; Noyori, R. *Chem. Commun.* **1999**, 1235.
[102] Rose, P. M.; Clifford, A. A.; Rayner, C. M. *Chem. Commun.* **2002**, 968.
[103] Jeong, N.; Hwang, S. H.; Lee, Y. W.; Lim, J. S. *J. Am. Chem. Soc.* **1997**, *119*, 10549.
[104] Wegner, A.; Leitner, W. *Chem. Commun.* **1999**, 1583.
[105] Cacchi, S.; Fabrizi, G.; Gasparrini, F.; Pace, P.; Villani, C. *Synlett* **2000**, 650.
[106] Isaacs, N. S.; Keating, N. *J. Chem. Soc., Chem. Commun.* **1992**, 876.
[107] Ikushima, Y.; Ito, S.; Asano, T.; Yokoyama, T.; Saito, N.; Hatakeda, K.; Goto, T. *J. Chem. Eng. Jpn.* **1990**, *23*, 96.
[108] Ikushima, Y.; Saito, N.; Arai, M. *J. Phys. Chem.* **1992**, *96*, 2293.
[109] Renslo, A. R.; Weinstein, R. D.; Tester, J. W.; Danheiser, R. L. *J. Org. Chem.* **1997**, *62*, 4530.
[110] Ikushima, Y.; Saito, N.; Arai, M. *Bull. Chem. Soc. Jpn.* **1991**, *64*, 282.
[111] Weinstein, R. D.; Renslo, A. R.; Danheiser, R. L.; Harris, J. G.; Tester, J. W. *J. Phys. Chem.* **1996**, *100*, 12337.
[112] Clifford, A. A.; Pople, K.; Gaskill, W. J.; Bartle, K. D.; Rayner, C. M. *J. Chem. Soc., Faraday Trans.* **1998**, *94*, 1451.
[113] Oakes, R. S.; Heppenstall, T. J.; Shezad, N.; Clifford, A. A.; Rayner, C. M. *Chem. Commun.* **1999**, 1459.
[114] Matsuo, J.; Tsuchiya, T.; Odashima, K.; Kobayashi, S. *Chem. Lett.* **2000**, 178.
[115] Chapuis, C.; Kucharska, A.; Rzepecki, P.; Jurczak, J. *Helv. Chim. Acta* **1998**, *81*, 2314.
[116] Fukuzawa, S.; Metoki, K.; Komuro, Y.; Funazukuri, T. *Synlett* **2002**, 134.
[117] Weinstein, R. D.; Renslo, A. R.; Danheiser, R. L.; Tester, J. W. *J. Phys. Chem. B* **1999**, *103*, 2878.
[118] Hitzler, M. G.; Smail, F. R.; Ross, S. K.; Poliakoff, M. *Chem. Commun.* **1998**, 359.
[119] Gray, W. K.; Smail, F. R.; Hitzler, M. G.; Ross, S. K.; Poliakoff, M. *J. Am. Chem. Soc.* **1999**, *121*, 10711.
[120] Mikami, K.; Matsukawa, S.; Kayaki, Y.; Ikariya, T. *Tetrahedron Lett.* **2000**, *41*, 1931.
[121] Komoto, I.; Kobayashi, S. *Chem. Commun.* **2001**, 1842.
[122] Clark, M. C.; Subramaniam, B. *Ind. Eng. Chem. Res.* **1998**, *37*, 1243.

[123] Santana, G. M.; Akgerman, A. *Ind. Eng. Chem. Res.* **2001**, *40*, 3879.
[124] Ginosar, D. M.; Thompson, D. N.; Coates, K.; Zalewski, D. J. *Ind. Eng. Chem. Res.* **2002**, *41*, 2864.
[125] Kuo, T.-W.; Tan, C.-S. *Ind. Eng. Chem. Res.* **2001**, *40*, 4724.
[126] Clark, M. R.; DeSimone, J. M. *Macromolecules* **1995**, *28*, 3002.
[127] Kendall, J. L.; Canelas, D. A.; Young, J. L.; DeSimone, J. M. *Chem. Rev.* **1999**, *99*, 543.
[128] Cooper, A. I. *J. Mater. Chem.* **2000**, *10*, 207.
[129] Wells, S. L.; DeSimone, J. *Angew. Chem. Int. Ed.* **2001**, *40*, 518.
[130] Mistele, C. D.; Thorp, H. H.; DeSimone, J. M. *J. Macromol. Sci., Pure Appl. Chem.* **1996**, *A33*, 953.
[131] Hamilton, J. G.; Rooney, J. J.; DeSimone, J. M.; Mistele, C. *Macromolecules* **1998**, *31*, 4387.
[132] Darensbourg, D. J.; Stafford, N. W.; Katsurao, T. *J. Mol. Catal. A* **1995**, *104*, L1.
[133] Super, M.; Berluche, E.; Costello, C.; Beckman, E. *Macromolecules* **1997**, *30*, 368.
[134] Mang, S.; Cooper, A. I.; Colclough, M. E.; Chauhan, N.; Holmes, A. B. *Macromolecules* **2000**, *33*, 303.
[135] Kapellen, K. K.; Mistele, C. D.; DeSimone, J. M. *Macromolecules* **1996**, *29*, 495.
[136] Hori, H.; Six, C.; Leitner, W. *Macromolecules* **1999**, *32*, 3178.
[137] de Vries, T. J.; Duchateau, R.; Vorstman, M. A. G.; Keurentjes, J. T. F. *Chem. Commun.* **2000**, 263.
[138] Kläui, W.; Bongards, J.; Reiss, G. J. *Angew. Chem. Int. Ed.* **2000**, *39*, 3894.
[139] Mesiano, A. J.; Beckman, E. J.; Russell, A. J. *Chem. Rev.* **1999**, *99*, 623.
[140] Matsumura, S.; Nakamura, T.; Yao, E.; Toshima, K. *Chem. Lett.* **1999**, 581.
[141] Mori, T.; Okahata, Y. *Chem. Commun.* **1998**, 2215.
[142] Matsuda, T.; Kanamaru, R.; Watanabe, K.; Harada, T.; Nakamura, K. *Tetrahedron Lett.* **2001**, *42*, 8319.
[143] Matsuda, T.; Harada, T.; Nakamura, K. *Chem. Commun.* **2000**, 1367.
[144] Matsuda, T.; Ohashi, Y.; Harada, T.; Yanagihara, R.; Nagasawa, T.; Nakamura, K. *Chem. Commun.* **2001**, 2194.
[145] Jessop, P. G.; Hsiao, Y.; Ikariya, T.; Noyori, R. *J. Am. Chem. Soc.* **1994**, *116*, 8851.
[146] Kayaki, Y.; Suzuki, T.; Ikariya, T. *Chem. Lett.* **2001**, 1016.
[147] Kawanami, H.; Ikushima, Y. *Chem. Commun.* **2000**, 2089.
[148] Yasuda, H.; He, L.-N.; Sakakura, T. *J. Catal.* **2002**, *209*, 547.
[149] Bhanage, B. M.; Fujita, S.; Ikushima, Y.; Arai, M. *Appl. Catal. A* **2001**, *219*, 259.
[150] Lu, X.-B.; He, R.; Bai, C.-X. *J. Mol. Catal. A* **2002**, *186*, 1.
[151] Lu, X.-B.; Feng, X.-J.; He, R. *Appl. Catal. A* **2002**, *234*, 25.
[152] Sakakura, T.; Saito, Y.; Okano, M.; Choi, J.-C.; Sako, T. *J. Org. Chem.* **1998**, *63*, 7095.
[153] Sakakura, T.; Choi, J.-C.; Saito, Y.; Matsuda, T.; Sako, T.; Oriyama, T. *J. Org. Chem.* **1999**, *64*, 4506.
[154] Choi, J.-C.; He, L.-N.; Yasuda, H.; Sakakura, T. *Green Chem.* **2002**, *4*, 230.
[155] Choi, J.-C.; Sakakura, T.; Sako, T. *J. Am. Chem. Soc.* **1999**, *121*, 3793.
[156] Zhao, T.; Han, Y.; Sun, Y. *Stud. Surf. Sci. Catal.* **2000**, *130A*, 461.
[157] Fujita, S.; Bhanage, B. M.; Ikushima, Y.; Arai, M. *Green Chem.* **2001**, *3*, 87.
[158] Yoshida, M.; Hara, N.; Okumura, S. *Chem. Commun.* **2000**, 151.
[159] Abla, M.; Choi, J.-C.; Sakakura, T. *Chem. Commun.* **2001**, 2238.
[160] Abla, M.; Choi, J.-C.; Sakakura, T. In *50th Symposium on Organometallic Chemistry, Japan*; Kinki Chemical Society, Japan: Osaka University, September, 2003.
[161] Rohr, M.; Geyer, C.; Wandeler, R.; Schneider, M. S.; Murphy, E. F.; Baiker, A. *Green Chem.* **2001**, *3*, 123.
[162] Kawanami, H.; Ikushima, Y. *Tetrahedron Lett.* **2002**, *43*, 3841.
[163] Tominaga, K.; Sasaki, Y. *Synlett* **2002**, 307.
[164] Choi, J.-C.; Kobayashi, Y.; Sakakura, T. *J. Org. Chem.* **2001**, *66*, 5262.
[165] Bitterwolf, T. E.; Kline, D. L.; Linehan, J. C.; Yonker, C. R.; Addleman, R. S. *Angew. Chem. Int. Ed. Engl.* **2001**, *40*, 2692.
[166] Choi, J.-C.; Sakakura, T. *J. Am. Chem. Soc.* **2003**, *125*, 7762.
[167] Carter, C. A. G.; Baker, R. T.; Nolan, S. L.; Tumas, W. *Chem. Commun.* **2000**, 347.
[168] He, L.-N.; Choi, J.-C.; Sakakura, T. *Tetrahedron Lett.* **2001**, *42*, 2169.

[169] Wittmann, K.; Wisniewski, W.; Mynott, R.; Leitner, W.; Kranemann, C. L.; Rische, T.; Eilbracht, P.; Kluwer, S.; Ernsting, J. M.; Elsevier, C. J. *Chem. Eur. J.* **2001**, *7*, 4584.
[170] Hadida, S.; Super, M. S.; Beckman, E. J.; Curran, D. P. *J. Am. Chem. Soc.* **1997**, *119*, 7406.
[171] DeSimone, J.; Selva, M.; Tundo, P. *J. Org. Chem.* **2001**, *66*, 4047.
[172] Hou, Z.; Han, B.; Zhang, X.; Zhang, H.; Liu, Z. *J. Phys. Chem. B* **2001**, *105*, 4510.
[173] Bhanage, B. M.; Ikushima, Y.; Shirai, M.; Arai, M. *Chem. Commun.* **1999**, 1277.
[174] Bhanage, B. M.; Ikushima, Y.; Shirai, M.; Arai, M. *Tetrahedron Lett.* **1999**, *40*, 6427.
[175] Jacobson, G. B.; Lee Jr., C. T.; Johnston, K. P.; Tumas, W. *J. Am. Chem. Soc.* **1999**, *121*, 11902.
[176] McCarthy, M.; Stemmer, H.; Leitner, W. *Green Chem.* **2002**, *4*, 501.
[177] Hâncu, D.; Beckman, E. J. *Green Chem.* **2001**, *3*, 80.
[178] Matsuda, T.; Watanabe, K.; Kamitanaka, T.; Harada, T.; Nakamura, K. *Chem. Commun.* **2003**, 1198.
[179] Liu, F.; Abrams, M. B.; Baker, R. T.; Tumas, W. *Chem. Commun.* **2001**, 433.
[180] Brown, R. A.; Pollet, P.; McKoon, E.; Eckert, C. A.; Liotta, C. L.; Jessop, P. G. *J. Am. Chem. Soc.* **2001**, *123*, 1254.
[181] Sellin, M. F.; Webb, P. B.; Cole-Hamilton, D. J. *Chem. Commun.* **2001**, 781.
[182] Reetz, M. T.; Wiesenhofer, W.; Franciò, G.; Leitner, W. *Chem. Commun.* **2002**, 992.
[183] Lozano, P.; de Diego, T.; Carrié, D.; Vaultier, M.; Iborra, J. L. *Chem. Commun.* **2002**, 692.
[184] Laszlo, J. A.; Compton, D. L. *Adv. Chem. Ser.* **2002**, *818*, 387.
[185] Meehan, N. J.; Sandee, A. J.; Reek, J. N. H.; Kamer, P. C. J.; van Leeuwen, P. W. N. M.; Poliakoff, M. *Chem. Commun.* **2000**, 1497.
[186] Sellin, M. F.; Cole-Hamilton, D. J. *J. Chem. Soc., Dalton Trans.* **2000**, 1681.
[187] Dharmidhikari, S.; Abraham, M. A. *J. Supercrit. Fluids* **2000**, *18*, 1.
[188] Ley, S. V.; Ramarao, C.; Gordon, R. S.; Holmes, A. B.; Morrison, A. J.; McConvey, I. F.; Shirley, I. M.; Smith, S. C.; Smith, M. D. *Chem. Commun.* **2002**, 1134.
[189] Stamp, L. M.; Mang, S. A.; Holmes, A. B.; Knights, K. A.; de Miguel, Y. R.; McConvey, I. F. *Chem. Commun.* **2001**, 2502.
[190] Cacchi, S.; Fabrizi, G.; Gasparrini, F.; Villani, C. *Synlett* **1999**, 345.
[191] Yeung, L. K.; Lee Jr., C. T.; Johnston, K. P.; Crooks, R. M. *Chem. Commun.* **2001**, 2290.
[192] Gordon, R. S.; Holmes, A. B. *Chem. Commun.* **2002**, 640.
[193] van den Broeke, L. J. P.; Goetheer, E. L. V.; Vekerk, A. W.; de Wolf, E.; Deelman, B.-J.; van Koten, G.; Keurentjes, J. T. F. *Angew. Chem. Int. Ed.* **2001**, *40*, 4473.
[194] He, L.-N.; Yasuda, H.; Sakakura, T. *Green Chem.* **2003**, *5*, 92.
[195] Kani, I.; Omary, M. A.; Rawashdeh-Omary, M. A.; Lopez-Castillo, Z. K.; Flores, R.; Akgerman, A.; Fackler Jr., J. P. *Tetrahedron Lett.* **2002**, *58*, 3923.
[196] Horváth, I. T.; Rábai, J. *Science* **1994**, *266*, 72.
[197] Sinou, D.; Pozzi, G.; Hope, E. G.; Stuart, A. M. *Tetrahedron Lett.* **1999**, *40*, 849.
[198] Eastoe, J.; Paul, A.; Nave, S.; Steyler, D. C.; Robinson, B. H.; Rumsey, E.; Thorpe, M.; Heenan, R. K. *J. Am. Chem. Soc.* **2001**, *123*, 988.
[199] Fink, R.; Beckman, E. J. *J. Supercrit. Fluids* **2000**, *18*, 101.
[200] Hu, Y.; Chen, W.; Xu, L.; Xiao, J. *Organometallics* **2001**, *20*, 3206.
[201] Ye, W.; DeSimone, J. M. *Ind. Eng. Chem. Res.* **2000**, *39*, 4564.
[202] Goetheer, E. L. V.; Baars, M. W. P. L.; van den Broeke, L. J. P.; Meijer, E. W.; Keurentjes, J. T. F. *Ind. Eng. Chem. Res.* **2000**, *39*, 4634.
[203] Jardine, F. H.; Osborn, J. A.; Wilkinson, G. *J. Chem. Soc. (A)* **1967**, 1574.
[204] Jacobson, G. B.; Lee Jr., C. T.; daRocha, S. R. P.; Johnston, K. P. *J. Org. Chem.* **1999**, *64*, 1207.
[205] Koukou, M. K.; Papayannakos, N.; Markatos, N. C.; Bracht, M.; Veen, H. M.; Roskam, A. *J. Membr. Sci.* **1999**, *155*, 241.
[206] Bhattacharyya, P.; Gudmunsen, D.; Hope, E. G.; Kemmit, R. D.; Paige, D. R.; Stuart, A. M. *J. Chem. Soc., Perkin Trans. 1* **1997**, 3609.

Index